ALSO BY DAVID QUAMMEN

NONFICTION

Yellowstone

The Chimp and the River

Ebola

Spillover

The Reluctant Mr. Darwin

Monster of God

The Song of the Dodo

ESSAYS

Natural Acts

The Boilerplate Rhino

Wild Thoughts from Wild Places

The Flight of the Iguana

FICTION

Blood Line

The Soul of Viktor Tronko

The Zolta Configuration

To Walk the Line

THE

TANGLED TREE

A Radical New History of Life

DAVID QUAMMEN

Simon & Schuster
New York London Toronto Sydney New Delhi

Simon & Schuster
1230 Avenue of the Americas
New York, NY 10020

First Simon & Schuster hardcover edition August 2018

SIMON & SCHUSTER and colophon are registered
trademarks of Simon & Schuster, Inc.

For information about special discounts for bulk purchases,
please contact Simon & Schuster Special Sales at 1-866-506-1949
or business@simonandschuster.com.

The Simon & Schuster Speakers Bureau can bring authors to
your live event. For more information or to book an event,
contact the Simon & Schuster Speakers Bureau at 1-866-248-3049
or visit our website at www.simonspeakers.com.

Interior design by Lewelin Polanco

Manufactured in the United States of America

1   3   5   7   9   10   8   6   4   2

Library of Congress Cataloging-in-Publication Data

Names: Quammen, David, 1948- author.
Title: The tangled tree : a radical new history of life / David Quammen.
Description: First Simon & Schuster hardcover edition. | New York : Simon
& Schuster, 2018. | Includes bibliographical references and index.
Identifiers: LCCN 2018004356| ISBN 9781476776620 |
ISBN 1476776628 | ISBN 9781476776644 (ebook)
Subjects: LCSH: Phylogeny--Molecular aspects.
Classification: LCC QH367.5 .Q36 2018 | DDC 591.3/8--dc23
LC record available at https://lccn.loc.gov/2018004356

ISBN 978-1-4767-7662-0
ISBN 978-1-4767-7664-4 (ebook)

*To Dennis Hutchinson and David Roe,
my attorneys of the soul*

Contents

THREE SURPRISES

An Introduction

Life in the universe, as far as we know, and no matter how vividly we may imagine otherwise, is a peculiar phenomenon confined to planet Earth. There's plenty of speculation and probabilistic noodling, but zero evidence, to the contrary. The mathematical odds and chemical circumstances do seem to suggest that life should exist elsewhere. But the reality of such alternate life, if any, is so far unavailable for inspection. It's a guess, whereas earthly life is fact. Some astounding discovery of extraterrestrial beings, announced tomorrow, or next year, or long after your time and mine, may disprove this impression of Earth's uniqueness. For now, though, it's what we have: life is a story that has unfolded here only, on a relatively small sphere of rock in an inconspicuous corner of one middling galaxy. It's a story that, to the best of our knowledge, has occurred just once.

The shape of this story, in its broad outlines as well as its finer details, is therefore a matter of some interest.

What happened, over the course of roughly four billion years, to bring life from its primordial origins into the fluorescence of diversity and complexity we see now? How did it happen? By what concatenation of accident and determination did it yield creatures so wondrous as humans—and blue whales, and tyrannosaurs, and giant sequoias? We know there have been crucial transitions in evolutionary history, improbable incidents of convergence, dead ends, mass extinctions, big events, and little ones with big consequences—including some fateful contingencies that have left behind evidence of their occurrence embedded subtly throughout the fossil record and the living world. Alter those few contingencies, as a thought experiment, and everything would be different. We wouldn't exist. Animals and plants wouldn't exist. Why did it happen as it did, and not some other way? Religions have their responses to such questions, but for science, the answers must be discovered and then supported with empirical evidence, not received in a holy trance.

This book is about a new method of telling that story, a new method of deducing it, and certain unexpected insights that have flowed from the new method. The method has a name: molecular phylogenetics. Wrinkle your nose at that fancy phrase, if you will, and I'll wrinkle with you, but, in fact, what it means is fairly simple: reading the deep history of life and the patterns of relatedness from the sequence of constituent units in certain long molecules, as those molecules exist today within living creatures. The molecules mainly in question are DNA, RNA, and a few select proteins. The constituent units are nucleotide bases and amino acids—more definition of those to come. The unexpected insights have fundamentally reshaped what we think we know about life's history and the functional parts of living beings, including ourselves. In particular, there have come three big surprises about who we are—we multicellular animals, more particularly we humans—and what we are, and how life on our planet has evolved.

One of those three surprises involves an anomalous form of creature, a whole category of life, previously unsuspected and now known as the archaea. (Their name gets uppercased when used as a formal taxonomic category: Archaea.) Another is a mode of hereditary change that was also unsuspected, now called horizontal gene transfer. The third is a revelation, or anyway a strong likelihood, about our own deepest ancestry. We ourselves—we humans—probably come from creatures that, as recently as forty years ago, were unknown to exist.

The discovery and identification of the archaea, which had long been mistaken for subgroups of bacteria, revealed that present-day life at the microbial scale is very different from what science had previously depicted, and that the early history of life was very different too. The recognition of horizontal gene transfer (HGT, in the alphabet soup of the experts) as a widespread phenomenon has overturned the traditional certitude that genes flow only vertically, from parents to offspring, and can't be traded sideways across species boundaries. The latest news on archaea is that all animals, all plants, all fungi, and all other complex creatures composed of cells bearing DNA within nuclei—that list includes us—have descended from these odd, ancient microbes. Maybe. It's a little like learning, with a jolt, that your great-great-great-grandfather came not from Lithuania but from Mars.

Taken together, these three surprises raise deep new uncertainties— and carry big implications about human identity, human individuality, human health. We are not precisely who we thought we were. We are composite creatures, and our ancestry seems to arise from a dark zone of the living world, a group of creatures about which science, until recent decades, was ignorant. Evolution is trickier, far more intricate, than we had realized. The tree of life is more tangled. Genes don't move just vertically. They can also pass laterally across species boundaries, across wider gaps, even between different kingdoms of life, and some have come sideways into our own lineage—the primate lineage—from unsuspected, nonprimate sources. It's the genetic equivalent of a blood transfusion or (different metaphor, preferred by some scientists) an infection that transforms identity. "Infective heredity." I'll say more about that in its place.

And meanwhile, speaking of infection: another result of this sideways gene movement involves the global medical challenge of antibiotic-resistant bacteria, a quiet crisis destined to become noisier. Dangerous bugs such as MRSA (methicillin-resistant *Staphylococcus aureus*, which kills more than eleven thousand people annually in the United States and many more thousands around the world) can abruptly acquire whole kits of drug-resistance genes, from entirely different kinds of bacteria, by horizontal gene transfer. That's why the problem of multiple-drug-resistant superbugs—unkillable bacteria—has spread around the world so quickly. By such revelations, both practical and profound, we're suddenly challenged to adjust our basic understandings of who we humans are, what has gone into the making of us, and how the living world works.

This whole radical reset of biological thinking arose from several points of origin in space and time. One among them, maybe the most crucial, deserves mentioning here: the time was autumn 1977; the place was

Urbana, Illinois, where a man named Carl Woese sat with his feet on his desk, before a blackboard filled with notes and figures, posed jauntily for a photographer from the *New York Times*. The accompanying *Times* story for which the photo was shot, announcing that Woese and his colleagues had discovered "a separate form of life" constituting a "third kingdom" of biological forms in addition to the recognized two, ran on November 3, 1977. It was front page, above the fold, shouldering aside items on the kidnapped heiress Patty Hearst and an arms embargo against the apartheid regime in South Africa. Big news, in other words, whether or not the average *Times* reader could grasp, from such a lean telling, just what was meant by "a separate form of life." That article marked the apex of Woese's fame, his Warhol moment: fifteen minutes of limelight, then back to the lab. Woese brought radical changes—to his own field, to the story of life—and yet he remains unknown to most people outside the rarefied corridors of molecular biology.

Carl Woese was a complicated man—fiercely dedicated and very private—who seized upon deep questions, cobbled together ingenious techniques to pursue those questions, flouted some of the rules of scientific decorum, made enemies, ignored niceties, said what he thought, focused obsessively on his own research program to the exclusion of most other concerns, and turned up at least one or two discoveries that shook the pillars of biological thought. To his close friends, he was an easy, funny guy; caustic but wry, with a love for jazz, a taste for beer and scotch, and an amateurish facility on piano. To his grad students and postdoctoral fellows and laboratory assistants, most of them, he was a good boss and an inspirational mentor, sometimes (but not always) generous, wise, and caring.

As a teacher in the narrower sense—a professor of microbiology at the University of Illinois—he was almost nonexistent as far as undergraduates were concerned. He didn't stand before large banks of eager, clueless students, patiently explaining the ABCs of bacteria. Lecturing wasn't his strength, or his interest, and he lacked eloquent forcefulness even when presenting his work at scientific meetings. He didn't like meetings. He didn't like travel. He didn't create a joyous, collegial culture within his lab, hosting seminars and Christmas parties to be captured in group photos, as many senior scientists do. He had his chosen young friends, and some of them remember good times, laughter, beery barbecues at the Woese home, just a short walk from the university campus. But those friends were the select few who, somehow, by charm or by luck, had gotten through his shell.

In later years, as he grew more widely acclaimed, receiving honors of all kinds short of the Nobel Prize, Woese seems also to have grown bitter. He considered himself an outsider. He was elected to the National Academy of Sciences, an august body, but tardily, at age sixty, and the delay annoyed him. He became, by some reports, distant from his family—a wife and two children, seldom mentioned in published accounts of his scientific labors. He was a brilliant crank, and his work triggered a drastic revision of one of the most basic concepts in biology: the idea of the tree of life, the great arboreal image of relatedness and diversification. For that reason, Woese's moment of triumph in Urbana, on November 3, 1977, has its place near the core of this book.

Other scientists and other discoveries are connected to Woese and his tree. A little-known British physician named Fred Griffith, for instance, in the mid-1920s, while researching pneumonia for the Ministry of Health, noticed an unexpected transformation among bacteria: one strain changing suddenly into another strain, presto, from harmless to deadly virulent. This was important in terms of public health (bacterial pneumonia was in those days a leading cause of death) but also, as even Griffith didn't realize, a clue to deeper truths in pure science.

The mechanism of Griffith's perplexing transformation remained obscure until 1944, when a quiet, fastidious researcher named Oswald Avery, at the Rockefeller Institute in New York, identified the substance, the "transforming principle," that can cause such sudden change from one bacterial identity to another. It was deoxyribonucleic acid. DNA. Less than a decade later, Joshua Lederberg and his colleagues showed that this sort of transformation, relabeled "infective heredity," is a routine and important process in bacteria—and, as later work would show, not just in bacteria. Meanwhile, the corn geneticist Barbara McClintock, discovering genes that bounce from one point to another on the chromosomes of her favorite plant, worked with very little support or recognition through the prime years of her career—and then accepted a Nobel Prize at age eighty-one.

Lynn Margulis, a Chicago-educated microbiologist unique in almost every way, shared at least one thing with McClintock: the frustrations of being dismissed by some colleagues as an eccentric and obdurate woman. In Margulis's case, it was for reviving an old idea that had long been considered wacky: endosymbiosis. What she meant by the term was, roughly, the cooperative integration of living creatures within living creatures. That is, not just tiny creatures within the bellies or noses of big creatures, but cells within cells. More specifically, Margulis argued that the cells constituting every creature in the more complex divisions of life—every human, every animal, every plant, every fungus—are chimerical things, assembled with captured bacteria inside nonbacterial receptacles. Those particular bacteria, over vast stretches of time, have become transmogrified into cellular organs. Imagine an oyster, transplanted into a cow, that becomes a functional bovine kidney. This seemed crazy when Margulis proposed it in 1967. But she was right about the matter, mostly.

Fred Sanger, Francis Crick, Linus Pauling, Tsutomu Watanabe, and other scientists played crucial parts in this chain of events too, sometimes by force of personality as well as by scientific brilliance. Slightly deeper in the past lie obscure figures such as Ferdinand Cohn, Edward Hitchcock, and Augustin Augier, as well as more famous ones, including Ernst Haeckel, August Weismann, and Carl Linnaeus. The ghost of Jean-Baptiste Lamarck rises here again to skulk along inescapably in the shadows of evolutionary thinking.

Such people, all contributors to a scientific upheaval, are of additional interest for the ways their works grew from their lives. They serve as good reminders that science itself, however precise and objective, is a human activity. It's a way of wondering as well as a way of knowing. It's a process, not a body of facts or laws. Like music, like poetry, like baseball, like grandmaster chess, it's something gloriously imperfect that people do. The smudgy fingerprints of our humanness are all over it.

Humans aren't the only important characters in this book. There are also a lot of other living creatures, whose unique histories and foibles illustrate points in the story I'm trying to tell. Many of them are microbes—those bacteria I've mentioned, those archaea, and other teeny things. Please don't be fooled by their smallness; their implications and impacts are big. And don't be daunted by their names, which are mostly expressed in scientific Latin: *Bacillus subtilis* and *Salmonella typhimurium* and *Methanobacterium ruminantium* and other monstrous tongue twisters. The reason I call them by those names is not because I like arcane language but because no other labels exist. Microbes generally don't get the courtesy of common names at the species level, casual monikers such as southern giraffe, olive bunting, monarch butterfly, and Komodo dragon. If the bacterium known as *Haemophilus influenzae* could be accurately called Fleming's nose-tickler, I promise you I would do it.

One other featured character, of the human sort, should be introduced here. He's a bearded American microbiologist with a penchant for philosophical musing, tucked away at a university in Nova Scotia. This man has linked Carl Woese, Lynn Margulis, and much of the new work in molecular phylogenetics into a pungent challenge against biology's central metaphor. His name is Ford Doolittle. He's tall, diffident in manner though not in thought, and enjoys causing a little intellectual discomfort. At the turn of the millennium, Doolittle published an essay titled "Uprooting the Tree of Life," which helped release a cascade of arguments. I caught wind of him through that essay and his related writings, notably those in which he discussed horizontal gene transfer and its implications. "Horizontal what?" was my earliest thought. Then I pilgrimed to Halifax and camped for days in his office. Doolittle is semiretired, still guiding graduate students, still well funded with a prestigious research grant, but no longer growing radioactive bacteria in a lab in order to deduce bits of their genomes (the totality of their DNA) from images on chest X-ray films. He's no longer pulling chopped molecules through electrophoretic gels, as he did in the pioneer days. He reads, he thinks, he writes, he draws. (He takes art photographs, mainly for his own amusement, and occasionally mounts a gallery show, but that's another realm of enterprise entirely.) In fact, part of what has made Ford Doolittle so influential is that, in addition to his qualifications in biology, he writes far better than most scientists—and he draws deftly, turning big concepts into graceful, cartoony shapes. Doolittle's father was a painter and an art professor. Young Ford considered an art career himself, though his father called that "a terrible way to make a living." Then, when he was fifteen years old, in 1957, the Soviets put Sputnik into space, persuading Ford and many other Americans that science and engineering were the more urgent, forceful pursuits. He went to Harvard College and studied

xvi 🕊 *Three Surprises: An Introduction*

biochemistry. The artistic impulse never left him. Nowadays, to illustrate his subversive thinking and his genial provocations, he draws trees that aren't trees.

Woese, Doolittle, Margulis, Lederberg, Avery, Griffith, and the others—they all have their roles in this story. But a more natural starting point is much earlier: London, 1837, with a very different scientist, in a very different situation.

PART ※ I

Darwin's Little Sketch

1

Beginning in July 1837, Charles Darwin kept a small notebook, which he labeled "B," devoted to the wildest idea he ever had. It wasn't just a private thing but a secret thing, a record of his most outrageous thoughts. The notebook was bound in brown leather, with a tab and a clasp; 280 pages of cream-colored paper, compact enough to fit in his jacket pocket. Portable, but no toss-away pad. Its quality of materials and construction reflected the fact that Darwin was an affluent young man, living in London as a naturalist of independent means. He had arrived back in England just nine months earlier from the voyage of HMS *Beagle*.

That journey, consuming almost five years of Darwin's life, on sea and land, mostly along the South American coastline and inland to the plains and mountains, though with notable other stops on the roundabout way home, would be the only major travel experience of his sheltered, privileged life. But it was enough. A mind-awakening and transformative opportunity, it had given him some large ideas that he wanted to pursue. It had opened his eyes to an astonishing phenomenon that demanded explanation. In a letter to his biology professor and friend John Stevens Henslow, back at Cambridge University, written from Sydney, Australia, Darwin mentioned his puzzling observations of the mockingbirds (not the finches) of the Galápagos Archipelago, a set of volcanic nubs in mid-Pacific. These gray, long-beaked birds differed from island to island but so subtly that they seemed to have diverged from one stock. Diverged? Three kinds of mockingbird? Varying slightly, this island to that? Yes: they appeared distinct but similar, in a way that suggested relatedness. If that impression were true, Darwin confided to Henslow, confessing an intellectual heresy, "such facts would undermine the stability of species."

The stability of species represented the bedrock of natural history. It was taken for granted, and important, not just among clergy and pious lay people but scientists too. That all the varied forms of creatures on Earth had been fashioned by God, in special acts of creation, and are therefore immutable, was an article of faith to the Anglican scientific establishment of Darwin's era. This tenet is known as the special-creation hypothesis, though at the time, it seemed less hypothesis than dogma. It had been embraced and supported by prominent naturalists and philosophers of the scientific culture within which Darwin had been educated at Cambridge. He was now home from his wildcat voyage, a youthful adventure with a bunch of rough English sailors, about which his stern father had been skeptical at the start. The experience had altered him—though not in the ways his father may have feared. He hadn't become a drunk or a libertine. He didn't curse like a bosun. Darwin's wanderlust, satisfied physically, was now intellectual. He intended to investigate, very discreetly, a radical alternative to scientific orthodoxy: that the forms of living creatures *weren't* eternally stable, as God had created them, but instead had changed over time, one into another—by some mechanism that Darwin didn't yet understand.

It was a risky proposition. But he was twenty-seven years old and deeply changed by what he had seen and, in a quiet way, very gutsy.

So he had set himself up in the big city, with lodgings on Great Marlborough Street, a convenient location for his visits to the British Museum. This was just a few doors down from the house where his elder brother, Erasmus, had already settled. Darwin joined scientific clubs, the Geological Society, the Zoological Society, but had no job. Didn't need one. The same formidable father who had first disapproved of the *Beagle* voyage—Dr. Robert Darwin, a wealthy physician up in the town of Shrewsbury—was now rather proud of his second son, the young naturalist well regarded within British scientific circles. Grumpy on the outside, generous within, Dr. Darwin had made supportive arrangements for both brothers. And Charles was single. He sauntered around London, he handled follow-up tasks on his specimens from the voyage, he worked on rewriting his *Beagle* diary into a travel book, and—very privately—he ruminated about that radical alternative to special creation. He read widely, scribbling facts and phrases into various notebooks. The "A" notebook was devoted to geology. The B notebook was first of a series on what, to himself only, he called "transmutation." You can guess what that meant. Darwin had begun thinking his way toward a theory of evolution.

He opened the B notebook, in July 1837, with a few phrases alluding to a book titled *Zoonomia; or the Laws of Organic Life*, published decades earlier by his own grandfather, another Erasmus Darwin. *Zoonomia* was a medical treatise (Erasmus was a physician), but it contained some provocative musings that sounded vaguely evolutionary. All warm-blooded animals "have arisen from one living filament," according to *Zoonomia*, and they possess "the faculty of continuing to improve" in ways that could be passed down across the generations, "world without end!" Improvement across generations? Heritable change throughout the history of the world? That was contrary to the special-creation hypothesis, but not too surprising from a gouty, libidinous freethinker and sometime poet such as old Erasmus. Darwin had read *Zoonomia* during his student days and shown little sign of giving his grandfather's daring ideas much credit. But now, on revisiting, he took them as a point of departure. Page one, entry one, in the B notebook: his grandfather's title, *Zoonomia*, followed by reading notes.

Then again, those wild suggestions didn't lead anywhere. Erasmus Darwin had offered no material mechanism for "the faculty of continuing to improve," and a material mechanism was what young Charles wanted, though he may not have fully realized that yet. As reflected in the B notebook, he now went from his grandfather's work to other readings, other speculations and questions, jotting down clipped phrases, often in bad grammar and punctuation. He wasn't writing to publish. These were messages to himself.

"Why is life short," he asked, omitting the question mark in his haste. Why is reproduction so important? Why do animals of a given kind tend to be constant in form across an entire country but to differ at least slightly on separate islands? He remembered the giant tortoises on the Galápagos, where his stopover had lasted only thirty-five days but catalyzed an up-heaval in his thinking. He remembered the mockingbirds too. And why had he seen two distinct kinds of "ostriches" (his label for big, flightless birds now known as rheas) on the Argentine Pampas, one living north of the Rio Negro, one south of it? Did creatures somehow become different when isolated? Put a pair of cats on an island, let them breed and inbreed there for generations, with a little pressure from enemies, and "who will dare say what result," Darwin wrote. He dared. The descendants might come to look different from other cats, might they not? He wanted to understand why.

Another important question: "Each species changes. does it pro-gress." Do the cats become *better* cats, or at least better cats for catting on that particular island? If so, how long would it take? How far would it go? What are the logical limits, if "every successive animal is branching up-wards" and with "different types of organization improving," new forms arising, old forms dying out? That one word, *branching*, was freighted with interesting implications: of directional growth, of divergence, of an arbo-real form. And these questions Darwin asked himself, they applied not just to cats and ostriches but also to armadillos and sloths in Argentina, to marsupials in Australia, to those huge Galápagos tortoises, and to the wolflike Falkland Islands fox, all peculiar in certain ways, all unique to their isolated places, but recognizably similar to their correlatives—other cats and tortoises and foxes, etcetera—elsewhere. Darwin had seen a lot. He was an acutely observant and reflective young man. He sensed that he had seen patterns, not just particulars. It almost seemed, he wrote, that there was a "law of adaptation" at work.

All this and more, facts and speculations, crammed into the first twenty-one pages of notebook B. The pages are mostly undated, so we can't know how many days or weeks passed in the opening burst of effort. Anyway, he didn't yet have his theory. Big ideas were coming at him like diving owls. He needed some order as much as he needed the jumble of tantalizing clues. Maybe he needed a metaphor. Then, on the bottom of page 21, Darwin wrote: "organized beings represent a tree."

2

W e don't know whether Darwin sat back after writing that
statement and breathed deep with a new sense of clarity, but
he might have. And he was entitled.

Then he scribbled on. The tree is "irregularly branched," he told the
B notebook, "some branches far more branched." Each branch diverges
into smaller branches, he wrote, and then twigs, "Hence Genera," the
next higher category above species, which would be the twiglets or termi-
nal buds. Some buds die away without yielding further growth—species
extinction, end of a line—while new buds appear, somehow. Although
the very idea of extinction had once been problematic among naturalists
and philosophers, doubted as a possibility or rejected outright on grounds
that God's acts of special creation couldn't be undone, Darwin recognized
that there's "nothing stranger in death of species" than in death of an individual. In fact, extinction was not just natural but necessary, making
space for new species as old ones die away. He wrote: "The tree of life
should perhaps be called the coral of life, base of branches dead," ancestral forms gone. Darwin knew something about coral, having seen reefs
at Keeling Atoll in the eastern Indian Ocean and elsewhere during the
Beagle voyage. They fascinated him; he concocted a theory of how reefs
are formed; and in 1842, five years after this notebook entry, he would

publish a book about coral reefs. Coral seemed apt—branching coral, not brain coral or table coral, was what he had in mind—because the lower limbs and base are lifeless calcitic skeleton, left behind like extinct forms of ancient lineages as the soft polyps advance upward like living species. But even he seems to have sensed that "the coral of life" didn't have the same memorable ring. He drew a feeble pen sketch, on page 26 in the B notebook, of a three-branched coral of life, with dotted lines depicting the inanimate lower sections. And then he let the coral idea slide, abandoning that metaphor.

The tree of life was better. It was already a venerable notion in 1837, and Darwin could adapt it to his purposes as an evolutionary theorist— easier than inventing a new trope from scratch. Of course, to make that adaptation was to alter its meaning radically. Never mind, he took the step. Ten notebook pages along, he sketched a much livelier and more complex figure in bold strokes, with a trunk rising into four major limbs and several minor ones, each major limb diverging into clusters of branches, one branch within each cluster labeled A, B, C, D. The branches B and C were near neighbors in the treetop, within adjacent clusters, indicating close relationships among the creatures on those branches. The

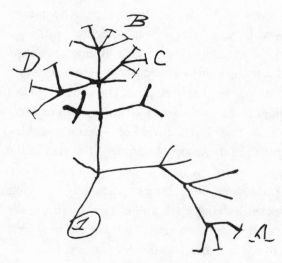

Darwin's 1837 sketch, redrawn by Patricia J. Wynne.

letter A was far away, on the opposite side of the tree's crown, signaling a more distant relationship—but still a relationship. The letters were place-holders, meant to represent living species, or maybe genera. *Felis, Canis, Vulpes, Gorilla.* We don't know exactly what he had in mind, and maybe it was nothing so specific. Anyway, this was a thunderous assertion, abstract but eloquent. You can look at the little sketch today, with its four labeled branches amid the limbs and the crown, and imagine the evolutionary divergence of all life from a common ancestor.

Just above the sketch, as though gesturing toward it bashfully, Darwin wrote: "I think."

3

Darwin didn't invent that phrase, "the tree of life," nor originate its iconic use, though he put it to new purpose in his theory. Like so many other metaphors embedded deep in our thinking, it came down murkily, modified and reechoed, from early versions in Aristotle and the Bible. (Why do these things always go back to Aristotle? Well, that's why he's Aristotle.) In the Bible, it's a grand bookend motif, invoked in Genesis 3 just as Adam and Eve are booted out of the Garden, and reappearing at the end of Revelation, on the very last page of the King James version—excellent placement for a launch into Western culture. There in Revelation 22, verses 1–2, the authorial prophet describes his ecstatic vision of the "water of life," flowing out like a pure river from the throne of God, and beside which grows "the tree of life," bearing fruit every month, plus leaves "for the healing of the nations." This tree possibly represents Christ, supplying his leafy and fruity blessings to the world; or maybe it's grace, or the Church. The passage is opaque, and differences in translations (one tree or many?) have confused things further. The point here is simply that the "tree of life" is an ancient poetic image, a resonant phrase, variously construable, with a long presence in Western thought.

In Aristotle's *History of Animals*, written during the fourth century BCE, the tree of life is not yet a tree. It's more like a ladder of nature or—as later

Latinized from his Greek—a *scala naturae*. According to Aristotle, the diversity of the natural world "proceeds" from lifeless things such as earth and fire to living creatures such as animals "little by little," in a progression so incremental that it's impossible to draw absolute lines between one form and another. This idea remained useful throughout the Middle Ages and beyond, turning up in woodcuts during the sixteenth century as a *Great Chain of Being* or a *Ladder of Ascent and Descent of the Intellect*, which typically rose step-by-step from inanimate substances such as stone or water, to plants and then beasts, then humans, then angels, and finally to God. By that point it was a "Stairway to Heaven," almost five centuries before Led Zeppelin.

The Swiss naturalist Charles Bonnet reverted to this linear, stair-step model as late as 1745, even while other Enlightenment thinkers and artists were allowing images of nature's diversity to burgeon sideways with limbs and branches. Bonnet's treatise on insects, published that year, included a foldout diagram of his "Idea of a Scale of Natural Beings," arranged in vertical ascent from fire, air, and water, through earth and various minerals, upward to mushrooms, lichens, plants, and then sea anemones, followed by tapeworms and snails and slugs, upward further to fish and then flying fish in particular, and then birds, above which came bats and flying squirrels, then four-legged mammals, monkeys, apes, and lastly man. See the logic? Flying fish are superior to other fish because they fly; bats and squirrels exist on a higher level than birds because bats and squirrels are mammals; orangutans and humans are the best of mammals, and humans are more best than anybody. Bonnet made his living as a lawyer but much preferred studying insects and plants. He was a lifelong citizen of the Republic of Geneva, his French ancestors having been chased out of France by religious persecution, and so maybe it's no accident that his ladder diagram culminated in people, not God.

The other notable absence from Bonnet's scale of natural beings, besides God, are microbes. He paid no attention to microorganisms, although the pioneering Dutch microscopist Antoni van Leeuwenhoek had discovered the existence of bacteria, protozoans, and other tiny "animalcules" about seventy years earlier. We all know Leeuwenhoek's name from our reading in high school of Paul de Kruif's *Microbe Hunters* (a terrible book full of concocted dialogue and bogus detail, but an influential doorway to the subject) or other storybook histories of science, though we might not remember that Leeuwenhoek was a draper in Delft who started making his own magnifying lenses in order to better inspect the thread-count of textiles. Then he turned the lenses onto other materials, out of sheer curiosity, and made astonishing discoveries: he found menageries of tiny creatures living in lake water, in rain water, in water from drain pipes, even in scrapings of crud from his own teeth.

Leeuwenhoek's revelatory observations of microbial life were reported in the journal of the Royal Society of London and became famous in scientific circles throughout Europe, but Charles Bonnet wasn't interested enough in those "very wee animals" to fit them into his rising scale—not even where they might dismissively have been slotted, somewhere between asbestos and truffles. That omission presages a lasting discomfort with placing microbes on the ladder of life or, harder still, arranging their diverse forms on the tree—and it's a discomfort to which I'll return, because it became acute in 1977.

The linear approach to depicting life's diversity was on the way out, notwithstanding Charles Bonnet's scale of nature, and being replaced by its more complicated and dimensional successor, the tree. By the late eighteenth century and the start of the nineteenth, natural philosophers (we'd call them scientists, but that word didn't yet exist) tried to classify and arrange living creatures into distinct groups and subgroups, reflecting their similarities and differences and some sort of organizing schema. The linear alignment, in order of what passed for increasing sublimity, the ladder raised toward God, was no longer satisfactory. There had been a knowledge explosion in Europe since the great age of sailing explorations began—knowledge of diverse animals, plants, and other creatures from all over the world—and scholars wanted to set that explosive abundance of new facts within hierarchical categories so that it could be easily accessed and used.

This wasn't evolutionary thinking; it was just data management. The knowledge would fill volumes (one man alone, the German naturalist Alexander von Humboldt, published a thirty-volume account of his travels in South America), making all the more necessary an overview, an organizing principle, that could be apprehended at a glance: an illustration. But the illustrators now needed two dimensions, not one, and so the ladder turned into a trunk, and the trunk sprouted limbs, and the limbs diverged into branches. This offered more scope, sideways as well as up and down, for arranging the varied abundance of known creatures.

The tree of life was an old symbol by then, an old phrase, dating back at least to those mentions in Genesis and Revelation. The tree had also served as a model for family histories—the genealogical tree or pedigree of a German duke, for instance. Now the secularized tree became useful for organizing biology. Among the first to embrace this convention was another Frenchman, Augustin Augier, who wrote in 1801 that "a figure like a genealogical tree appears to be the most proper to grasp the order and gradation" of what concerned Augier: the diversity of plants.

Augier was an obscure citizen of the French Republic, living in Lyon, working on botany part-time; his real profession was unknown, his biographical details lost, even to a historian of Lyonnais botanists writing a hundred years later. Augier disappeared. But he left behind a book, a little octavo volume, in which he proposed a new classification of plants, "according to the order that Nature appears to have followed." That is, a "natural order," as opposed to an artificial classification system based merely on human whim or convenience. And in the book was a figure representing that system: his *arbre botanique* (botanical tree). Its trunk and limbs look almost as orderly and stiff as a menorah, but its sideways branching and copious leafing suggest a rife multiplicity of plant forms.

Again, this didn't imply any heretical ideas about origins. Augier was no evolutionist before his time. His natural order wasn't meant to suggest that all plants had descended from common ancestors by some sort of material process of transformation. God was their maker, shaping the varied forms individually: "It appears, and one can hardly doubt it, that the Creator, when making flowers, followed certain proportions and progressions in the number of their different parts." Augier's contribution, as he saw it, was discovering those proportions and progressions—design principles that had satisfied the Deity's neat sense of pattern—and using them after the fact to organize botanical knowledge into a tidy system.

Augier wasn't the first naturalist to hanker for a natural order of nature's diversity. Aristotle had classified animals as "bloodless" and "blooded." In the first century of our era a Greek physician named Dioscorides, attached to the Roman army, gathered lore on more than five hundred kinds of plants, arranging them in a compendium mainly on the basis of their medicinal, edible, and perfumatory uses. That book, in various reprints and translations, served as a trusted botany text for the next fifteen hundred years. Toward the end of its run, around the time of the Renaissance, as people traveled more widely and paid closer attention

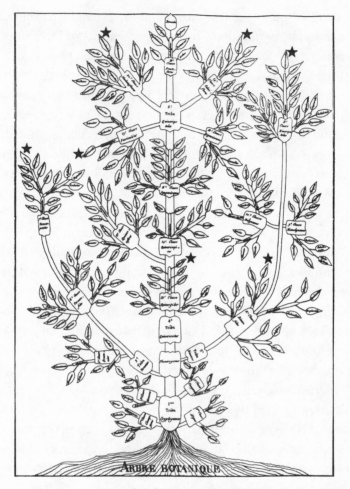

Augier's *Arbre Botanique*, 1801.

to the empirical details of nature, old Dioscorides gave way to newer illustrated herbals. These were essentially field guides to botany, graced with
better illustrations based on improvements in drawing and woodcut techniques, but still organized for convenience of use, not natural order. In the
sixteenth century, Leonhart Fuchs produced one of those books, an herbal
cataloging hundreds of plants, beautifully illustrated and arranged in al-
phabetical order. Two centuries later, the great systematizer Carl Linnaeus described a genus of plants with purplish red flowers, naming it
Fuchsia in honor of Leonhart Fuchs (and hence we got also the color, fuchsia). Linnaeus himself, a Swede who traveled widely as a young man and

then took up a professorial life in Uppsala, emerged from this herbalist tradition but went beyond it.

Linnaeus's *Systema Naturae*, as first published in 1735, was a unique and peculiar thing: a big folio volume of barely more than a dozen pages, like a coffee-table atlas, in which he outlined a classification system for all the members of what he considered the three kingdoms of nature: plants, animals, and minerals. Notwithstanding the inclusion of minerals, what matters to us is how Linnaeus viewed the kingdoms of life.

His treatment of animals, presented on one double-page spread, was organized into six columns, each topped with a name for one of his classes: Quadrupedia, Aves, Amphibia, Pisces, Insecta, Vermes. Quadrupedia was divided into several four-limbed orders, including Anthropomorpha (mainly primates), Ferae (doggish forms such as wolves and foxes, plus cat forms such as lions and leopards, in addition to bears), and others. His Amphibia encompassed reptiles as well as amphibians, and his Vermes was a catchall group containing not just worms, leeches, and flukes but also slugs, sea cucumbers, starfish, barnacles, and other sea animals. He divided each order further into genera (with some recognizable names such as Leo, Ursus, Hippopotamus, and Homo), and each genus into species. Apart from the six classes, Linnaeus also gave a half column to what he called Paradoxa: a wild-card assemblage of mythic chimeras and befuddling but real creatures, including the unicorn, the satyr, the phoenix, the dragon, and a certain giant tadpole (*Pseudis paradoxa*, under its modern label) that, strangely, paradoxically, shrinks during metamorphosis into a much smaller frog. Across the top of the chart ran large letters: CAROLI LINNAEI REGNUM ANIMALE. His animal kingdom. It was a provisional effort, grand in scope, integrated, but not especially original, to make sense of faunal diversity based on what was known and believed at the time. Then again, animals weren't Linnaeus's specialty.

Plants were. His classification of the vegetable kingdom was more innovative, more comprehensive, and more orderly. It became known as the "sexual system" because he recognized that flowers are sexual structures, and he used their male and female organs—their stamens and pistils, those delicate little stems sticking up to present and receive pollen—for characterizing his groups. Linnaeus defined twenty-three classes, into which he placed all the flowering plants, based on the number, size, and arrangement of their stamens. Then he broke each class into orders, based on their pistils. To the classes, he gave names such as Monandria, Diandria, and Triandria (one husband, two husbands, three husbands), and, within each class, ordinal names such as Monogynia, Digynia, and Tryginia (numbers of wives, yes, you get the idea), thereby evoking all sorts of polygamous and polyandrous ménages that must have caused lewd smirks and disapproving scowls among his contemporaries. A plant of the Monogynia order within the Tetrandria class, for instance: one wife with four husbands. Linnaeus himself seems to have enjoyed the sexy subtext. And it didn't prevent his botanical schema from becoming the accepted system of plant classification throughout Europe.

Our man Augustin Augier, coming along a half century later with his botanical tree of classification, seems to have seen himself challenging Linnaeus's overly neat sexual system. "Stamen number is a striking character," Augier conceded, but "not when it comes to the examination of plants"—that is, not always unambiguous and therefore not reliable as a basis for organizing the great jumble of botanical life. He nodded respectfully to Linnaeus—also to the French botanist Joseph Pitton de Tournefort, who had sorted plants into roughly seven hundred genera based on their flowers, their fruits, and other bits of their anatomy—and offered his own system, using multiple characters for different levels of sorting and to resolve the ambiguities and fine gradations. "This figure, which I call a *botanical tree*, shows the agreements which the different series of plants maintain amongst each other, although detaching themselves from the trunk; just as a genealogical tree shows the order in which different branches of the same family came from the stem to which they owe their origin." All discrete, yet all connected: bits of the same tree.

But they weren't connected, in Augier's mind, by descent from shared ancestors. Despite the hint he gave to himself in his language about family trees—all branches divergent from "the stem to which they owe their origin"—there is no evidence in Augier's writing or his tree figure that he had embraced, or even imagined, the idea of evolution.

4

That idea was coming soon, and, with its arrival, the tree of life would change meaning. The change was drastic, soul shaking to many people who lived through it, because it reflected a challenge to faith, and it met strong resistance. Jean-Baptiste Lamarck, France's great early evolutionist, and Edward Hitchcock, an American who prided himself a "Christian geologist," are the two scientists whose works—and whose graphic illustrations—best reflect how tree thinking shifted during the decades before Darwin unveiled his theory of evolution.

Lamarck was a protean figure: a soldier from a family of soldiering minor nobility who transformed himself into a botanist, then into a professor of zoology at the Muséum National d'Histoire Naturelle in Paris, to which he was appointed in 1793, on the eve of the Reign of Terror. His title at the museum put him in charge "of insects, of worms, and microscopic animals," three categories of life he had never studied, but he adapted fast, and even invented the word *invertebrates* to cover them. He abandoned plants and studied his invertebrates through the grimmest days of the French Revolution, earning a measly salary but at least keeping his head, as other scientists such as Antoine-Laurent Lavoisier went to the guillotine. Lamarck had probably helped his standing among the revolutionaries back in 1790 while employed at what was then the Jardin du Roi, when he urged dropping the royal label and renaming that institution the Jardin des Plantes. Clearly, he had good political instincts. He held the conventional view of species—that they were fixed forever and created by God—until 1797, but then his views changed, possibly as a result of his study of fossil and living mollusks, which seemed to show patterns of gradual transformation. He came out as an evolutionist on May 11, 1800, in his first lecture for the year's course on invertebrates. After that, he published three major works on evolutionary zoology, the most influential being his *Philosophie Zoologique* in 1809.

Lamarck outlived four wives and three of his seven children, living beyond the revolution, through the Napoleonic era and most of the Bourbon Restoration, a handsome man with a downturned mouth, balding slowly across his pate, blind for his final ten years, his faithful daughter, Cornelie, giving her life to him and reading him French novels. He died at eighty-five and was eulogized by important colleagues such as Geoffroy St. Hilaire, after which things didn't go so well: his remains were interred at the Montparnasse Cemetery in a common trench, not a permanent individual plot, and because such burial trenches were regularly recycled, his bones may have ended up in the Paris catacombs, along with those of thousands of paupers and other neglected folk. There was no Lamarck grave to visit. He became, according to one biographer, rather quickly "forgotten and unknown." His fame would return, if not immediately, but still it was a cold finish for the world's first serious evolutionary theorist.

Lamarck nowadays is commonly associated with what his name came to represent: Lamarckism, an easy but imprecise label for the idea of the inheritance of acquired characteristics. Many people are vaguely aware of him as a predecessor to Darwin; he is seen as a forerunner whose theory was provocative but wrong, refuted by later evidence because it depended, as Darwin's did not, on that illusory notion of acquired traits being heritable. (The real facts aren't so simple. For instance, Darwin himself included the inheritance of acquired characteristics as a force in evolution, under the label "use and disuse.") The most familiar example of such inherited adjustments, which Lamarck himself offered, is the giraffe. The proto-giraffe on the dry plains of Africa stretches to reach high foliage, its neck lengthens (supposedly) from the effort, its front legs lengthen too, and therefore (again supposedly) its offspring are born with longer necks and front legs. Lamarckism, in that cartoonish form, has been easy to despise but harder to kill off entirely.

It came back into fashion during the late nineteenth century, when the general idea of evolution gained acceptance but the crucial details of Darwin's particular theory, offering natural selection as the primary mechanism, were widely rejected. Natural selection just seemed too mechanistic, too stark and unguided, and many evolutionists found it unpalatable. This situation went on for decades—the world accepting Darwin's idea of evolution but not his explanation of how it occurs— though only historians remember that now. Lamarckism became neo-Lamarckism and seemed a less nihilistic alternative. It has continued to linger as a dubious but ineradicable notion—embodied in that single tenet, the inheritance of acquired characteristics—enjoying small surges of reconsideration even down to the present day.

But that single tenet was never Lamarck in totality. He had other ideas, some even worse. He believed in spontaneous generation. He disbelieved in extinction, at least as a natural process. He argued that "subtle fluids," surging through the bodies of living creatures, helped reshape them adaptively.

In one of his earlier botanical works, before the shift to animals and the epiphany about evolution, Lamarck had arranged plants in what he called "the true order of gradation": from least perfect and complete to most, ascending along an old-fashioned ladder of life. He matched that with a separate ladder for animals, a "counterpart" arrangement,

TABLEAU

Servant à montrer l'origine des différens animaux.

Vers.

Infusoires.
Polypes.
Radiaires.

Insectes.
Arachnides.
Crustacés.

Annelides.
Cirrhipèdes.
Mollusques.

Poissons.
Reptiles.

Oiseaux.

Monotrèmes.

M. Amphibies.

M. Cétacés.

M. Ongulés.

M. Onguiculés.

Lamarck's tree of dots, 1809.

showing an ascending series of forms: from worms, through insects, through fish and amphibians and birds, to mammals. Neither of those ladders hinted at divergence from common ancestors or transformation. But in the 1809 book *Philosophie Zoologique*, he included a different sort of figure, subtle yet dramatic, depicting animal diversity. It was a branched diagram, descending down the page, with major animal groups connected by dotted lines, like one of those connect-the-dots games for kids on the paper placemats at a pancake house. Connect the dots and discover that the secret shape is . . . an airplane! Or . . . an elephant! Or . . . George Washington's head! In Lamarck's dotted figure, the secret shape was a tree.

Birds sat perched on a branch divergent from reptiles. Insects had diverged from the main trunk before it yielded mollusks. Walruses and other marine mammals lay farther along that trunk, beyond which still other branches led to whales, then to hoofed mammals, and finally to all other mammals. Wrong though it was about the particulars, and despite being upside down, this figure marked an important transition in scientific thought. Scholars tell us that it was the earliest evolutionary tree.

5

E dward Hitchcock stands as a counterpoint to Lamarck, with
that first evolutionary tree, in that Hitchcock offered a last *pre-*
evolutionary tree in the decades before Darwin changed everything. In fact, Hitchcock presented two separate trees of life, one for
animals, one for plants, in his 1840 book *Elementary Geology*, which became a successful and often-reprinted text in the mid-nineteenth century. Hitchcock's trees were also innovative—among the first based on
deep knowledge of fossils, not just close observation of living creatures.
He called his illustration a "Paleontological Chart," and what it shows is
diversification of the animal and plant kingdoms charted against geological time, from the Cambrian period (beginning about 540 million years
ago) to the present.

Hitchcock's trees weren't classically tree shaped, spreading outward
into a canopy like a maple or an oak. Each of the two, the one for animals and the one for plants, looks more like a windbreak of tightly placed
Lombardy poplars grown to maturity along a roadway. The base of each
windbreak is a thick, solid trunk from which rise slender stems, fluffy
with foliage but without much branching as they ascend. Vertical, parallel, they seem independent: crustaceans, worms, bivalves, vertebrates.
The vertebrate stem does branch into several shafts. The shaft leading up

to modern mammals culminates in the word *Man*, atop which sits a regal crown adorned by a cross.

The crowned "Man," with its cross, tells us what we need to know about Hitchcock's sense of hierarchy in the living world. He grounded his geology firmly within the tradition known as natural theology, meaning science purposed to illuminate the power and wisdom of God as creator of all, with humans as the culmination of that divine creativity. He was a devout, driven New England Yankee, and his "Paleontological Chart" reflected his view of humans as the apogee of creation, as well as his findings in geology.

Hitchcock was born to a poor family in Deerfield, Massachusetts, his father a Revolutionary War veteran and a hatter by trade, with debts and three sons, who found just enough money to see his boys through primary school and some time at the local academy. After that, as Hitchcock recalled, "nothing was before me but a life of manual labor." He balked at the idea of apprenticing as a hatter, to his father, or in any other trade. Instead he worked on a farm—it was rented land, cropped by one of his brothers—for a period that stretched on so long, or what felt like so long, that later he claimed not to remember how many years. With his free time, especially rainy days and evenings, young Edward studied science and the classics. Ambitious and hungry, he thought he was preparing himself for Harvard. Under the influence of an uncle, he took up astronomy. Then came the great comet of 1811, a celestial passerby that reached its peak of brightness in the north sky during autumn that year, when Hitchcock was eighteen. He borrowed some instruments from Deerfield Academy and spent night after night measuring its progress. "I gave myself to this labor so assiduously that my health failed," he wrote later.

The health crisis brought on a religious conversion: from Unitarianism, into which he had drifted, back to the Congregationalism of his father. That passed for a drastic rethink in Edward Hitchcock's life. In lieu of Harvard, he returned to Deerfield and somehow got hired, at age twenty-three, as principal of the academy. Then he studied for the ministry, was ordained, and became pastor of a Congregationalist church in Conway, Massachusetts, just up the road from Deerfield. Throughout these years and for the rest of his life, Hitchcock remained an invalid in self-image if not bodily, obsessed with his own fragility, continually

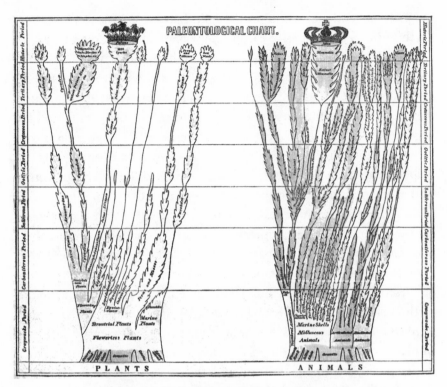

Hitchcock's "Paleontological Chart," 1857 version.

complaining that he felt death nearby, although he lived to be seventy. One scholar, having looked into his life and work, called him "a hypochondriac of the first rank."

Conveniently for his scientific career, he was "dismissed" from the Conway pastorate in autumn 1825 on the grounds of impaired health and imminent death if (according to his own worried judgment) he didn't stop preaching, circuiting the parish, and running revivals. Amherst College, recently founded, hired him to teach chemistry and natural history, and he stayed there the rest of his life, serving later as professor of natural theology and geology, and for one nine-year stretch also as president. The early years of Hitchcock's career at Amherst spanned the period when Charles Lyell, in England, published his multivolume *Principles of Geology*, a radical work that challenged Scripture-based interpretations of the geological record, including Hitchcock's own.

The conventional school of thought, known as catastrophism, saw Earth's history as a series of cataclysmic upheavals cast down like thunderbolts by the Creator, such as the bolt that brought forty days and forty nights of rain, documented in Genesis as Noah's flood. These catastrophes were considered directional and purposeful, in the sense that God used them as occasions for purging the planet of some creatures (dinosaurs, begone) and adding new creations (mammals, arise). Lyell's alternative view was uniformitarianism, insisting that the processes and events that shaped Earth in the past were physical, such as erosion and deposition, as well as the occasional volcanic eruption—things that continue to occur in the present at roughly the same rate they did in the past. Those forces caused extinctions, among other effects. Second thoughts by God about what fauna and flora should exist, according to Lyell, did not enter into it.

Hitchcock read Lyell's work promptly as the first three volumes came out, from 1830 to 1833, and found it all discomforting. He was no young-earth creationist himself; he acknowledged that volcanism and erosion were continuing processes. But he worried that Lyell's view of the planet would "exclude a Deity from its creation and government." In an article on deluges, comparing the biblical with the geological records, Hitchcock wrote cattily: "We know nothing of Mr. Lyell's religious creed. But there is something in such an ambiguous mode of treating of scriptural subjects that reminds us of infidel cunning and duplicity." Lyell was a dutiful Anglican, not an infidel, at least when he authored *Principles of Geology*, but Hitchcock seems to have sensed, maybe better than Lyell himself, that his work would nudge some readers toward godless, materialistic ideas.

One of those so nudged was Charles Darwin, who read Lyell's three volumes aboard the *Beagle* and followed their influence, not just toward uniformitarianism in geology but eventually (because Lyell described Lamarck's ideas, without endorsing them) toward a theory of evolution. So although Hitchcock was wrong about Lyell's supposed "cunning and duplicity," he was right about *Principles of Geology* taking readers—one crucial reader, anyway—onto a slippery slope.

In 1840, seven years after Lyell's third volume appeared, Hitchcock published his own *Elementary Geology*, and with it that Paleontological Chart of Lombardy poplars, included as a hand-colored, foldout figure presenting his two nonevolutionary trees of life. The chart showed changes in Earth's flora and fauna over geological time, with this or that group of plants or animals waxing or waning in diversity and abundance, but not much branching of one from another. The cause of those changes, Hitchcock explained in his text, was God's direct agency, adding and subtracting creatures, improving and perfecting the world as a long-term project. The major groups were present all along, according to this slightly tortured schema, but new species manifesting "a higher organization" had been inserted along the way, until at last Earth was ready for "more perfect" kinds of creatures, "the most generally perfect of all with man at their head." The gradual introduction of "higher races," he wrote, "is perfectly explained by the changing condition of the earth which being adapted for more perfect races Divine Wisdom introduced them." These were special creations by the Deity, appropriate as environments changed. God wasn't rethinking the planet's fauna and flora, just adjusting them to newly available niches. If that doesn't quite make sense, don't blame Charles Lyell or me.

Hitchcock's *Elementary Geology* was a hit. Between 1840 and the late 1850s, it went through thirty editions, to which he made minor revisions of language and data. Throughout all those editions, the trees figure remained—unchanged except for color adjustments. Then something happened. As a consequence of that something, or else by improbable coincidence, the thirty-first edition of Hitchcock's book, in 1860, contained a notable difference. An omission. No trees.

What happened was that in 1859 Charles Darwin published *On the Origin of Species*. His book also contained a tree, but one with dangerous new meaning.

6

By that point, Darwin had incubated his theory in secret for half a lifetime. After sketching his little tree into the B notebook in 1837, he had continued reading, gathering facts, pondering patterns, trying out phrases, brainstorming fervidly for another sixteen months in a series of such notebooks, labeled "C" and "D" and "E," like a man pushing puzzle pieces around on a table. Then suddenly, in November 1838, as recorded in the E notebook, he solved the puzzle of *how* species must evolve. Combining three pieces in his mind, he hit upon an explanatory mechanism for evolution.

The first piece was hereditary continuity. Offspring tend to resemble their parents and grandparents, providing a stable background of similarity throughout time. The second factor, a countertrend to the first, was that variation does occur. Offspring don't *precisely* resemble their parents. Brown eyes, blue eyes, taller, shorter, differences of hair color or nose shape among humans; wing markings in a butterfly, beak size in a bird, length of neck in a giraffe. Reproduction is inexact. Likewise, siblings, as well as parents and offspring, differ from one another. Darwin saw that these two pieces, heredity and variation, stand together in some sort of dynamic tension.

The third puzzle piece, which he had begun considering just recently, having been alerted to it by his eclectic reading, was that population growth always tends to outrun the available means of subsistence. Earth is always getting too full of life. One female cat may give birth to five kittens; one rabbit may deliver eight bunnies; one salmon may lay a thousand eggs. If all those offspring were to survive, and reproduce in their turns, there would soon be a very great lot of cats and bunnies and salmon. Whatever the litter size, whatever the lifetime fecundity, whatever the kind of organism, including humans, we all tend to multiply by geometric progression, not just by arithmetic increase—that is, more like 2, 4, 8, 16 than like 2, 3, 4, 5. Meanwhile, living space and food supply don't increase nearly so quickly, if at all. Habitat doesn't replicate itself. Places get crowded. Creatures go hungry. They struggle. The result is competition and deprivation and misery, winners and losers, unsuccessful efforts to breed and, for the less fortunate individuals, early death. Many are called, but few are chosen. The book that awakened Darwin to this reality was *An Essay on the Principle of Population*, by a severely logical clergyman and scholar named Thomas Malthus.

Malthus's gloomy treatise was first published in 1798. It went through six editions in the next three decades and influenced British policy on welfare. (It argued against the relatively easy charity of the contemporary Poor Laws, which were soon changed.) Darwin read it in early autumn 1838–"for amusement," as he recalled later. Seldom is amusement more productive. He came away with the population piece, combined that with his two other pieces, and scribbled an entry in his D notebook about "the warring of the species as inference from Malthus." Yes, this "warring" applied not just to humans, Darwin realized, but also to other creatures. Competition was fierce, and opportunities were finite. "One may say there is a force like a hundred thousand wedges," Darwin wrote, all trying to "force every kind of adapted structure" into the gaps in the economy of nature. "The final cause of all this wedgings," he added, "must be to sort out proper structure & adapt it to change." By "final cause," he essentially meant final result: the struggle yielded well-adapted forms. That was the essence, though still inchoate and crudely stated.

Darwin seemed to leave Malthus behind as he finished the D notebook, but returned to him soon in the next. That one, labeled E, begun in October 1838, was bound in rust-brown leather, with a metal clasp. It's one of the true relics in the history of biology. In its earlier pages, Darwin ruminated further about "the grand crush of population" and alluded repeatedly to what he now called "my theory." He was growing more confident and clear. Then, on or soon after November 27, with his usual clipped grammar and eccentric punctuation, he wrote:

Three principles, will account for all

1. Grandchildren, like, grandfathers
2. Tendency to small change . . . especially with physical change
3. Great fertility in proportion to support of parents

Inheritance, variation, overpopulation. He saw how they fit. Put those three together and turn the crank: you'll get differential survival, based on something or other. Based on what? Based on which variations turn out to be most advantageous. And those variations will tend to be inherited. The result will be gradual transmutation of heritable forms, and adaptation to circumstances, by a process of selective culling. Eventually he gave the crank a name: natural selection.

Twenty years passed after the E notebook entry. The world heard nothing about natural selection.

7

It was a perplexingly long delay, almost two decades, between the writing of those four lines in his secret E notebook and the first public announcement of Darwin's theory. Longer still, twenty-one years, to publication of the theory in book form—*On the Origin of Species* appeared in November 1859. The reasons for that delay, which were both scientific and personal, both anxious and tactical, have been minutely examined in other works (including some of mine). We can skip over them here except to note that, when Darwin finally went public with his theory, it was because a younger naturalist had forced his hand by coming forward with the same idea.

Alfred Russel Wallace, after four years of fieldwork in the Amazon and four more in the Malay Archipelago, had hit upon the notion of natural selection (framed in his own language, not that pair of words) and written it up in a short paper. As recounted by Wallace long afterward, the idea came during a layover in his collecting travels through the northern Moluccas. He suffered a bout of fever (maybe malarial), and, amidst it, he had this extraordinary insight. Variation plus overpopulation, minus the unsuccessful variants, would yield heritable adaptation. When the fever broke, and the sweat dried, and the dreamy brainstorm still seemed cogent, Wallace composed his manuscript and then tried to get it considered.

But he was a poor man's son, working his way through the tropics by selling decorative specimens—bird skins, butterflies, pretty beetles—not a gentleman traveler as Darwin had been on the *Beagle*. Wallace wasn't well educated or well connected. He knew almost nobody in the higher circles of British or European science, and almost nobody in those circles knew him—not face-to-face and not as a peer, anyway. He was a collector of dried creatures for pay, a natural-history tradesman. There was class stratification in science as in every other part of Victorian British society. But he had published a few earlier papers in a respectable journal, and one of those papers had drawn favorable attention from Charles Lyell, the great geologist. Oh, and Wallace knew one other famous man, not personally but as a sort of pen pal, who had spoken generously to him in a letter: Charles Darwin.

It was now February 1858. Hardly anyone at that point recognized Darwin for what he was—an evolutionary theorist, in secret—and though Lyell was among that small group who did, as a close friend and confidant, Alfred Wallace certainly wasn't. Charles Darwin to him was just a conventionally eminent naturalist, author of the *Beagle* chronicle and other safe books, including several on the taxonomy of barnacles. But a Dutch mail boat would soon stop at the port of Ternate, in the Moluccas, where Wallace had fetched up. He was excited by his own discovery, if it was a discovery, and eager to share this dangerous hypothesis with the scientific world. So he packed up his paper with a cover letter and mailed the packet to Mr. Darwin, hoping that Darwin might find it worthy. If so, maybe Darwin would share it with Mr. Lyell, who might help get it published.

The packet reached Darwin, probably on June 18, 1858, and hit him like a galloping ox. He felt crushed, scooped, ruined—but also honor-bound to grant Wallace's request, passing the paper on toward publication. It would mean, Darwin knew, letting the younger man take all the credit for this epochal idea he himself had incubated for twenty years but was not yet quite ready to publish. Despite that, he did send the Wallace paper along to Lyell—communicating yelps of his own anguish along with it. Lyell took not just the paper but also the hint. Along with another of Darwin's close scientific allies, the botanist Joseph Hooker, Lyell talked Darwin back from despair, suggested a posture of sensible fairness rather than self-abnegating honor, and brokered a compromise of shared credit. The result was a clumsily conjoined presentation—a pastiche of Wallace's paper plus excerpts from Darwin's unpublished writings—before a British scientific club, the Linnean Society, in the summer of 1858. Lyell and Hooker offered an introductory note, and then simply watched and listened. Proxies read the works aloud, with neither of the authors present. (Darwin was at home, where his youngest son had just died of scarlet fever; Wallace was still out in the far boonies of the Malay Archipelago.) This joint presentation made almost no impression on anyone, not even the few dozen Linnean members in attendance, because the night was hot, the language was obscure, the logic was elliptical, and the big meaning didn't jump forth.

Seventeen months later, Darwin published *On the Origin of Species*. That 1859 book, not the 1858 paper or excerpts, launched the Darwinian revolution. It was only an abridged and hasty abstract of the much longer (and more tedious) book on natural selection that Darwin had been writing for years, but *The Origin* was just enough, in the right form, at the right time. It presented the theory as "one long argument," not just a bare syllogism, and with oodles of data but not many footnotes. It was plainspoken, and readable by any literate person. It became a bestseller and went into multiple editions. It converted a generation of scientists to the idea of evolution (though not to natural selection as the prime mechanism). It was translated and embraced in other countries, especially Germany. That's why Darwin is still history's most venerated biologist and Alfred Russel Wallace is a cherished underdog, famous for being eclipsed, to the relatively small subset of people who have heard of him.

The crux of the "one long argument" comes in chapter 4 of *The Origin*, titled "Natural Selection," in which Darwin describes the central mechanism of his theory. It's the same combination of three principles that he had scratched into his notebook two decades earlier, plus the turned crank. "Natural selection," he wrote in the book, "leads to divergence of character and to much extinction of the less improved and intermediate forms of life." Lineages change over time, he stated. You could see that in the fossil record. Different creatures adapt to different niches, different ways of life, and thereby diversify into distinct forms and behaviors. Transitional stages disappear. Then he wrote: "The affinities of all the beings of the same class have sometimes been represented by a great tree. I believe this simile largely speaks the truth."

8

Darwin explored the tree simile in one extended paragraph, ending that chapter of *The Origin*. "The green and budding twigs may represent existing species," he wrote. From there he worked backward: woody twigs and small branches as recently extinct forms; competition between branches for space and for light; big limbs dividing into branches, then those into lesser branches; all ascending and spreading from a single great trunk. "As buds give rise by growth to fresh buds," Darwin wrote, and those buds grow to be twigs, and those twigs grow to be branches, some vigorous, some feeble, some thriving, some dying, "so by generation I believe it has been with the great Tree of Life, which fills with its dead and broken branches the crust of the earth, and covers the surface with its ever branching and beautiful ramifications." There's a nice word: *ramifications*.

It's especially good in this context because, while the literal definition is "a structure formed of branches," from the Latin *ramus*, of course the looser definition is "implications." Darwin's tree certainly had implications.

Furthermore his book, like Edward Hitchcock's, included a treelike illustration. This was the *only* illustration, the only graphic image of any sort, in the first edition of *The Origin*. It appeared between pages 116 and 117, amid his discussion of how lineages diverge over time. A foldout, again like Hitchcock's, but published in simple black and white. It was a schematic figure, not an artfully drawn tree, not even so lively as the little sketch in his notebook long ago. Darwin called it a diagram. It showed hypothetical lineages, proceeding upward through evolutionary time and diverging—that is, dotted lines, rising vertically and branching laterally. Darwin was no artist, but, even lacking such talent, he could have laid out this diagram with a pencil and a ruler. In its draft version, as sent to the lithographer, he probably had. But it made the arboreal point.

Each increment of vertical distance on the ruled page, Darwin explained, stood for a thousand generations of inheritance. Deep time. Eleven major lineages began the ascent. Eight of those came to dead ends—meaning, they went extinct. Trilobites, ammonites, ichthyosaurs, and plesiosaurs had all suffered such ends, leaving no descendants of any sort. One lineage rose through the eons without splitting, without tilting, like a beanstalk—meaning that it persisted through time, unchanged. That's much the way horseshoe crabs, sometimes called living fossils, have survived relatively unchanged (at least externally, so far as fossilization can show) over 450 million years. The other two lineages, dominating the

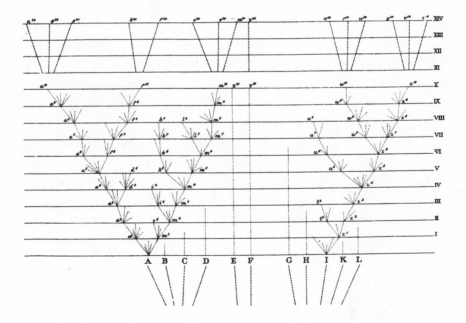

Darwin's diagram of divergence, from *On the Origin of Species*, 1859.

diagram, branched often and spread horizontally—as well as climbed vertically. Their branching and horizontal spread represented the exploration of different niches by newly evolved forms. So there it all was: evolution and the origins of diversity.

Back in Massachusetts, Edward Hitchcock read Darwin's book, and it stuck in his craw. This wasn't his first exposure to the idea of transmutation (he knew of Lamarck's work and some other wild speculations), but it was the latest statement of that idea, the most concrete and logical, and therefore the most dangerously persuasive. Like some other pious scientists who chose to see God's hand acting directly in the fossil record—Louis Agassiz at Harvard, François Jules Pictet in Geneva, and Adam Sedgwick, who had been Darwin's mentor in geology at Cambridge—Hitchcock wasn't pleased.

Into the 1860 edition of his *Elementary Geology*, he inserted his rejoinder to Darwin's book, based mainly on proof by authority. He noted that Pictet saw no evidence for transmutation in the fossil record of fishes. Agassiz said that the resemblances among animals derive from—where?—the mind of the Creator. "It is well to take heed to the opinions of such masters in science," Hitchcock wrote, "when so many, with Darwin at their head, are inclined to adopt the doctrine of gradual transmutation in species."

That was mild but firm, a dismissive shrug. Hitchcock would ignore Charles Darwin and encourage his readers to do likewise. More telling, more defensive, was his other response: he removed the trees figure from his own book. No more Paleontological Chart. It seems never to have appeared in another edition of *Elementary Geology*.

Darwin and Darwin's followers owned the tree image now. It would remain the best graphic representation of life's history, evolution through time, the origins of diversity and adaptation, until the late twentieth century. And then rather suddenly a small group of scientists would discover: oops, no, it's wrong.

PART * II

A Separate Form of Life

9

~~~~ ~~~

Molecular phylogenetics, the study of evolutionary relatedness using molecules as evidence, began with a suggestion by Francis Crick, in 1958, offered passingly in an important paper devoted to something else. That was characteristic of Crick—so brilliant and recklessly imaginative that he sometimes influenced the course of biology even with his elbows.

You know Crick's name from the most famous triumph of his life: solving the structure of the DNA molecule, with his young American partner James Watson, in 1953, for which he and Watson and one other scientist would eventually, in 1962, receive the Nobel Prize. Crick wasn't wasting his time, in 1958, mooning about dreams of glory in Stockholm. He was still interested in DNA, but he had moved on from the sheer structural question to other big problems. He had bent his mind intensely, but with his usual sense of merry play, to the challenge of deciphering the genetic code.

The code, as you've heard many times but might need reminding, is written in an alphabet of four letters, each letter representing a component—a nucleotide base, in chemistry lingo—of the DNA double helix. The four letters are: A (for adenine), C (cytosine), G (guanine), and T (thymine). DNA's full moniker is deoxyribonucleic acid, of course, and it's worth understanding why. The two helical strands of the double helix, twining around a central axis in parallel with each other, are composed of units called nucleotides, linked in a chain, each nucleotide containing a base (that's the A, C, G, or T), a sugar (that's the deoxyribose), and a phosphate group (that's the acidic part). The sugar end of one nucleotide bonds to the phosphate end of the next, forming the two long helical strands. I just called them parallel, but to be more precise, those strands are *anti*parallel to each other, since the sugar-phosphate binding gives them directionality—a front end and a back end—and the front end of one strand aligns with the back end of the other. The nucleotide bases, linked crossways by hydrogen bonds, hold the strands together. The base A pairs with T, the base C pairs with G, forming a stable structure, like the steps in a spiral staircase. This is the nifty arrangement that Watson and Crick deduced.

It's not *just* a stable structure, though. It's a wondrously efficient one for storing, copying, and applying heritable data. When the two strands are peeled apart, the sequence of bases along one of the strands (the template strand) represents genetic information ready to be duplicated or used. Watson and Crick noted that capacity with exquisite coyness in their 1953 paper. The paper was lapidary, only a page long, as published in the journal *Nature*, and included a sketch. Near the end, having proposed their double helix structure and the matchup of bases, always A with T and C with G, they wrote: "It has not escaped our notice that the specific pairing we have postulated immediately suggests a possible copying mechanism for the genetic material."

But copying that material, for hereditary continuity, was one thing. Translating it into living organisms was another. Translated how? By what steps does the information in DNA become physically animate?

This mystery leads first to proteins. There are four kinds of molecule essential to living processes—carbohydrates, lipids, nucleic acids, and proteins—often collectively called the molecules of life. Proteins might be the most versatile, serving a wide range of structural, catalyzing, and transporting functions. Their piecemeal production, and the controls on the process of building and using them, are encoded in DNA. Every protein consists of a linear chain of amino acids, folded upon itself into an elaborate secondary structure. Although about five hundred amino acids are known to chemistry, only twenty of those serve as the fundamental components of life, from which virtually all proteins are assembled. But what sequences of the four bases determine which amino acids shall be added to a chain? What combination of letters specifies leucine? What combination produces cysteine? What arrangement of A, C, G, and T delivers its meaning as glutamine? What spells tyrosine? This fundamental matter—how do bases designate aminos?—became known as "the coding problem," to which Francis Crick addressed himself in the late 1950s. Solving it was a crucial step toward understanding how organisms grow, live, and replicate.

There were questions within questions. Do the bases work in combinations? If so, how many? Two-base clusters, selected variously from the group of four and in specified order (CT, CG, AA, and so on) would allow only sixteen combinations, not enough to code twenty amino acids. Then maybe clusters of three or more? If three (such as CTC, CGA, AAA), do those triplets overlap one another, or do they function separately, like three-letter words divided by commas? If there are commas, are there periods too? Four letters, in every possible combination of three, yield sixty-four variants. Are all sixty-four possible triplets used? If so, that implies some redundancy; different triplets coding for the same amino acid. Does the code include a way of saying "Stop"? If not, where does one gene end and another begin? Crick and others were keen to know.

Crick himself had also started thinking beyond that problem, to the question of how proteins are physically assembled from the coded information, with one amino acid brought into line after another. How does the template strand find or attract its amino acids? How do those units become linked? He wanted to learn not just the language of life—its letters, words, grammar—but also the mechanics of how it gets spoken: its equivalent of lungs, larynx, lips, and tongue.

Crick was back in England by the mid-1950s, after a sojourn in the United States, and based again at the Cavendish Laboratory in Cambridge, where he had worked with Jim Watson. He had a contract with the Medical Research Council (MRC), a government agency with some mandate for fundamental as well as medical research. Solving the DNA structure, though it had brought scientific fame to Crick and Watson and would eventually bring the Nobel Prize, provided no immediate cure for Crick's dicey financial situation, all the more acute since the birth of his and his wife Odile's third child. He had to work for pay: a modest salary from the

MRC and whatever small change the occasional radio broadcast or popular article might bring. Now he was sharing his office, his pub lunches, his fevered conversations, and his blackboard with another scientist, Sydney Brenner, rather than with Watson. One colleague at the Cavendish, upon early acquaintance with Crick, concluded that "his method of working was to talk loudly all the time." When not talking, or listening to Brenner, he spent his time reading scientific papers, rethinking the results of other researchers, combing through such bodies of knowledge for clues to the mysteries that engaged him. He was not an experimentalist, generating data. He was a theoretician—probably the century's best and most intuitive in the biological sciences.

Sometime in 1957 Crick gathered his thoughts and his informed guesses on this problem—about how DNA gets translated into proteins—and in September he addressed the annual symposium of the Society for Experimental Biology, convened that year at University College London. His talk "commanded the meeting," according to one historian, and "permanently altered the logic of biology." The published version appeared a year later, in the society's journal, under the simple title "On Protein Synthesis." Another historian, Matt Ridley, in his short biography of Crick, called it "probably his most remarkable paper," comparable to Isaac Newton's *Principia* and Ludwig Wittgenstein's *Tractatus*. It was a commanding presentation of insights and speculations about how proteins are built from DNA instructions. It noted the important but still-fuzzy hypothesis that RNA (ribonucleic acid), the *other* nucleic acid, which seemed to exist in DNA's shadow, is somehow involved. Might RNA play a role in manufacturing proteins, possibly by helping express the order (coded by DNA) in which amino acids are linked one to another? Amid such ruminations, Crick threw off another idea, almost parenthetically: ah, by the way, these long molecules could also provide evidence for evolutionary trees.

As published in the paper: "Biologists should realize that before long we shall have a subject which might be called 'protein taxonomy'—the study of the amino acid sequences of the proteins of an organism and the comparison of them between species."

He didn't use the words "molecular phylogenetics," but that's what he was getting at: deducing evolutionary histories from the evidence of long molecules. Comparing slightly different versions of essentially the same protein (such as hemoglobin, which transports oxygen through the blood of vertebrates), as found in one creature and another, could allow you to draw inferences about degrees of relatedness between them. Those inferences would be based on assuming that the variant hemoglobins had evolved from a common ancestral molecule and that, over time, in divergent lineages, small differences in the amino sequences would have crept in, by accident if not by selective advantage. The degree of such differences between one hemoglobin and another should correlate with the amount of time elapsed since those lineages diverged. From such data, Crick suggested, you might draw phylogenetic trees. Humans have one variant of hemoglobin, horses have another. How different? How long since we shared an ancestor with horses? It could be argued, Crick added, that protein sequences also represent the most precise observable register of the physical identity of an organism, and that "vast amounts of evolutionary information may be hidden away within them."

Having tossed off this fertile suggestion, Crick returned in the rest of the paper to his real subject: how proteins are manufactured in cells. That was his way. A passing thought, with the heft of a beer truck. Essentially he had said: Look, *I'm* not pursuing this protein taxonomy business, but *somebody* should.

10

Somebody did, though not immediately. Seven years passed, during which several other scientists began noodling along various routes that would lead to a similar idea. Two of them were Linus Pauling and Emile Zuckerkandl, who gave their own fancy name to the enterprise—they called it "chemical paleogenetics"—and they converged on it by very different trajectories.

Zuckerkandl was a young Viennese biologist whose family had escaped Nazi Europe via Paris and Algiers. He got to America, did a master's degree at the University of Illinois (long before Carl Woese would arrive there), then returned to Paris after the war for a doctorate. He found work at a marine laboratory on the west coast of France and studied the molting cycles of crabs, which involve a molecule analogous to hemoglobin. His interest drifted from crustacean physiology to questions at the molecular level, and he hankered to return to America. In 1957 Zuckerkandl finagled a chance to meet Pauling, who by then was a celebrated chemist with the first of his two Nobel Prizes already won. The prize had given Pauling some latitude to expand his own range of concerns, from lab chemistry at the California Institute of Technology to the wider world, and some leverage in pursuing those concerns. He had two in particular: genetic diseases such as sickle cell anemia and the threats posed by thermonuclear weapons, including radioactive fallout from testing. By the late 1950s, Pauling was raising his voice. He initiated a petition against atmospheric nuclear testing that more than eleven thousand scientists signed. He had become, along with Bertrand Russell, the provocative British philosopher, also a Nobel winner, one of the world's most august peaceniks.

Pauling's initial encounter with Zuckerkandl coincided with his increasing interest in genetics, evolution, and mutation—most pointedly, the mutations that might be caused by radiation released in weapons tests. His interest in disease led in the same direction, because sickle cell anemia is a problem that results from mutations in one of the genes for hemoglobin. Pauling found Zuckerkandl impressive enough that he offered the younger man a postdoctoral fellowship in chemistry at Caltech. Then, when Zuckerkandl arrived in Pasadena, intending to continue work on the crab-molting molecule, Pauling discouraged that project and said, "Why don't you work on hemoglobin?"

Pauling suggested further that he take up a newly invented technique—still primitive but promising—that employed electrophoresis (separating molecules by their sizes, using electrical charge) and other methods to "fingerprint" such proteins, distinguishing one variant from another. Comparing protein molecules that way, Pauling figured, might allow researchers to draw some evolutionary conclusions. So Zuckerkandl went to work, learning the technique and applying it to hemoglobin in variant forms. Before long, he could see the close similarity between human hemoglobin and chimpanzee hemoglobin, and that human hemoglobin was less similar to hemoglobin found in orangutans. He could also tell a pig from a shark just by looking at the molecular fingerprints. Of course, there were easier ways to tell a pig from a shark, but never mind. Although it wasn't such a precise methodology as he might have wished, this sort of molecular comparison was a start.

Over the next half dozen years, Zuckerkandl's work thrived, and he published a series of papers with Pauling. Some of those were invited contributions to celebratory volumes, *Festschriften*, in honor of eminent scientists, generally on some occasion such as retirement or a big, round birthday. Such invitations came often because of Pauling's own eminence, and he recruited Zuckerkandl as coauthor to do much of the thinking and most of the writing. In the meantime, Pauling won his second Nobel, this time the Peace Prize in recognition of his efforts against nuclear weapons proliferation and testing. That one didn't add to his scientific reputation (in fact, he resigned from his Caltech professorship because university administrators and trustees disapproved of his peace activism), but it certainly helped amplify his public voice. He was a busy man, much in demand. The invitations—to speak, to visit, to contribute scientific papers for ceremonial volumes—continued. Because such papers didn't normally go through the peer-review filter, they could be a little more bold and speculative than a typical journal article. One of them, written in 1963 to honor a Russian scientist on his seventieth birthday, was titled "Molecules as Documents of Evolutionary History." Two years later, it was reprinted in English in the *Journal of Theoretical Biology*, giving it much broader reach and influence. Pauling and Zuckerkandl were wading into the same pond where Francis Crick had dipped his toe.

Their 1963 paper made an important distinction between molecules that carry genetic information—such as DNA or the proteins it encodes—and other molecules, such as vitamins, that cycle through a living creature and out the other end. Information molecules have histories that can be deduced; they have ancestors from which the variant forms, in this creature or that, have descended. Scrutiny of such molecules, wrote Zuckerkandl and Pauling, can tell us three things: how much time has passed since the lineages split, what the ancestral molecules must have looked like, and what were the lines of descent. The first of those three kinds of information became known as the molecular clock, although Zuckerkandl and Pauling hadn't yet named it. The third kind implied trees.

Zuckerkandl continued reworking and developing these ideas, with Pauling as his coauthor and sponsor. In September 1964, before a distinguished and argumentative symposium audience at Rutgers University, he delivered a long paper that became the definitive version of their shared ideas and that, despite Zuckerkandl having done most of the writing, has been called the "most influential of Pauling's later career." In this paper, the two authors offered their memorable metaphor: if the minor changes in molecular variants are proportional to elapsed time over the eons, they said, what you have is "a molecular evolutionary clock."

It was tentative, a hypothesis. The hypothesis was disputed at the Rutgers symposium and would be controversial in coming years, but it captured attention, it focused thought, and it promised a whole new way of measuring life's history, if it was right. The molecular clock has since been called "one of the simplest and most powerful concepts in the field of evolution," and also "one of the most contentious." Crick himself later judged it "a very important idea" that turned out to be "much truer than people thought at the time."

Emile Zuckerkandl, meanwhile, moved back to France. Along with Pauling and just a few others, he had helped launch a new scientific enterprise, and when a *Journal of Molecular Evolution* came into being, in 1971, he was its first editor in chief. His name isn't familiar to the wider world, as Pauling's is, but if you say "Zuckerkandl and Pauling" to a molecular biologist today, he or she will think "molecular clock." Fitting as that may be, it overlooks the other important point: the other metaphor embedded in the long Rutgers paper, where Zuckerkandl wrote that "branching of molecular phylogenetic trees should in principle be definable in terms of molecular information alone." This was a whole new way of sketching those trees, which rose and spread their branches as the clock ticked.

11

C arl Woese came to the University of Illinois, in Urbana, in 1964, the same year Zuckerkandl delivered the paper at Rutgers. The enterprise that would become molecular phylogenetics—back then bruited under other names, such as Crick's protein taxonomy, and Pauling and Zuckerkandl's chemical paleogenetics—had begun to attract interest. Woese saw its deepest possibilities more clearly than anyone else. Molecular sequence information, he realized, could be used to read the shape of the past.

Woese was thirty-six years old and was hired with immediate tenure, which gave him some latitude to undertake risky, laborious research projects without need to worry about quick publications. His professorship was in the Department of Microbiology, though he had trained as a biophysicist, not a microbiologist, and had spent little time if any peering through microscopes at bacteria and other tiny bugs. He was more interested in molecular biology, then still in its early phase. It was a thrilling new branch of science, its methods just being invented, its cardinal principles just taking shape, and he wanted to be part of that. But the molecular clock wasn't Woese's topic, and the prospect of a molecular tree of life hadn't yet captured his imagination. He was focused instead on the genetic code—and not just what he called the cryptographic aspect:

the matter of which bases in which combinations specified which amino acids for building proteins. He wanted to go deeper in time and meaning; he wanted to understand how the code had evolved.

He was well aware that Francis Crick and others, including the eclectic Russian physicist George Gamow, had been working on the cryptographic aspect as a theoretical problem, treating it like an abstract intellectual game. That problem had been illuminated, but not solved, since Crick's 1958 paper, by a new recognition of RNA's role, as a messenger molecule somehow carrying DNA instructions to the site in a cell where proteins are built. But what was the structure of RNA, and how did it play that role? Gamow and the others were puzzled, and to them the puzzle was a thrilling game. They had even formed an elite, semifacetious little club—limited to twenty members, reflecting the twenty amino acids of life—for the private exchange of ideas about how coding and protein synthesis might work. They called it the RNA Tie Club—*RNA* because that molecule was still the mysterious intermediary; *Tie* because such neckwear evoked, and mocked, the clubby bond of an old school tie. As tokens of club membership, these scientists had embroidered neckties, all alike. They had individual tiepins, each representing one amino acid. They embraced their respective amino identities, at least jocularly: Serine and Lysine and Arginine, etcetera. Cute. Woese wasn't a member.

The cryptographic riddle, so intriguing to Gamow and Crick and the others, was this: How could the four bases of DNA—represented by those four cardinal letters, A, C, G, and T—be combined in groups of at least three, with or without commas, to produce the twenty different amino acids? Woese addressed it alone. He knew that a team led by Marshall Nirenberg, a young biochemist at the US National Institutes of Health, had made better progress with an experimental approach than the RNA Tie Club was making with collegial theorizing. But he wanted to go deeper.

"I differed from the whole lot of them," Woese wrote decades later, "in perceiving the nature of the code as inseparable from the problem of the nature and origin of the decoding mechanism." The decoding mechanism? By that, he meant whatever organ or molecule translated the DNA information into real, physical proteins. Its origin? To him, at that time, this was *the* central biological concern. He wanted to understand not just how that decoding mechanism worked but also how it had come into being roughly four billion years ago. He recognized, more clearly than anyone else, that life could not have progressed beyond its simplest primordial forms without a translation system for applying the information in DNA.

No statement from Woese is more telling of his character, his cantankerous self-image as a scientific outsider, than the beginning of that sentence just quoted: "I differed from the whole lot of them . . ." He was a loner by disposition. He took a separate path. Not in the club. No RNA tie. He published a few papers in *Nature* on the coding question, and a comment in *Science*—all under his sole authorship, suggesting ideas, critiquing what others had done. He offered his own view in full, an evolutionary view, in a 1967 book, *The Genetic Code*, which was visionary, ambitious, closely reasoned, and mostly wrong. But in science, wrong doesn't mean useless. Trying to imagine the origins of the genetic code brought Woese around, almost reluctantly, to the tree of life.

He needed some such universal diagram, Woese realized, as a framework for understanding the evolution of that one crucial system at life's core—the translation system, turning DNA-coded information into proteins. Deep biology required deep history. This conundrum has been nicely expressed by Jan Sapp, a plant geneticist who became a historian of biology and came to know Woese well: "A universal tree would therefore hold the secret to its own existence." History illuminating biology and vice versa. Evolutionary biology *is* history, after all. But there was a problem. For microbiology—bacteria and other single-celled creatures—a tree didn't exist. The known trees didn't encompass such organisms, or portray their diversity, to any satisfactory extent. Animals could be compared with one another on the basis of their physical appearance and behavior, as Linnaeus and Darwin had compared them; plants could be compared; fungi could be compared. They could be arranged in treelike patterns that reflected their relationships as deduced from such external, visible evidence. But that was impossible with microbes because, even under a high-powered microscope, so many of them looked so much alike.

There were a few basic shapes—rods, spheres, filaments, spirals—and those had served (reliably or not) to define major groups of bacteria. But at the finer level, the level of what we would think of as species, bacterial classification into a natural system, showing evolutionary relationships, was difficult. Arguably impossible. Even some of the experts had given up. It couldn't be done on the basis of appearance and behavior. It couldn't be done by way of physiological characteristics (which, in microbes, are what pass for behavior). It couldn't be done at all, unless someone invented a new method.

"A slight diversion in my research program would be necessary," Woese recollected later—a wry comment, because by then the diversion had lasted two decades.

12

On June 24, 1969, Woese in Urbana wrote a revealing letter to Francis Crick in Cambridge. He had struck up an acquaintance with Crick about eight years earlier when Woese was an obscure young biologist at the General Electric Research Laboratory in Schenectady, New York, and Crick was already world renowned for the DNA structure discovery. It had begun as a tenuous exchange of courtesies, through the mail—Woese requesting, and receiving, a reprint of one of Crick's papers on coding—but by 1969, they were friendly enough that he could be more personal and ask a larger favor. "Dear Francis," he wrote, "I'm about to make what for me is an important and nearly irreversible decision," adding that he would be grateful for Crick's thoughts and his moral support.

What he hoped to do, Woese confided, was to "unravel the course of events" leading to the origin of the simplest cells—the cells that microbiologists called prokaryotes, by which they meant bacteria. Eukaryotes constituted the other big category, the other domain, and all forms of cellular life (that is, not including viruses) were classified as one or the other. Prokaryotes (*pro* being the Greek for "before," *karyon* the Greek for "nut" or "kernel") are cells without nuclei. Eukaryotes (*eu* for "true") are the more complicated creatures, including multicellular animals, and plants, and fungi, plus certain single-celled but complex organisms such as amoebae, whose cells contain nuclei (hence the name, meaning "true kernel"). Prokaryotes ("before kernel") seem to have existed on Earth before eukaryotes. Although bacteria are still around and still vastly successful, dominating many parts of the planet, they were thought in 1969 to be the closest living approximations of early life-forms. Investigating their origins, Woese told Crick, would require extending the current understanding of evolution "backward in time by a billion years or so," to that point when cellular life was just taking shape from . . . something else, something unknown and precellular.

Oh, just a billion years further back? Woese was always an ambitious thinker. "There is a possibility, though not a certainty," Woese told Crick, "that this can be done using the cell's 'internal fossil record.'" What he meant by that term was merely the evidence of long molecules, the linear sequences of units in DNA, RNA, and proteins. Comparing such sequences—variations on the same molecule, as seen in different creatures—would allow him to deduce the "ancient ancestor sequences" from which those molecules had diverged, in one lineage and another. And from such deductions, such ancestral forms, Woese hoped to glean some understanding of how creatures had evolved in the very deep past. He was talking about molecular phylogenetics, still without using that phrase, and he hoped by this technique to look back at least three billion years.

But which molecules would be the most telling? Which would represent the best internal fossil record? Frederick Sanger, a humble but visionary biochemist in England, had sequenced the amino acids of bovine insulin, and insulins are a fairly old family of molecules in animals and other eukaryotes, but they don't go back nearly as far as Woese wanted. Other scientists had sequenced a protein called cytochrome c, also crucial in cell biochemistry among many creatures. But those didn't satisfy Woese. He wanted something more basic, more universal—something that went *all* the way back, or nearly all the way, to the beginnings of life.

"The obvious choice of molecules here lies in the components of the translation apparatus," he told Crick. "What more ancient lineages are there?" By "translation apparatus," Woese meant the decoding mechanism, the system that turns DNA information into proteins—the same system that Crick had groped toward understanding in his 1958 paper "On Protein Synthesis." Investigating the translation apparatus would in turn bring Woese around toward his starting point: his desire to learn how the genetic code itself might have evolved. Now, eleven years after Crick's protein paper, the system was much better understood.

The components Woese had in mind were pieces of a tiny molecular mechanism common to all forms of cellular life. It's called the ribosome. Nearly every cell contains ribosomes in abundance, like flakes of pepper in a stew, and they stay busy with the task of translating genetic information into proteins. Hemoglobin, for instance, that crucial oxygen-transporting protein. Architectural instructions for building hemoglobin molecules are encoded in the DNA, but where is hemoglobin actually produced? In the ribosomes. They are the core elements of what Woese called the translation apparatus.

Crick hadn't used that phrase, "translation apparatus," in his paper. He hadn't even used the word *ribosomes,* but he touched upon them vaguely under their previous name, *microsomal particles.* These particles had only recently been discovered (in 1956, by a Romanian cell biologist using an electron microscope) and at first no one knew what they did. Then they became recognized as the sites where proteins are built, but a big question remained: how? Some researchers suspected that ribosomes might actually *contain* the recipes for proteins, extruding them as an almost autonomous process. That notion collapsed in 1960, almost with a single flash of insight, when Crick's brilliant colleague Sydney Brenner, during a lively meeting at Cambridge University, hit upon a better idea. Matt Ridley has described the moment in his biography of Crick:

> Then suddenly Brenner let out a "yelp." He began talking fast. Crick began talking back just as fast. Everybody else in the room watched in amazement. Brenner had seen the answer, and Crick had seen him see it. The ribosome did not contain the recipe for the protein; it was a tape reader. It could make any protein so long as it was fed the right tape of "messenger" RNA.

This was back in the days before digital recording, remember, when sound was recorded on magnetic tape. The "tape" in Brenner's metaphor was a strand of RNA—that particular sort called messenger RNA (one of several forms of RNA that perform various functions) because it carries messages from the cell's DNA genome to the ribosomes. A ribosome consists of two subunits, one large, one small, fitted together and performing complementary functions. The small subunit reads the RNA message. The large subunit uses that information to join the appropriate amino acids into a chain, constituting the protein. The ribosomes and the messenger RNA, plus a few other pieces, constitute what Woese called the translation apparatus. By 1969, when Woese wrote to Crick, their crucial roles were appreciated.

Every living cell, including bacteria, including the cells of our own bodies, including those of plants and of fungi and of every other cellular organism, contains many ribosomes. They function as assembly mechanisms, taking in genetic information, plus raw material in the form of amino acids, and producing those larger physical products: proteins. In plainer words: ribosomes turn genes into living bodies. Because the proteins they produce become three-dimensional molecules, a better metaphor than Brenner's tape-reader, for our own day, might be this: the ribosome is a 3-D printer.

Ribosomes are among the smallest of identifiable structures within a cell, but what they lack in size they make up for in abundance and consequence. A single mammalian cell might contain as many as ten million ribosomes; a single cell of the bacterium *Escherichia coli*, better know as *E. coli*, might get by with just tens of thousands. Each ribosome might crank out protein at the rate of two hundred amino acids per minute, altogether producing a sizzle of constructive activity within the cell. And this activity, because it's so basic to life itself, life in all forms, has presumably been going on for almost four billion years. Few people, in 1969, saw the implications of that ancient, universal role of ribosomes more keenly than Carl Woese. What he saw was that these little flecks—or some molecule within them—might contain evidence about how life worked, and how it diversified, at the very beginning.

Another of Woese's penetrating insights, back at this early moment, was to focus on a particular portion of ribosomes: their *structural* RNA. Usually we think of RNA in the role I mentioned above—as an information-bearing molecule, single stranded rather than double helical like DNA, carrying the coded genetic instructions to the ribosomes for application. Transient in space (through the cell) and transient in time (used and discarded). But that's only one kind of RNA, messenger RNA, performing one function. There's more. RNA can serve as a building block as well as a message. Ribosomes, for instance, are composed of structural RNA molecules and proteins, just as an espresso machine might be made of both steel and plastic. "I feel," Woese confided to Crick in the letter, "that the RNA components of the machine hold more promise than (most of the) protein components." Those RNA components held more promise for plumbing deep history, he reckoned, because they were so old and, probably, so little changed over time.

Woese saw the secret truth that RNA—not just a molecule, but a family of versatile, complex, underappreciated molecules—is really more interesting and dynamic than its famed counterpart, DNA. And this is where that family enters the story and begins taking its position near the center. Woese had decided he would use ribosomal RNA as the ultimate molecular fossil record.

"What I propose to do is not elegant science by my definition," he confided to Crick. Scientific elegance lay in generating the minimum of data needed to answer a question. His approach would be more of a slog.

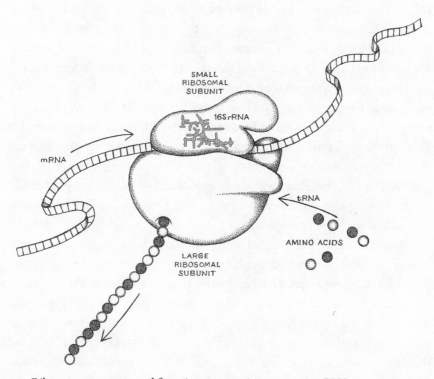

Ribosome structure and function: converting messenger RNA to protein.

He would need a large laboratory set up for reading at least portions of the ribosomal RNA. That itself was a stretch at the time. (The sequencing of very long molecules—DNA, RNA, or proteins—is so easily done nowadays, so elegantly automated, that we can scarcely appreciate the challenge Woese faced. Work that would eventually take him and his lab members arduous months, during the early 1970s, can now be done by a smart undergraduate, using expensive machines, in an afternoon.) Back in 1969, Woese couldn't hope to sequence the entirety of a long molecule, let alone a whole genome. He could expect only glimpses—short excerpts, read from fragments of ribosomal RNA molecules—and even that much could be achieved only laboriously, clumsily, at great cost in time and effort. He planned to sequence what he could from one creature and another and then make comparisons, working backward to an inferred view of life in its earliest forms and dynamics. Ribosomal RNA would be his rabbit hole to the beginning of evolution.

Gearing up the laboratory would be step one. Given his low level of administrative skill, he admitted to Crick, that much would be difficult. But besides lab equipment and money and administration, Woese perceived one other necessity. "Here is where I'd be particularly grateful for your advice and help," he told Crick. He hoped to enlist "some energetic young product of Fred Sanger's lab, whose scientific capacities complement mine." By that, he meant: for this great sequencing effort, Woese would need a helper who knew how to sequence.

13

Fred Sanger's pioneering work was the standard at that time for efforts at sequencing RNA. Building on ideas from earlier researchers, Sanger had developed techniques for cutting a long molecule into short pieces, then separating those pieces by electrophoresis, pulling them apart within a column of gel. The gel column served as a racetrack for fragments of different sizes. With an electrical force applied, each fragment would be attracted toward one end and would migrate through the gel at its own speed, dependent on its molecular size and its electrical charge. As their differing speeds spread them apart, the fragments would show as a characteristic oval spot in a two-dimensional pattern, as captured on film. Each oval could then be read as a short squib of code, using other means of cutting and pulling. This was an advance on the same general method that Pauling had recommended to Zuckerkandl for distinguishing variant forms of a molecule by "fingerprinting."

Fred Sanger had two things, but perhaps not much else, in common with Linus Pauling: chemistry and a pair of Nobel Prizes. Unlike Pauling, he was a quiet, unassuming man, from a Quaker upbringing in the English Midlands, who won both his Nobels in the branch of science he and Pauling shared—he was the only person awarded twice for chemistry. He received the first prize in 1958, at age forty, for work on the molecular structure of a protein—specifically, bovine insulin. To solve that structure, Sanger adapted some relatively primitive methods from other researchers, in an ingenious way, allowing him to determine which sequences of amino acids compose the two long branches of the insulin molecule. This was a Nobel-worthy achievement for what it said not just about blood-sugar regulation in cows but also about proteins in general: that they're not amorphous things but have, each protein, a determined chemical composition. From proteins, Sanger turned to sequencing RNA, then DNA, and won his second Nobel in 1980 for the culminating phase of his DNA work. Soon after, at age sixty-five, he retired from science and turned his energies to gardening. He had a nice little home in a village near Cambridge.

"My work had sort of come to a climax," he said later, and he didn't care to morph into an administrator. He declined a knighthood, having no desire to be addressed as "Sir Fred" by friends and strangers. "A knighthood makes you different, doesn't it," he said, "and I don't want to be different." But that Cincinnatus retirement lay long in the future when Carl Woese, in his 1969 letter to Crick, daydreamed of getting a Sanger protégé to help him.

One of Sanger's grad students had already come to Urbana, in fact, as a postdoc in the lab of another scientist within Woese's department. That postdoc was David Bishop, brought over to assist Sol Spiegelman in sequencing viral RNA. Spiegelman had recruited Woese to the University of Illinois, rescuing him from obscurity at General Electric, in 1964. One year after Bishop's arrival, Spiegelman left Illinois, returning to Columbia University in New York City, where his career had begun, and eventually taking Bishop with him. That might have yanked the Sanger techniques beyond Woese's grasp. But in the interim months, Woese found a promising doctoral student named Mitchell Sogin and assigned him to learn what he could from Bishop before Bishop left. Molecular biology was in its formative phase, and though results could be announced in journal papers, the gritty details of lab methodology were often passed person to person, like the gift of stone tools or fire.

Mitch Sogin was a bright Chicago kid who had come down to the University of Illinois as an undergraduate on a swimming scholarship, planning to do premed. The swimming ended, the allure of medicine faded, but Sogin stuck around to earn a master's degree in industrial microbiology within the Department of Food Science, part of the College of Agriculture. He worked on bacteria—specifically, the germination of bacterial spores, a matter of some practical interest to the food industry, given the implications for human health. Carl Woese, inhabiting a different department, almost a different universe, happened to have a lingering interest in spore germination from studies earlier in his career. For that slim reason, someone sent young Sogin to meet him. They clicked.

"And so I would go down and talk to him," Mitch Sogin told me, almost fifty years later. "I liked him."

Sogin was seventy at the time of our conversation, with a face that looked youthful but was now framed by thick, white hair. Behind his glasses, with his diffident smile, he resembled a professorial Paul Simon. We sat in his third-floor office in an old redbrick building on Water Street in Woods Hole, Massachusetts, headquarters of the Marine Biological Laboratory, a venerable research institution, where Sogin held the position of senior scientist and director of a center for comparative molecular biology and evolution. He seemed slightly bemused to have ended up there at Woods Hole, studying microbial communities of the oceans, microbial communities of the human gut, and microbial stowaways on space vehicles bound for Mars, as I nudged him to recall his early encounters with Carl Woese, back in 1968.

At that dicey moment in history, Sogin found himself, by age and geography, at the top of the rolls for his local Selective Service board. He hadn't been drafted yet, but it seemed imminent, and this was before the first lottery made draft boards less arbitrary. "I had to make a sudden decision whether to stay in school or whether to go to Vietnam." The war was at its ugliest; the Tet offensive in February that year had curdled the thinking of many young American males (including Mitch Sogin and me), and, unfair as it was, you could still get a deferment for graduate school. "Decided to stay in school," Sogin told me. "It was simple." He began work toward a doctorate under the mentoring of Woese. His topic was ribosomal RNA.

Woese had noticed something about Mitch Sogin during their early interactions: the kid was not just smart but also handy around equipment. Some combination of talents—dexterity, mechanical aptitude, precision, patience, a bit of the plumber, a bit of the electrician—made him good not just at experimental work but also at creating the tools for such work. Sol

Spiegelman had ordered and paid for a collection of apparatus to be used for RNA sequencing by the Sanger method; but now Spiegelman was off to Columbia, leaving behind the tools.

"So Carl inherited that equipment. But he had no one that knew how to use it." No one, that is, until Sogin joined his lab. "I was essentially responsible for importing all the technology"—importing it from Spiegelman's lab, and other sources, into the Woese operation. Sogin learned as much as possible from Bishop about Fred Sanger's techniques before Bishop decamped to New York, and then Sogin became Woese's handyman as well as his doctoral student, assembling and maintaining an array of hardware to enable the sequencing of ribosomal RNA.

Woese himself was not an experimentalist. He was a theorist, a thinker, like Francis Crick. "He never used any of the equipment in his own lab," Sogin said. None of it—unless you count the light boxes for reading films. Sogin himself had built these fluorescent light boxes, on which the film images of RNA fragments, cast by radioactive phosphorus onto large X-ray negatives, could be examined. He had converted an entire wall of bookshelves, using translucent plastic sheeting and more fluorescent bulbs, into a single big, vertical light box, like a bulletin board. They called it the light board. Viewed over a box or taped up on the light board, every new film would show a pattern of dark ovals, like a herd of giant amoebae racing across a bright plain. This was the fingerprint of an RNA molecule. Recollections from his lab members at the time, as well as a few old photographs, portray Carl Woese gazing intently at those fingerprints, hour upon hour.

"It was routine work, boring, but demanding full concentration," Woese himself recalled later. Each spot represented a small string of bases, usually at least three letters but no more than about twenty. Each film, each fingerprint, represented ribosomal RNA from a different creature. The sum of the patterns, taking form in Carl Woese's brain, represented a new draft of the tree of life.

14

The mechanics of this effort in Woese's lab, during Mitch Sogin's time and for much of the next decade, were intricate, laborious, and a little spooky. They involved explosive liquids, high voltages, radioactive phosphorus, at least one form of pathogenic bacteria, and a loosely improvised set of safety procedures. Every boy's dream. Courageous young grad students, postdocs, and technical assistants, under a driven leader, were pushing their science toward points where no one, not even Fred Sanger or Linus Pauling, had gone before. The US Occupational Safety and Health Administration (OSHA), though recently founded, was none the wiser.

The fundamental goal was to sequence variants of a molecule from the deepest core of all cellular life, compare those variants, and deduce the history of evolutionary relationships since the beginning. Woese had already settled on that one universal element of cellular anatomy, the ribosome, the machine that turns genetic information into proteins, but there remained a crucial decision: *Which* ribosomal molecule should he study? Ribosomes comprise two subunits, as I've mentioned—a small one snuggled beside a larger one, like an auricle and a ventricle of the heart, each constructed of both RNA and proteins. The RNA fractions include several distinct molecules of different lengths. At first, Woese targeted a short RNA molecule from the large subunit, known as 5S ("five-S") for obscure reasons that I don't ask you to contemplate. Just remember 5, a smallish number. That molecule proved unsatisfactory because its very shortness limited the amount of information it contained. The alphabet of nucleotides composing RNA is slightly different from that of DNA—it's A, C, G, and U (for uracil) in place of T (for thymine)—and there was just not enough of the A-C-G-U alphabet in any little 5S sequence to distinguish different creatures from one another. So he switched to a longer molecule in the small subunit, and at the risk of causing your eyes to roll back in your head, I'm going to tell you its name. Why? Because it's important, and once you've got it, you own it: 16S rRNA. There. Not so bad?

In English we say: "sixteen-S ribosomal RNA." It's a structural component of every bacterium on Earth, and bacteria were what Woese studied initially.

There's a close variant, 18S rRNA, in the ribosomes of more complex creatures, such as animals and plants and fungi. This 16S molecule and its 18S variant, therefore, could serve as the reference standard, the great clue, for deducing divergence and relatedness among all cellular organisms. It was, arguably, the single most reliable piece of evidence, molecular or otherwise, for drawing a tree of life. And that recognition, though it never made the front page of the *New York Times*, was Carl Woese's single greatest contribution to biology in the twentieth and twenty-first centuries.

The immediate goal for Woese, back in the early 1970s, was to extract ribosomal RNA from different organisms, to learn as much as possible about the genetic sequence of the chosen rRNA molecule from each organism, and to make comparisons from which he could gauge degrees of relatedness. He started with bacteria, because many kinds of bacteria are easy to grow in a lab, and their collective history is very ancient. Looking at bacteria from numerous different families allowed him the prospect of seeing contrasts, even in such a slowly evolving molecule as 16S rRNA. He and his team proceeded by extracting ribosomal RNA from the bacterial cells, purifying samples of the 16S molecules in each, and cutting those molecules into variously sized fragments with enzymes. Then they separated the fragments by electrophoresis, using an electrical field and a racetrack of soaked paper or gel.

In electrophoresis, a solution of mixed fragments is added to the racetrack, the power is turned on, and the electrical force pulls small fragments along faster than large ones, causing them to separate as distinct bands or ovals along the track. In Woese's effort, each fragment comprised just a few of those A, C, G, U bases—maybe three, maybe five, maybe eight, maybe as many as twenty, but always a minuscule fraction of the full molecule. Those small fragments could then be pulled again, this time in a sideways direction, and their exact sequence would begin to come clear, based on the chemical and electrical differences among A, C, G, and U. Small fragments were easier to sequence by this method than one mammoth chain. AAG was easier to discern, as you might imagine, than AAUUUUUCAUUCG.

There were several stages of work. The primary run began the process of separating the fragments from one another. The secondary run, in a sideways dimension, revealed more about each fragment, which grew discretely recognizable as it raced not just down the racetrack but also now across. Those fragments, because of their radioactive content, showed as ovals burned onto the X-ray films. The oval-marked films would let an expert interpreter such as Woese infer the sequences—that is, to sort the As, Cs, Gs, and Us from one another and determine their order in each fragment. Once illuminated that way, a fragment became more like a word than like a shadowy amoeba. It had its own spelling. What was the spelling of this little word, this fragment, or that one? Was it CAAG? Or was it CAUG? Was it something a little longer and quite different—maybe CUAUGG? The answers were important because from those words, added up into paragraphs, Woese would deduce the degree of relatedness of the creatures from which they had come.

If the sequences were still ambiguous after a secondary run, as they often were, at least for longer fragments, then those were cut further, using other enzymes, and a third run was made. Rarely there might be a fourth run, but that was usually impracticable (as well as unnecessary) because the short half-life of the radioactive phosphorus that had been fed into these bacteria meant that its radiation faded quickly, and, after two weeks, the bits wouldn't burn their images onto film. With experience, Woese developed a good sense of how to cut the fragments and get it all done in three runs at most.

Mitch Sogin and his successors did the culturing of microbes, the extraction of RNA, the cutting, and the electrophoresis. They added improvements to the methodology—different enzymes for cutting, modifications of the electrophoresis—and by 1973, the Woese lab had become the foremost user of Sanger-type RNA-sequencing technology in the world. While the grad students and technicians produced finger-prints, Woese spent his time staring at the spots. Was this effort tedious in practice as well as profound in its potential results? Yes. "There were days," he wrote later, "when I would walk home from work saying to myself, 'Woese, you have destroyed your mind again today.'" The years between 1968 and 1977 were lonely and long. Today sequencing is a snap, but Woese was ahead of his time, gathering data like a man crawling across desert gravel on his hands and knees. He couldn't have done it without a strong sense of purpose.

Being his assistant or his student called on a certain gravelly fortitude too. Mitch Sogin described the deliveries of radioactive phosphorus (an isotope designated as P-32, with a half-life of fourteen days), which by 1972 amounted to a sizable quantity arriving every other Monday. The P-32 came as liquid within a lead "pig," a shipping container designed to protect the shipper, though not whoever opened it. Sogin would draw out a measured amount of the liquid and add it to whatever bacterial culture he intended to process next. "I was growing stuff with P-32. It was crazy," he said, tossing that off as a casual memory. "I don't know why I'm alive today." Because the bacteria were cultured in growth media lacking other phosphorus, a vital nutrient, they would avidly seize the P-32 and incorpo-rate it into their own molecules. Sogin would then extract and purify the ribosomal RNA, "all the while not contaminating the laboratory." That was the hope, anyway. For separating 16S from the other ribosomal frac-tions, he used "home-built electrophoresis units," cylinders of acrylamide gel through which the different molecular fragments would migrate at different speeds. (Acrylamide is a water-soluble thickener, sometimes used in industry as well as in science.) Then he would freeze the gel and attempt to slice it, like bologna, with a very precise knife. The slicing was difficult: slices would fall off when they shouldn't, he had to work the material at just the right temperature, and "this was pretty radioactive stuff." Sogin then cut the 16S molecules into fragments with an enzyme, and those fragments would run a race of their own, not through cylinders of gel but along a racetrack of special, absorbent paper.

One end of the long paper strip went into a receptacle known as a Sanger tank (as developed by Fred Sanger), containing a liquid buffer. The strip passed over a rack, beyond which its far end dropped into another Sanger tank, and both tanks were wired to an apparatus that provided the electrical pull. At the bottom of the tanks were high-voltage platinum electrodes, covered by three inches of liquid buffer and then at least fifteen inches of Varsol, a solvent not unlike paint thinner, intended to cool the paper strip. "Varsol is both volatile and explosive," Sogin said. The power source delivered around 3,500 volts and plenty of amps, he recalled—"certainly enough to kill you." Also enough, with an errant spark into the Varsol, to blow you up.

This whole panoply of dangerous, intricate machinery dwelt within a shielding hood that could be closed behind large sliding doors, floor to ceiling, in a nook off the main lab known as the electrophoresis room. Set up the system, close the doors, turn on the juice, hope for the best. "I was too stupid to be afraid of anything," Sogin told me. "Too naïve. Too young. Immortal." He was also lucky. Nobody got hurt.

Around the time Sogin finished his doctorate and prepared to leave, Woese hired a young woman named Linda Bonen, a walk-in from a different building, to take on some of the technical work. Raised in rural Ontario, she had come down to the University of Illinois and gotten a master's degree in biophysics. Woese trained her for this new lab work himself—how to chop the RNA into fragments, how to run the electrophoresis in two dimensions, how to prepare the films, even a bit about how to interpret them, deducing which spot on a film represented which fragment, which little blurt of letters. Was it UCUCG, or was it UUUCG? Tricky to tell. But here's GAAGU, obviously different. Woese coached her patiently on the tasks and their meaning.

"He was very good about bringing me along," Bonen recalled four decades later, when I visited her at the University of Ottawa, where she was by then a biology professor herself, gray haired, deeply expert in molecular genetics, gentle mannered as a schoolteacher. "The end product would be a 'catalog' for microbe X," she said, meaning simply a list of the different fragments found within the 16S rRNA molecules of that creature. A catalog. If the fragments resembled words, these catalogs were the paragraphs. Comparing one catalog with another revealed the degree of similarity between any two organisms, by a very precise standard, and more dissimilarity could be taken to reflect more distance in evolutionary time. Where had the great limbs diverged from the trunk, the big branches from the limbs—and why there, and why then, and leading to what creatures? Beyond the mind-numbing methodology of data collection, those were the questions Woese hoped to answer.

What was he like, I asked Bonen, as a boss and a teacher?

"Well, he never came across as a boss," she said. "He was very soft-spoken and quiet, reserved. I'm sure you've . . ." She hesitated. "Did you know him yourself? Did you ever meet him?"

Never. I didn't explain to her, but the reason was simple: Woese died, in late 2012, an old man taken down hard and fast by pancreatic cancer, just before I picked up the trail.

"To everybody, he was Carl," she said. "He was not a boss."

Bonen showed me a photograph, a memento from her personal files: the youngish Carl Woese in his lab, bathed by yellow-green light, jaw set firmly, gazing up at a pattern of dark spots. Short brown hair, striped sport shirt, handsome and jaunty enough to have stood onstage amid the Beach Boys. Almost apologetically, she said: "That's the only good picture I have." This was all different from what I had expected. My mental image was of the later man: the shy, crotchety, and august Dr. Carl Woese.

He was shy, yes, Bonen said. But "august," no, that was wrong, not a word she would ever . . . and here again her voice fell away. Then she added: "I only knew him in a short period of time."

15

Ken Luehrsen, soon after Linda Bonen's short period, had a different sort of experience in the Woese lab. He was an undergraduate at Illinois when he first encountered Woese as one of the instructors for a seminar in developmental biology, well outside Woese's field of expertise. The logic behind this mismatch, according to Luehrsen, was that "other professors just liked to hear Carl's take on things so they might incorporate some of his ideas into their own research." Woese was notoriously brilliant, full of ideas, but jealous of his expended effort. "Undoubtedly, Carl found an opportunity to get a teaching credit where he didn't have to do too much work." In a seminar, students would be assigned to make presentations explicating this journal paper or that one, and Woese could easily moderate the discussion. He hated classroom teaching of the more arduous sort—preparing and delivering lectures, God forbid—because "he felt it took him away from his real love: understanding the origin and evolution of life."

After the seminar acquaintance, Luehrsen went to this formidable figure and asked to do an honors project under his guidance. Woese not only accepted him but also, to Luehrsen's surprise, "he plopped me down in his office," a very small room containing two desks, both covered with chaotic stacks of papers, and said (either seriously or as a tease) that it was so he "could keep an eye on me." Luehrsen was befuddled. Should he really be there? Should he scram whenever the phone rang and give Woese his privacy? His discomfort eased when he saw that Woese himself spent little time in that office and most of his time in the lab, "reading 16S rRNA fingerprints at his light board."

After Woese's death, Ken Luehrsen wrote a short memoir describing the man's work, his temperament, and their interactions so long ago, for publication with other Woese tributes in a scientific journal. He brought it all to mind again when I tracked him down in San Carlos, California, on the edge of Silicon Valley, where he was now a senior scientist and biotech inventor in the late afternoon of his career, consulting for a small company lodged behind glass doors in an office park. By that time, he held many patents in biotechnology, for methodologies to create antibodies and other molecular products, and lived comfortably in an old counter-culture enclave across the peninsula, a place known as Half Moon Bay, from which he could commute to the action. He worked when he felt like it. At this firm, he was the grizzled elder, surrounded by smart young colleagues seated in carrels, for whom "Woese" was at most a dimly recognizable name, like "Darwin" or "Fibonacci." Tall and thin, with a goatee, relaxed and a little sardonic, Luehrsen suggested we escape downtown for sushi—after which we talked for most of the afternoon.

"I may have been a junior at the time," he said about his first acquaintance with Woese. "I didn't know anything." Despite Luehrsen's ignorance, the great man invested some effort in him; a private tutorial was less abhorrent to Woese than lecturing at banks of indifferent faces. "He explained to me what he was doing. I maybe understood a quarter of it." But the youngster paid close attention and caught on fast. "I think he saw somebody who was interested, and I was a pretty hard worker."

It was 1974 when Luehrsen joined the Woese lab as an undergraduate assistant, paired with a graduate student and assigned the unenviable job of extracting radioactive rRNA from bacterial cultures. They would dump ten millicuries (a large dose) of P-32 into this culture or that and, after overnight incubation to let the bacteria suck it up, spin the mixture in a centrifuge to gather the hot bacteria into a little pellet. After dissolving the pellet in a buffer, they would squash that brew through the laboratory version of a French press, not too unlike the one you might use for coffee. This served to rip open the bacterial cells and set their innards adrift. Luehrsen and his partner would then pull out the ribosomal RNA by chemical extraction, after which the different fractions—the 16S molecules versus the others, including that shorter one, known as 5S—were separated using Mitch Sogin's home-built cylinders of acrylamide gel. In addition to acrylamide (today recognized as a probable carcinogen), they were working with phenol, chloroform, ethanol, and the radioactive phosphorus. "What a mess that often was! The Geiger counter was always screaming," Luehrsen wrote in his memoir.

One of the bacteria he cultured and squashed was *Clostridium perfringens*, the microbe responsible for gas gangrene, an ugly form of necrosis that takes hold in muscle tissue made vulnerable by wounds, especially the sort that lay open among injured soldiers on battlefields. When he realized this, Luehrsen complained, but Woese "just chuckled and said not to worry" in the absence of an open wound. He had been to medical school for "two years and two days," Woese said, and he could assure Luehrsen that *Clostridium perfringens* was unlikely to give him gangrene. Luehrsen took the episode as a lesson—not a lesson to trust Woese but to rely on his own perspicacity more—and never probed the matter of why Woese had quit medical school two days into his third-year rotation in pediatrics.

After graduating from Illinois in 1975, Ken Luehrsen stayed to work toward a PhD under Woese's supervision, just as Woese shifted the lab's focus, slightly but critically, in a way that would lead toward his most startling discovery. So far, they had targeted their molecular analyses on common bacteria and a few other single-celled organisms such as yeast—easy to obtain, easy to grow in the lab. But that was just a preliminary effort as they refined their methods. "One of the things he wanted to do was to look at unusual bacteria," Luehrsen told me. Woese hoped this might give a view "deep into evolution," where he could see "deep divergences" between one big branch of life and another. So he struck up a collaboration with a colleague in the Microbiology Department, Ralph Wolfe, one of the world's leading experts in culturing a group known as the methanogens.

Methanogens: their name derives from an odd aspect of their biochemistry, producing methane as a byproduct while metabolizing hydrogen and carbon dioxide in environments lacking oxygen. To say it more plainly, these bugs generate swamp gas in muddy wetlands, from which it bubbles up, and similar gas in the bellies of cows, whence it emerges by belch and fart. Certain methanogens also thrive beneath the Greenland ice cap, deep in the oceans, and in other extreme environments, such as hot desert soils. Despite these shared metabolic traits, Ralph Wolfe advised Woese, there was an odd discontinuity among the assemblage of methanogens—discontinuity in terms of their shapes. Some were cocci (spherical), some were bacilli (rod shaped). Since the cocci and the bacilli were considered two distinct kinds of bacteria, microbiologists had been puzzled about how to classify the methanogens—together by metabolism or separately by shape. That conundrum captured Woese's interest.

Having told me this much, and more, Ken Luehrsen finished our conversation and sent me away with some gifts. One was a black-and-white print of a photo he took in the mid-1970s, a snapshot, showing Woese at his light board, engrossed before a pattern of dark spots, with a handful of felt-tip pens for color coding what he saw, a pencil for data registry behind his right ear. Luehrsen's other gift was a single yellowing sheet—not a copy, the original—from his own notebook of the time. It was a catalog of

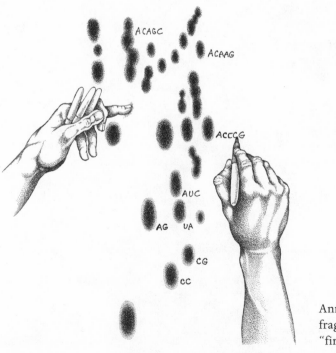

Annotating RNA fragments on a "fingerprint" film.

fragments from an organism, more of those telling blurts of the four coding letters, neatly recorded in two columns. UCUCG. CAAG. GGGAAU, and dozens more. At the top, also hand lettered, an abbreviation indicated the name of the organism as it was known at the time: *Methanobacterium ruminantium*. Later, I realized that, notwithstanding the name, this was no bacterium. Luehrsen had given me the genetic rap sheet on a separate form of life.

16

How do you classify the methanogens? Where do they fit on the tree of life? To what other little bugs are they most closely related? Those questions, which Woese and his colleagues were asking themselves in the mid-1970s, fell within the scope of an important discipline with a dry name: bacterial taxonomy. That's the enterprise of sorting bacteria into nested groups: species, genera, families, etcetera. You name something *Methanobacterium ruminantium*, and then where do you put it?

This may sound like an exercise in arcana, a marginal activity of risible triviality beside which stamp collecting looks like an adventure sport. Bacteria are tiny, relatively simple, invisible. But if being invisible made things unimportant, gravity and microwaves would be unimportant too. It's useful to recall that most life-forms on Earth are microbial, that they determine the conditions of existence for the rest of us, and that even the human body contains at least as many microbial cells (those tiny passengers that live in your gut, on your skin, in the follicles of your eyelashes, and elsewhere) as human cells. Your environment is highly microbial too. Your food. The air you breathe. Microbes run the world, and a very large portion of those microbes are bacteria. Some of them serve as helpful partners of humanity. Some are benign. Some are rapacious, ready to poison your blood, fill your lungs, kill you. So it's no small matter, telling one bacterium from another.

Scientists once believed it might be possible to do this from visual evidence obtained through a microscope. They even presumed that the concept of species, as understood for animals and plants and fungi, could be applied to bacteria. These were useful simplifications in their era—like the simplifications of Newtonian physics, before correction by Einstein—but that era was a long time ago.

The early hero in the field was a man named Ferdinand Julius Cohn, a botanist and microbiologist at the University of Breslau (now Wrocław, Poland) during the late nineteenth century. Cohn is an appealing figure, and only partly because his important contributions have been overshadowed by those of better-remembered contemporaries whose accomplishments were more practical and dramatic: Louis Pasteur, Robert Koch, Joseph Lister. They worked on disease, agriculture, and wine. Cohn worked mainly on describing and classifying microscopic organisms. No one makes Hollywood movies about bacterial taxonomists.

Cohn wasn't the first researcher to classify bacteria, making distinctions between kinds, trying to place the whole group in its proper position on the tree of life. But his effort was more hardheaded and percipient than the others, and he did much to bring bacteriology out of a fog of confusions that had lingered for more than a century, ever since startled observers such as Leeuwenhoek had noticed these little creatures through simple microscopes. Several insights and adjustments of method helped him make progress. Microscopy improved, with better lenses and precision instruments in which they were mounted. Cohn's lab started culturing bacteria on solid media such as slices of cooked potato, not in liquid nutrient, the old way. That allowed Cohn to choose, cultivate, and consider different strains separately. Also, he recognized that physiological and behavioral characteristics as well as structural ones could be useful for distinguishing bacterial species: How do they grow in different media? How do they move? By this time, too, Cohn had embraced Darwin's theory of evolution, and so it made sense to him that bacterial strains might change and adapt over time. This was incremental change, very different from the sort of utter transformation—one bacterial form suddenly morphing into another—that some scientists imagined to occur. Cohn didn't buy transformation. He saw bacteria as fundamentally stable in their identities. Finally, he published his system, dividing them into four tribes: spherical, rod shaped, filamentous, and spiral, each of which got an imposing Latinate name. Within the tribes, he drew finer distinctions, separating them into genera and species.

Not everyone in the field accepted Cohn's classification of bacterial species or his conviction about their stable identities, and the idea of shape-shifting bacteria lingered for more than a decade. The longer judgment of science historians was good to him, as a man and a scientist, noting his "reserve" against self-promotion, his modesty, his eloquent lecturing, and his success in "disentangling almost everything that was correct and important out of a mass of confused statements on what at that time was a most difficult subject to study." Besides arguing for the reality of bacterial species and sketching a way to classify them, Cohn did much, along with Pasteur, to kill the resilient delusion that new life-forms arise by spontaneous generation. They don't, he showed. When bacteria seem to appear out of nowhere, it's because they have arrived from somewhere: contamination, floating through the air, reawakening spores. Cohn's work was "entirely modern in its character and expression," according to an authoritative chronicler of the field, writing in 1938, "and its perusal makes one feel like passing from ancient history to modern times." But what looked modern in 1938, of course, doesn't look modern now.

Even the devoutly empirical Ferdinand Cohn made mistakes. For one: after all his research, he still believed, as many of his colleagues did, that bacteria belong to the kingdom of plants. So his tree of life, by later standards, was badly wrong. For another: the premise of radical transformation, one bacterial form to another, turns out to be vastly more complicated than he could imagine.

17

C haos" was the name of the group into which Linnaeus, the great systematizer, in the 1774 edition of his *Systema Naturae*, had lumped Leeuwenhoek's bacteria and other little creatures. That was a durable judgment. Even well into the twentieth century, decades after Ferdinand Cohn, experts were still arguing about whether bacterial taxonomy was a meaningful enterprise or hopelessly chaotic.

Beginning in 1923, the standard source for identifying bacteria was a thick compendium, *Bergey's Manual of Determinative Bacteriology*, edited by the bacteriologist David Hendricks Bergey. But as microbiology progressed, it became clear that the *Bergey's* system was vague, inconsistent, and, on some fundamentals, inaccurate. It didn't offer a tree of bacterial life. It was only a glorified field guide. Still, other researchers who critiqued *Bergey's Manual*, and then tried to improve on it, found the critiquing much easier than the improving. The task of bacterial classification was just so difficult. There was almost no fossil record of bacterial ancestors. There weren't enough differences of external shape and internal anatomy, even as seen through powerful microscopes, to support fine distinctions. Physiological characters could also be misleading, if they reflected parallel adaptations rather than shared ancestry. What did that leave for a classifier to use? (Hint: Carl Woese would offer an answer, but not until 1977.) This conundrum came to a head in 1962, when two of the world's leading microbiologists, C. B. van Niel and Roger Stanier, essentially threw up their hands in despair.

Van Niel was a Dutchman, educated in Delft, who in 1928 decamped to California, where he taught at a marine biological station that was part of Stanford University. His particular interests were bacterial physiology and taxonomy. Roger Stanier was a younger Canadian who became van Niel's student, then his special protégé, then his collaborator. In 1941, when Stanier was still just twenty-five years old, he and van Niel coauthored an influential paper on bacterial classification.

That paper stood as definitive for a generation—until both authors renounced it. Stanier himself later admitted some embarrassment about it, all the more so because he had arm-twisted van Niel to sign on as coauthor—student and teacher together, although the work was mainly Stanier's. What the paper contained, besides a pointed critique of *Bergey's Manual*, was a shiny new proposal for classifying bacteria—not just a checklist or a field guide but a "natural" system reflecting their evolutionary relationships. That system divided the familiar bacteria into four major groups (as Ferdinand Cohn had done) and placed them in a kingdom of simple creatures along with just one other group: the blue-green algae.

Algae? Yes, the blue-green algae, as they were then called, had long been an ambiguous group, because they seemed to straddle the line between bacteria and plants. (This was partly what allowed Cohn to believe that all bacteria *were* plants—the blurry lines around blue-green algae.) *Algae* was a catchall term for a loose assemblage of creatures that photosynthesize, including these tiny blue-green creatures, but that didn't mean all algae shared a single common ancestor. Did they? Stanier and van Niel said no. By their new definition of things, blue-green algae were more similar to bacteria than to other algae, and these two groups should be lumped together in a kingdom of their own, apart from everything else. Eventually they labeled such cells *procaryotic*—meaning "before kernel," as I've mentioned—and set them in contrast to *eucaryotic* cells, comprising all else. (Their spellings were later corrected, from more accurate transliteration of the Greek roots, to *prokaryotic* and *eukaryotic*.) The kernel in question was a cell nucleus. Just as a bacterium doesn't have one, neither do the creatures that were then known as blue-green algae (and are now classified as cyanobacteria). Advances in microscopy since the end of World War II, including electron microscopy, had given micro-biologists a better view of those distinctions and others, making possible a fresh analysis of what a bacterium is—and what it isn't. Stanier and van Niel offered that fresh analysis along with the prokaryote category in a new paper, published in 1962, titled "The Concept of a Bacterium." By their lights, the "abiding intellectual scandal of bacteriology" was that no such concept had ever been clearly delineated. What *was* a bacterium? Um, hard to say.

They tried to correct that by placing bacteria and blue-green algae to-gether as prokaryotes, and setting them in contrast to the alternative cat-egory, eukaryote, which encompasses all other forms of cellular life. The chief distinguishing features of a prokaryote, according to Stanier and van Niel, were: (1) no cell nucleus, (2) cell division by simple fission, rather than the elaborate process of chromosome pairing known as mitosis, and (3) a cell wall strengthened by a certain sort of latticework molecule with a fancy name, peptidoglycan. I know, it looks like the moniker of a flying reptile from the Jurassic. Forget about it for now, and when peptidoglycan comes back as an important clue toward understanding the deepest struc-ture of the tree of life, and the twig on the branch on the limb from which we humans have sprouted, I'll remind you.

The dichotomy between prokaryotes and eukaryotes, creatures without cell nuclei and those with, relatively simple beings and relatively complex, became a fundamental organizing principle of biology. Stanier and his two coauthors of a textbook would later say that it "probably rep-resents the greatest single evolutionary discontinuity to be found in the present-day living world." It was also a salubrious reminder to humans of our inescapable linkage to other creatures, including some very humble ones. We are, at the most basic level of classification, eukaryotes. So are amoebae. So are yeasts. So are jellyfish, sea cucumbers, the little parasites that cause malaria, and rhododendrons. To an average person, the gap between an amoeba and a bacterium may seem narrow (partly because most of us have never, or at least not since high school biology, looked through a microscope at either), but the prokaryote-eukaryote distinction reveals it as oceanic. You could think of the living world—and, beginning from Stanier and van Niel's 1962 paper, biologists *did* think of the living world—as divided into proks and euks.

Besides putting that idea into play, "The Concept of a Bacterium" is notable for having signaled surrender, by Stanier and van Niel, in the battle of bacterial taxonomy. About this they were candid, confessional, and brusque. Ever since Leeuwenhoek, microbiologists had been seeking the best way to classify bacteria. Ever since Darwin, they had been arguing about how one bacterium was related to another. Enough was enough. "Any good biologist finds it intellectually distressing to devote his life to the study of a group that cannot be readily and satisfactorily defined." C. B. van Niel himself had devoted forty years. He and Stanier now alluded to the "elaborate taxonomic proposal" they had published back in 1941, "which neither of us cares any longer to defend." Never mind *that*. They admitted having "become sceptical about the value" of any such formal systems, or the effort spent to develop them, although they still affirmed the importance of figuring out just what the devil bacteria *are*.

This skepticism, this taxonomist's despair, had been wiggling up inside van Niel for a long time. Two decades earlier, even as he was signing onto that first elaborate proposal, he had confessed his gloom to Stanier in a letter: "Many, many years ago I often went around with a sense of futility of all our (my) efforts. It made me sick to go around in the laboratory (this was in Delft) and talk and think about names and relations of microorganisms." Was any of it real? Was there any value to putting bacteria into labeled boxes? "During those periods I would go home after a day at the lab, and wish that I might be employed somewhere as a high-school teacher." Not that he would enjoy such teaching, he realized, but at least "it would give me some assurances that what I was doing was considered worth-while." Nowadays we might see that as a signal of bipolar disorder, but it's just as likely that van Niel simply viewed bacterial taxonomy with great clarity.

Under their revised spellings, *prokaryote* and *eukaryote*, those two became enshrined for a generation as the most fundamental categories of life. Eukaryotes had cell nuclei. Prokaryotes did not. That dichotomy seemed to represent, as Stanier and his coauthors had written, the greatest single evolutionary divide in the living world. There were two basic kinds of creature, the proks and the euks, and there was nothing between.

What makes this worth knowing is that Carl Woese proved it wrong.

18

As of early 1976, with Ken Luehrsen and others still helping, Woese had done his unique form of catalog analysis on samples from roughly thirty species, using differences in ribosomal RNA molecules to measure their relatedness. Most were prokaryotes, but he also looked at a few eukaryotes (which carried that slightly different molecule in their ribosomes, 18S rRNA instead of 16S), including yeast, for purposes of gross comparison. He could tell a prok from a euk just by inspecting the spots on a sheet of film. And he was eager to see those "unusual bacteria," the methanogens, about which Ralph Wolfe had alerted him.

The tricky thing about methanogens was that, since oxygen poisoned them, they were hard to grow in a laboratory. But Wolfe's lab team included an ingenious doctoral student, Bill Balch, who had solved that problem by devising a way to culture methanogens in pressurized aluminum tubes with black rubber stoppers, and using syringes to move things in and out. Balch gave the methanogens an atmosphere of hydrogen and carbon dioxide instead of oxygen, plus a liquid growth medium, and they thrived. Woese sent his own postdoc, a rangy young man named George Fox, trained in chemical engineering, to work with Balch on growing some of these methanogens and tagging them with radioactive phosphorus. Fox, Ken Luehrsen, and other members of the Woese lab then combined their efforts on the rest of the process: extracting the radioactive RNA, purifying it to get concentrations of 16S and 5S molecules, chopping those molecules into pieces, running the electrophoresis to separate the fragments, and printing the spots onto films. Their first methanogen carried a formal name so long (*Methanobacterium thermoautotrophicum*) that even Woese himself dismissed that as "a fourteen-syllable monstrosity" and preferred using a shorter label, denoting the particular laboratory strain: delta H. Examining its primary fingerprint on his light board, Woese noticed something odd.

He was practiced enough by now at reading such fingerprints that he could immediately recognize a certain pair of small fragments, common to all bacteria, that "screamed out" their membership in the prokaryotes. He looked for them on the primary film from delta H. They were missing. Intrigued but patient, he waited for the secondary fingerprint, with the fragments pulled sideways to reveal more detail. He got that from his technician several days later. On June 11, 1976, he taped the primary film up on his light board again, with the secondary now in front of him on the light table, and began trying to interpret what he saw. He intended, as usual in this stage of the process, to use the secondary film as a guide for inferring the base sequences of the fragments in the primary pattern. Apart from his board and his table, the room was dark. His face, we can imagine, reflected an eerie glow. Quickly he noticed more oddities.

The two missing fragments were still missing, but it wasn't just that. Woese turned to a different part of the pattern, expecting to see another familiar fragment—a "signature" sequence in all prokaryotes. Not there. Instead, he found a strange fragment, a longish sequence that shouldn't have been present at all. "What was going on?" he later recalled wondering. This methanogen rRNA just "was not feeling" prokaryotic. And the more fragments he sequenced, the less prokaryotic it felt. By this time, he knew the sequences of ribosomal RNA in bacteria so intimately that his "feel" for the molecule was a persuasive standard of normality. And something in this particular creature, delta H, was abnormal. Some bacterial fragments were appearing where expected, as expected, yes. But some others looked eukaryotic, suggesting a completely distinct form of life: a yeast, a protozoan, *what?* And still others were just weird. What *was* this RNA? he wondered, and what manner of organism did it represent? It couldn't be from a prokaryote. It wasn't eukaryotic. It wasn't from Mars, because it contained too many familiar stretches of RNA code. "Then it dawned on me," he wrote. There was "something out there"—out there in the teeming ecosystems of planet Earth, he meant—other than prokaryotes and eukaryotes. A third form of life, separate.

Woese called this, whimsically, his "out-of-biology experience." It would be the watershed moment of his scientific life.

19

After his death in December 2012, Woese's files of scientific correspondence, manuscripts, journal articles, and other materials went to the University of Illinois Archives to be indexed, curated, and preserved. The archives are held in several different locations, one of which is the Archives Research Center, a sort of annex, housed in an old, barnlike building of red brick on Orchard Street near the south edge of the campus. A sign in front identifies this, confusingly but historically, as the Horticulture Field Laboratory; a bank of yew bushes and a riot of hostas guard the entrance. Inside, filed neatly in thirty-four boxes that can be accessed by request, are the Carl Woese Papers. I was working there at a table one hot July afternoon, reading through letters, looking for clues about the human side of this peculiar man, when John Franch arrived, wearing a dark T-shirt and a ball cap. Franch is the assistant archivist who was sent to clean out Woese's lab after the funeral, and who knows the material found there better than anyone else. He had heard about my interest and wanted to show me something.

He led me toward the back of the building, where the roof arches high, and unlocked a door. This was one of the "vaults," he told me, that formerly served for storing fruit—apples in particular—from the horticultural research orchards from which Orchard Street got its name. At one point, there were 125 varieties of apple grown just behind the building, and they came in by the basket and the crate to be stored here or pressed for cider and vinegar. Beyond the door, we entered an air-conditioned room, empty of apples now but lined along its left side with tall metal shelves, along its right side with tables. The shelves held hundreds of large, flat yellow boxes—the original packaging of Kodak medical X-ray films—representing the library of Woese's RNA sequencing fingerprints. Each box was labeled along its edge with a date and the organism whose fragments were depicted.

Across the room, some films lay on the tables, where Franch had been working over them. He showed me three large sheets, carefully taped together, forming a triptych of images. I stared at the patterns of dark spots: amoebae galloping on a plain. To me, they made no particular sense. But to Woese, they had spoken eloquently of identity, relationship, evolution. If something was odd, he would have seen it.

This is delta H, Franch said.

20

Immediately after his epiphany, Woese shared it with George Fox, the postdoc he had assigned to work with Bill Balch on growing the methanogens. As recalled later by Fox, Woese "burst into my room in the adjoining lab" with the announcement that they had something unique. From there he proceeded throughout the lab, among his young students and assistants, "proclaiming that we had found a new form of life. He then pointed out," by Fox's memory, tart and amused, "that this was of course contingent on my having not screwed up the 16S rRNA isolation." Being cautious, they repeated the whole process with delta H and got the same result. So no, Fox hadn't screwed up.

"George was always skeptical," Woese himself wrote later about their reactions to the discovery, adding that he valued such skepticism as good scientific instinct. Fox's doctorate in chemical engineering suited him well to offset epiphanic leaps, even by the boss, with empirical caution. In fact, their shared instinct for skepticism about such a startling result helps explain why these two men worked so well together. But the anomalies in the fingerprints persuaded Fox too. By his account, they seemed to "jump off the page," and he agreed that those differences suggested a third, very distinct form of life.

Still, Woese and Fox both knew that convincing other scientists of such an epochal discovery would be difficult. More data were needed. So the Woese lab went back to work, with Balch's methodology and help, on culturing and fingerprinting still another methanogen. Woese and his colleagues worked quietly, for the time being. By the end of 1976, they had five additional genetic catalogs from five more methanogenic microbes, all quite different from one another but sharing signs of a much greater, much deeper, and shared difference from anything else known to exist.

21

Bacteria are versatile and diverse. That's an understatement. Wide-spread—another understatement. They are hard to categorize, hard to identify, hard to sort into related groups, as even Stanier and van Niel finally admitted. They are nearly ubiquitous across most parts of Earth's expanse, in both natural and human-made environments, floating through the air, coating surfaces everywhere, awash in the oceans, even present in rocks deep underground. Your skin as I've said is covered with them. Your gut is teeming. Your human cells may be out-numbered by them at a three-to-one ratio in your body. Bacteria live also in mudholes and hot springs and puddles and deserts, atop mountains, deep in mines and caves, on the tabletops at your favorite restaurant, and in the mouths of you and your dog.

A species called *Bacillus infernus* has been cultured from core samples of Triassic siltstone, buried strata at least 140 million years old, drilled up from almost two miles beneath eastern Virginia. Under the Pacific Ocean, 35,755 feet deep in the Mariana Trench, lie sediments that have also yielded living bacteria. In Antarctica, a body of water known as Sub-glacial Lake Whillans, lidded by half a mile's thickness of ice and super-cooled to just below zero, harbors a robust community of bacteria. They thrive there in the darkness and cold, eating sulphur and iron compounds from crushed rock.

Then again, some like it hot. Those are called thermophiles. Among the most famous of thermophilic bacteria is *Thermus aquaticus*, first cultured from a sample collected in Yellowstone National Park by the microbiologist Thomas Brock and a student, Hudson Freeze, in 1966. Brock and Freeze had found it in a steaming, multicolored pool called Mushroom Spring, in Yellowstone's Norris Geyser Basin, at a temperature of about 156 degrees Fahrenheit. Functioning in such heat, *Thermus aquaticus* contains a specialized enzyme for copying its DNA, one that performs well at high temperatures, which became a key element in the polymerase chain reaction technique for amplifying DNA. That technique, widely useful in many aspects of genetic research and biotech engineering, earned its chief developer (but not Thomas Brock) a Nobel Prize.

Other heat-loving bacteria can be found around hydrothermal vents on the sea bottom, where they help anchor the food chains, producing their own organic material from dissolved sulfur compounds vented out with the hot water, and being fed upon by little crustaceans and other animals. A giant tube worm, one of those gaudy red creatures that waggle around such vents, with no mouth, no digestive tract, gets its nutrition from bacteria growing within its tissues.

By one estimate, the total mass of bacteria exceeds the total mass of all plants and animals on Earth. They have been around, in one form or another, for at least three and a half billion years, strongly affecting the biochemical conditions in which most other living creatures have evolved. That we don't see bacteria is simply because our eyes are not calibrated to the appropriate scale. There may be more than a billion bacterial cells in an average ounce of soil, and five million in a teaspoon of fresh water, but we can't hear their crackle or their fizz. A single kind of marine bacteria known as *Prochlorococcus marinus*, which drifts free in the world's tropical oceans and photosynthesizes like a plant, may be the most abundant creature on Earth. One source places its standing population at three octillion individuals, a number that looks like this: 3,000,000, 000,000,000,000,000,000,000.

They vary in shape and in size—interestingly in shape, drastically in size. A bacterial cell, on average, is about one-tenth as big as an animal cell. At the upper end of the range is *Thiomargarita namibiensis*, an odd thing discovered on the sea floor near Namibia, its cells ballooning up to three-quarters of a millimeter in diameter, stuffed with pearly globules of sulfur. At the lower end of the range is *Mycoplasma hominis*, a tiny bacterium with a tiny genome and no cell wall, which manages nonetheless to invade human cells and cause urogenital infections.

Bacterial shapes, as I've mentioned, range through rods, spheres, filaments, and spirals, with variations that in some cases represent adaptations for movement or penetration. It turns out that their geometries, notwithstanding the efforts and convictions of Ferdinand Cohn, are unreliable guides to their phylogeny. Shape can be adaptive, but adaptations can be convergent as well as ancestral. Roundness may be good as a hedge against desiccation. Elongation as a rod or filament seems to help with swimming, and a flagella definitely does. Filamentous bacteria that are star shaped in cross section, recently discovered in a wonderfully named substance called "mine-slime," deep in a South African platinum mine, may profit from all their surface area by way of enhanced absorption in nutrient-poor environments. The twisting motion of spirochetes, such as the ones that cause syphilis and Lyme disease, evidently allows them to wiggle through obstacles that other bacteria can't easily cross, such as human organ linings, mucous membranes, and the barrier between our circulatory system and our central nervous system—a fateful degree of access. Even the less dynamic shapes, the short rods known as bacilli, the spheres known as cocci, and the rods slightly curved like commas, serve well enough the bacteria responsible for a long list of diseases: anthrax, pneumonia, cholera, dysentery, hemoglobinuria, blepharitis, strep throat, scarlet fever, and acne, among others.

Although many bacteria live as solitary cells, taking their chances and meeting their needs independently, others aggregate into pairs, clusters, little scrums, chains, and colonies. The coccoid cells of *Neisseria gonorrhoeae*, which cause gonorrhea, lump together by twos, forming bilobed units resembling coffee beans. The genus *Staphylococcus* gets its name from Greek words for "granule" (*kókkos*, the spheroid aspect) and "a bunch of grapes" (*staphylè*), because staph cells tend to bunch. Most of the forty staph species are harmless, but *Staphylococcus aureus* can inflict skin infections, sinus infections, wound infections, blood infections, meningitis, toxic shock syndrome, plus other nasty conditions, and if you're so unlucky as to pick up a dose of those little grapes in one of their antibiotic-resistant forms, such as MRSA, a monstrous product of horizontal gene transfer (as I've mentioned, and to which I'll return), you could be in a world of hurt. Cells of *Streptococcus* species, including those that cause impetigo and rheumatic fever, stick together like beads on a chain.

Bacteria can also form stubborn, complex films on certain surfaces—the rocks of a sea floor, the glass wall of an aquarium, the metal ball of your new artificial hip—where they may cooperate together in exuding a slimy extracellular substance that helps nurture them collectively, maintain the stability of their little environment, serve as a sort of communications matrix among them, and even protect them from antibiotics. These living slicks, known as biofilms, can be thinner than tissue paper or as thick as a good dump of snow, and may incorporate multiple species. The little rods of *Acinetobacter baumannii* are infamous for their ability to lay down persistent biofilms on dry, seemingly clean surfaces in hospitals.

Cyanobacteria, including that monumentally abundant *Prochlorococcus*, convert light to energy and deliver, as byproduct, a large share of Earth's free atmospheric oxygen. Purple bacteria photosynthesize too, but do it by drawing upon sulfur or hydrogen instead of water as fuel for the process, and they don't produce oxygen. Lithotrophic bacteria, the rock eaters, deriving their energy from iron, sulfur, and other inorganic compounds, exist in more ingenious variants than you care to know. Japanese researchers have recently discovered a new bacterium, *Ideonella sakaiensis*, that digests plastic. Certain enterprising ocean bacteria, such as *Marinobacter salarius*, have risen to the challenge of degrading hydrocarbons from the Deepwater Horizon oil spill. Other bacteria are quite capable, in the presence of oxygen or without it, of feasting on garbage, sewage, various inorganic compounds, plants, fungi, and animal tissue, including human flesh. Lactic acid bacteria, which may be rod shaped or spherical, turn up in milk products, busy at their task of carbohydrate fermentation and resistant to the acid they create. Many of them also like beer.

Not all such particulars were known to Carl Woese in 1977 as he examined the fingerprints from his first few methanogens. But the vast scope, ubiquity, and multifariousness of bacteria certainly were. The terrain of bacteriology was known even better to Ralph Wolfe, who had trained in the classic fundamentals under van Niel and others. Woese's reaction to his own preliminary results must have seemed all the more radical, then, all the more shocking, as he shared it not just with George Fox and members of his own lab but also with Wolfe, just after they repeated the rRNA analysis of delta H, the first methanogen. "Carl's voice was full of disbelief," Wolfe wrote in a memoir, "when he said, 'Wolfe, these things aren't even bacteria.'"

22

Ralph Wolfe told me the same story, with some elaboration, thirty-nine years later when I called on him in Urbana. By then, he was an emeritus professor of microbiology, ninety-three years old, a frail and slender gentleman with a quick smile, still maintaining his office and coming to it, as though retirement were not an entirely satisfying option. On the wall behind his desk hung a replica of Alessandro Volta's *pistola*, a gunlike device invented by Volta in the late 1770s for testing the flammability of swamp gases, including methane. On the desk itself were papers and books and a computer.

Woese's lab back in the day had been in Morrill Hall, on South Goodwin Avenue, and Wolfe's was in an adjacent building, connected by a walkway. Woese would occasionally trundle over on various business. "He came down the hall and happened to see me," Wolfe recalled, "and says, 'Wolfe, these things aren't even bacteria!'" Wolfe laughed gently and, for my benefit, continued reenacting the scene.

" 'Of course they are, Carl.' " They look like bacteria in the microscope, Wolfe had told him. But Woese wasn't using a microscope. He never did. He was using ribosomal RNA fingerprints.

" 'Well, they're not related to anything I've seen.'" Coming back to the present, Wolfe said: "That was the pivotal statement that changed everything."

23

W   e went into fast-forward mode," Woese recalled in his ac-
          count of these events. By the end of 1976, his team had done
          fingerprints and catalogs on five additional methanogens,
with more in the pipeline. And sure enough, he wrote, none of the new
catalogs was prokaryotic, not in the prevailing sense of that word, which
meant bacteria and only bacteria. None of the organisms was eukaryotic,
either. But "they were all of a kind!"—a third kind, something else, something anomalous, something hitherto unsuspected to exist. Woese started
thinking that he would need to declare a new kingdom of life—create a
new name, invent a huge new category—to recognize their uniqueness
and contain them. It wasn't really a new kingdom, of course. It was a
newly *discovered* natural grouping of life-forms, which had existed apart
for a long time, unrecognized, and which might be called a "kingdom"
or an "urkingdom" or a "domain," according to preferred human convention.

Woese believed that this discovery, still unannounced, offered "a rare
opportunity to put the theory of evolution to serious predictive test." He
meant Darwin's theory of evolution, as opposed to any others—the one
that recognized hereditary continuity plus a degree of random variation
over long stretches of time, and explained the shaping of that variation, to

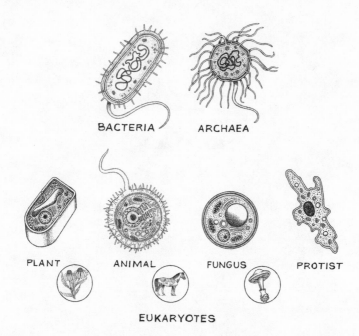

BACTERIA   ARCHAEA

PLANT   ANIMAL   FUNGUS   PROTIST

EUKARYOTES

Three domains and (within the eukaryotes) four kingdoms, four types of cell.

yield adaptation and diversity, mainly by way of natural selection. If Woese's preliminary findings were correct, he noted, those findings should serve as a guide for predicting roughly what further data and discoveries would appear. From the premise that 16S rRNA represented a very slow-ticking molecular clock, with a minimum of selected variation, he deduced that his newly found kingdom must represent a very old division. *Very* old—having originated near the beginning of cellular life, maybe three and a half billion years ago. Now he would try to sketch its boundaries and its characteristics. As he and his team added more microbes to its membership—more methanogens and maybe other creatures too, each known by its catalog of RNA fragments—Woese expected two things: that this unnamed kingdom would remain dramatically distinct from the rest of the living world and that it would nonetheless encompass great diversity. "Testing these two main evolutionary predictions," he wrote, "drove our work from that point on."

In August the team published a carefully limited paper, just a hint of what was coming, in the *Journal of Molecular Evolution*, the same journal at which Emile Zuckerkandl continued to serve as editor. It was a logical match of subject and outlet because Zuckerkandl, back in his days as Linus Pauling's sidekick, had helped articulate the very premise that Carl Woese was now putting dramatically to use: that the branching of lineages "should in principle be definable in terms of molecular information alone." The molecular information at issue in this case consisted of ribosomal RNA sequences from the first two methanogens Woese's team had characterized. One of those methanogens was a strain of *M. ruminantium*, isolated from rumen fluid (from the paunch of a cow) donated by a friendly contact in the university's Department of Dairy Science. The other was delta H, the conveniently nicknamed strain of the fourteen-syllable monstrosity, *M. thermoautotrophicum*, known to live at high temperatures and metabolize hydrogen. I asked Ralph Wolfe where they had gotten their starter sample of that exotic beast, delta H.

"It was isolated here from the sewage." More specifically, from a sewage sludge digester.

"In Urbana?"

"Yeah."

The first author on this discreet paper was Bill Balch, Wolfe's graduate student, who had earned his authorship priority by developing the sealed-tube technique of growing and labeling methanogens. "It was because of that technique," Wolfe told me, "that we could now do these experiments with Carl. Because everything was sealed, and you could now inject the P-32 into the culture." P-32, remember, was the radioactive phosphorus. "Whereas the previous techniques, you had to keep opening the stopper and flushing it out, and it would have been a radioactive nightmare to do it that way." Balch's system allowed for injecting the P-32 by syringe through the black rubber stopper. Balch grew the microbes, George Fox extracted the RNA, and Woese's trusted lab technician at the time, a young woman named Linda Magrum (she had replaced the earlier Linda in that role, Linda Bonen), prepared the fingerprint films for Woese to analyze. All three of them, plus Ralph Wolfe himself, appeared as coauthors, with Woese's name last, reflecting his role as senior author. Besides describing the methodology, this paper noted drily that the two methanogens didn't look much like "typical" bacteria. It mentioned that the divergence might represent "the most ancient phylogenetic event yet detected"—a big claim, vague enough as stated to pass almost unnoticed.

In October the team published a second paper, in a more far-reaching journal, the *Proceedings of the National Academy of Sciences* (known as *PNAS*). This time George Fox was first author, and the data covered ten species of methanogen, each one assessed for similarity to the other nine and to three species of what the authors still cautiously called "typical bacteria." Fox had created a simple measurement system by which the catalog of one microbe could be compared with the catalog of another, yielding a decimal number—a coefficient—representing degree of similarity. Comparing each of these thirteen microbes with all the others gave an overall picture of which were how closely similar to which others. The data could be arranged in a rectangular table, names down the left margin, names again across the top, numbers at each cross point, as in a chart showing the various mileage distances between all pairs in a list of cities. Instead of mileage: a similarity coefficient. From those numbers, and the premise that similarity reflected relatedness, Fox generated a dendrogram, a branching figure, showing nodes of divergence between major lineages and a branch for each organism. Although they printed this dendrogram sideways—like a bracket for the NCAA basketball tournament—rather than vertically, it was, in fact, a tree: the first of the new trees of life in the era of Carl Woese. There would be many more.

This one showed the "typical bacteria" occupying one major limb. The ten methanogens all branched from a second major limb. "These organisms," said the paper, "appear to be only distantly related to typical bacteria." Again the five authors were saying less than what they believed. The phrase "typical bacteria" was an interim delicacy that would soon disappear.

A third paper, the most bold and dramatic, appeared in *PNAS* a month later under the authorship of Woese and Fox alone. Its title hinted only obliquely at its intent: to reorganize "the primary kingdoms" of life. Again using Fox's similarity coefficients, it compared methanogens against one another and against "typical bacteria," and each of those also against several eukaryotic organisms, including a plant and a fungus. Its conclusion was radical: there are three major limbs on the tree of life, not two. The prokaryote-eukaryote dichotomy, as proposed by Stanier and van Niel, as generally accepted throughout biology, is invalid. "There exists a third kingdom," Woese and Fox wrote, "and it includes—but may not be limited to—the methanogens. It isn't the bacteria, and it isn't the eukaryotes, they explained. It's a separate form of life.

The two authors gave their kingdom a tentative name: archaebacteria. *Archae-* seemed apt, suggesting archaic, because the methanogens appeared so ancient, and their metabolism might have been well suited to early environments on Earth, about four billion years ago, before the onset of an oxygen-rich atmosphere. Woese had made that very point in an interview with the *Washington Post*. "These organisms love an atmosphere of hydrogen and carbon dioxide," he said (or at least, so he was quoted). "Just like the primitive earth was thought to be," he said, adding, "No oxygen and very warm." But the other half of that compound label, archae*bacteria*, tended to blur the central point of the discovery: that, as Woese had announced to Wolfe, these things aren't even bacterial forms of life. They're quite different. Wolfe himself told Woese that *archaebacteria* was a terrible choice. If they're not bacteria, why retain that word at all? The provisional name stuck for about a dozen years, before being emended to something better, something that stood by itself: the archaea.

24

eorge Fox was no longer a rangy young man when I sat with
him in a nondescript pizza parlor near the campus in Urbana,
after the opening session of a Carl Woese memorial symposium, and watched him eat a nondescript little pizza. Fox is a man who
prefers simple, plain food, and he had cringed when I ordered pepperoni
and mushrooms on my own. At age sixty-nine, he carried the full body
and slight jowls of a lifetime spent in laboratories and classrooms; wire-
rim spectacles had replaced the dark horn-rimmed glasses he had worn
in the 1970s photos, and his brown hair was graying at the temples, but
his eyes still shined brightly blue as he recalled the days and years with
Woese. Now a professor at the University of Houston, Fox had flown up
for the Woese meeting, which was hosted by the Carl R. Woese Institute
for Genomic Biology (its name reflecting the fact that Woese has become
a venerated brand at the University of Illinois). Fox would give one of the
invited talks.

He had spent his academic career at three institutions: Houston, for
almost three decades; preceded by Illinois, as a postdoc with Woese; and
before that it was Syracuse University, as an undergraduate and PhD student. The circumstances of Fox's arrival in Urbana were haphazard, beginning from a coincidence in Syracuse, where Woese himself grew up.

There at the university, Fox belonged to a professional engineering fraternity, Theta Tau, of which Carl Woese's father—also named Carl Woese—was a founder, and so Fox was required to know the name. As he shifted interest from chemical engineering to theoretical biology, he noticed and became fascinated by some of the early work of Carl Woese the son. In particular, there was a paper on what Woese called a "ratchet" mechanism of protein production by ribosomes—a risky proposal, a wild and interesting idea (later proven wrong in its details), published in 1970. So Fox wrote to this ratchet guy asking for a postdoc fellowship, and Woese seemed to see the Syracuse connection as karma. He had a position to fill, yes, with the departure of Mitch Sogin, the ultimate handyman grad student, and he offered that to Fox.

"We did not discuss salary," Fox said over his pizza and Coke. "He never sent me a letter offering the position. It was all completely verbal." On such assurance, Fox got married and showed up in Urbana that autumn with his wife. Arriving unannounced, he encountered a man at the lab door, an unprepossessing figure in jeans and a drab shirt, with a chain holding a huge bunch of keys. "He looked like the goddamn janitor." Fox gave his name and prepared to talk his way in. "No!? Welcome!" It was Woese.

"He sat me down in his office and . . ." Fox hesitated. "You got a piece of paper?" On a yellow sheet from my legal pad, he began sketching the layout of the lab. He drew a long rectangle and subdivided it. There were three major rooms, he explained, and the middle room, here, held the light table, where Carl usually worked. Linda Magrum and Ken Luehrsen were here, in the left room. Over here on the right side of the center room was Carl's little personal office and the electrophoresis room. The radiation room and the darkroom were across the hall, and then storage, three more spaces barely bigger than closets. Woese gave Fox a table in his office, Fox said, with a door that stayed open, "so he could see me." Like the young Luehrsen, only more so, as a postdoc, Fox was on probation.

At the beginning, Woese assigned him to assembling sequences from 5S rRNA, the shortest and least informative of the ribosomal RNA molecules, as a way of getting up to speed on what the lab was doing. That project yielded some unexpected results, impelling Woese to try to make Fox an experimentalist. But it wasn't his forte, and he knew that. He wanted to do the sort of "theoretical stuff," the deep evolutionary analysis of molecular data—what would now be called bioinformatics—that Woese himself did. Reading the code, drawing conclusions that went back three billion years and more. Woese, on the other hand, wanted him to generate data. "I was under a lot of pressure," Fox recalled—the pressure of Woese's expectations versus his own interests and skills. "What I had to do was, every other day, come up with a novel insight, so that he would continue to allow me to work on the sequence comparison project." Failing that impossible standard, he was banished back to the lab, set to the tasks of growing hot cells and extracting their ribosomal RNA. But Fox continued, in flashes, to show his value to Woese as a thinker. Gradually he proved himself, not just sufficiently to work on sequence comparisons but well enough to become Woese's trusted partner, as well as the sole coauthor on the culminating paper in 1977, with its announcement of a third kingdom of life.

25

ondering how that announcement was greeted by the scientific community at the time, I had put the question to Ralph Wolfe, several months before the pizza with George Fox. "It was a disaster," Wolfe said mildly. Then he explained, with the sympathy of friendship, why Woese's declaration of a third kingdom—the substance of the claim, and the manner in which Woese made it—had sounded discordantly to many of their peers. The crux of the problem was a press release.

Woese's lab had been supported by the National Science Foundation and the National Aeronautics and Space Administration, the latter under its exobiology program (devoted to extraterrestrial biology, in case there is any), presumably because grant administrators felt that his research on early evolution might help illuminate the question of life on other planets. As the first *PNAS* paper in the methanogens-aren't-bacteria series moved toward publication, Woese acceded to a suggestion from the federal agencies and allowed a public announcement of his findings from Washington, rather than just letting the article drop in the journal's November issue and speak for itself—which was how science, in those days, was customarily done. Ralph Wolfe knew nothing about this, despite his close connection with the work, until one day when a mutual acquaintance let slip that the press release would appear tomorrow. "What press release?" Wolfe asked.

The cat was out. It was an indelicate situation. "A few minutes later," Wolfe told me, "Carl was in my office, explaining."

Wolfe showed no dudgeon as he recounted this. The human comedy is various, not always funny; Woese's lapse was just a miscommunication between friends, a misstep by a colleague he held in high regard. To understand what went wrong, you had to consider an insult Woese had suffered years earlier, a hurt he had carried long afterward. "He presented a paper in Paris," Wolfe said. It was on the ratchet model, the same clever but incorrect idea that later caught George Fox's interest. Woese had conceived this brainstorm—a conceptual construct for how ribosomes work in manufacturing proteins—and called it a Reciprocating Ratchet Mechanism, by which RNA cranks through the ribosome structure, adding amino acids to the protein chain, a notch forward, and then a reload, and then another notch forward, but never a notch back.

"He didn't present any evidence for it," Wolfe said. "He just presented this as a concept." The audience at the Paris meeting may have included luminaries such as Jacques Monod, François Jacob, and Francis Crick, whom he knew a bit better than the others. "It was the last paper before lunch," Wolfe said, "and nobody asked any questions. They all got up, and left, and went to lunch. And this hurt Carl. It was almost a mortal wound. He was just so offended by the behavior of these scientists. He told me that 'I resolve next time they will not ignore me.' And so this was the rationale behind his press release."

The press release went out from Washington, presumably with an embargo to the date of journal publication. On November 2, 1977, the third kingdom became an open topic for all comers. The following day, based on that alert and three hours with Woese in his office, a reporter for the *Times* told the story on page 1, beneath the photo I've already mentioned—of Woese with his Adidas on a messy desk—and a headline emphasizing the ancientness theme: "Scientists Discover a Form of Life That Predates Higher Organisms." The article, by a veteran *Times* man named Richard D. Lyons, began:

Scientists studying the evolution of primitive organisms reported today the existence of a separate form of life that is hard to find in nature. They described it as a "third kingdom" of living material, composed of ancestral cells that abhor oxygen, digest carbon dioxide and produce methane.

That was relatively accurate compared with coverage in some other news outlets. The *Washington Post* did less well than the *Times*, reporting that Woese claimed to have found the "first form of life on earth," which suggested that a dawn organism, the very earliest living creature, self-assembled somehow about four billion years ago, had survived to occupy sewage in twentieth-century Urbana. Wrong. The *Chicago Tribune* was worse still, proposing that *Methanobacterium thermoautotrophicum* (misspelled) had left no fossil record because it "evolved and went into hiding" at a time before rocks had yet formed. Which rocks? "Utter nonsense," Wolfe said. The *Tribune* story even carried a dizzy headline asserting "Martianlike Bugs May Be Oldest Life." And from there the coverage spooled outward, via United Press International and other echo chambers, to small-town papers such as the *Lebanon Daily News* in Pennsylvania, under similar headlines tooting about "Oldest Life Form" rather than the distinctness between methanogens and all ("typical") bacteria. At very least, the stories bruiting "Oldest Life Form" were missing an essential point presented by Woese and Fox. A headline about "Weirdest Life Form" might have captured that better.

The problem, according to Ralph Wolfe, was not just announcing scientific results by press release but also that Carl Woese himself lacked facility as a verbal explainer. He had never developed the skills to give a good lecture. He stood before audiences—when he did so at all, which wasn't often—and thought deeply, groped for words, and started and stopped, generally failing to inspire or persuade. Then suddenly that November of 1977, for a very few days, he had the world's attention.

"When reporters called him up and tried to find out what this was all about," Wolfe told me, "he couldn't communicate with them. Because they didn't understand his vocabulary. Finally, he said, 'This is a third form of life.' Well, wow! Rockets took off, and they wrote the most unscientific nonsense you can imagine." The press-release approach backfired, the popular news accounts overshadowed the careful *PNAS* paper, and many scientists who didn't know Woese concluded, according to Wolfe, that "he was a nut."

Wolfe himself heard from colleagues immediately. Among his phone calls on the morning of November 3, 1977, "the most civil and free of four-letter words" was from Salvador Luria, one of the early giants of molecular biology, a Nobel Prize winner in 1969 and a professor there at Illinois during Wolfe's earlier years, who called now from the Massachusetts Institute of Technology (MIT), saying: "Ralph, you must dissociate yourself from this nonsense, or you're going to ruin your career." Luria had seen the newspaper coverage but not yet read the *PNAS* article, with the supporting data, to which Wolfe referred him. He never called back. But the broader damage was done. After Luria's call and others, Wolfe recollected in his memoir, "I wanted to crawl under something and hide."

To me, he added: "We had a whole bunch of calls, all negative, people outraged at this nonsense. The scientific community just totally rejected the thing. As a result, this whole concept was set back by at least a decade or fifteen years." Wolfe himself felt badly burned by the events, his professional reputation in peril. There arose a wall of resistance—cast up by visceral objection to science by press release—against recognizing the archaea as a separate form of life. "Of course, Carl was very bitter all through the eighties and well into the nineties," Wolfe said. "He was bitter that the scientific community rejected his third form. His phylogeny and taxonomy." As it had been for Stanier and van Niel, and still earlier for Ferdinand Cohn, bacterial taxonomy was a hot issue again. This time the evidence was molecular, and the deeper story was of evolution on its broadest scale.

26

It's hard to know in retrospect, and perhaps tempting to overestimate, just how severely Carl Woese was doubted, dismissed, and ridiculed during the decade following 1977. Certainly there was some of that, especially in America. But the resistance to his big claim softened somewhat after still another article, coauthored again with Ralph Wolfe and Bill Balch, offered many kinds of evidence (in addition to the 16S rRNA data) for considering methanogens a separate form of life. And in Germany, on the other hand, his idea of the newfound kingdom met a warm reception.

Researchers there—three in particular—had been developing some parallel observations. The first was Otto Kandler, a botanist and microbiologist from Munich, with an interest in cell walls, who happened to visit Urbana earlier in 1977, before the papers were published, and met Woese through Ralph Wolfe. "Ralph marched him into my office to hear the official word from George and myself," according to Woese's later memory of encountering Kandler. "I think he smiled." With a smile or not, Otto Kandler easily accepted the premise that methanogens were profoundly unique, because he had suspected it himself. His own work had shown him something even Woese and Wolfe didn't know: that the cell walls of at least one methanogen were starkly anomalous. They contained no

peptidoglycan. Remember that stuff, *peptidoglycan*—the latticework molecule, a strengthener of cell walls, that Stanier and van Niel had cited as
one of the defining characters of all prokaryotes? It didn't exist, zero, in
the cell walls of a certain methanogen Kandler was studying. Furthermore, he told Woese, it seemed absent also from some other untypical
bacteria, which lived amid high concentrations of salt. They were known,
for that affinity, as halophiles. Salt lovers.

The tip from Kandler about anomalous cell walls triggered a memory
in George Fox. He had once been taught, in a microbiology course, that
all bacteria have peptidoglycan walls—all except the extreme halophiles.
Reminded of that by the German, Fox went to the library to verify it,
and, in the process, he found another clue to the defining characters for
inclusion in this third kingdom. Here we get technical again, but I'll keep
it simple: weird lipids.

Lipids are a group of molecules that includes fats, fatty acids, waxes,
some vitamins, cholesterol, and other substances useful in living creatures for purposes such as energy storage, biochemical signaling, and
as the structural basis of membranes. Fox, rummaging in his library,
learned that halophiles contain lipids unlike those in other bacteria. They
were structured differently, with radically different chemical bonds. Carl
Woese now had another omigod moment: *Omigod, these salt lovers are full
of weird lipids, just like our methanogen.* The fact of such weird lipids in
halophiles had been reported by other researchers a dozen years earlier—
as Fox found in the library—but no one had drawn any conclusions. It
was merely a little anomaly. But for Woese, in his ferment of discovery, it
clicked into the larger pattern. "In my whole career I had never paid attention to lipids, and here we were with lipids on the brain!"

And not just the lipids he found in halophiles. Fox also turned up the
fact that two other kinds of extremity-loving bugs, known by their genus
names as *Thermoplasma* and *Sulfolobus*, also had weird lipids of the same
sort. Those two groups preferred environments that were very hot and
very acidic, such as hot springs in areas of volcanic activity. In the technical lingo, they were thermophilic and acidophilic. Perverse little beasts,
by our standards. Both had recently been isolated—one from a coal refuse
pile, the other from a hot spring in Yellowstone—and characterized in the
lab of Thomas Brock, the codiscoverer of *Thermus aquaticus*. Alerted to

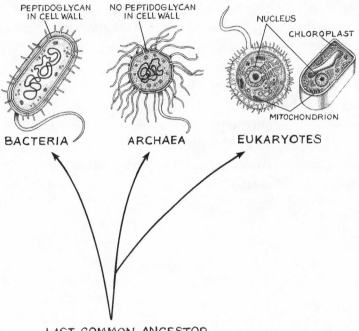

PEPTIDOGLYCAN
IN CELL WALL

NO PEPTIDOGLYCAN
IN CELL WALL

NUCLEUS

CHLOROPLAST

MITOCHONDRION

BACTERIA   ARCHAEA   EUKARYOTES

LAST COMMON ANCESTOR

The three domains of life: Bacteria, Archaea, Eukaryotes.

the weird-lipids connection by Fox, Woese got hold of samples and began trying to grow them and catalog them.

In light of all this, Woese suddenly became very keen to fingerprint some salt lovers. He reckoned that "if unusual cell walls meant anything, perhaps the extreme halophiles would turn out to be members of our new 'far out' group." George Fox, by this time, had left for the University of Houston. With Fox gone and his other lab people already busy, Woese couldn't wait for another student or collaborator to come along, so he started the wet work himself. Fortunately for him, growing halophiles is relatively easy. "I donned my acid-eaten lab coat (which had hung on the back of my office door for over a decade) and went back to the bench." He grew the cultures in quantity from samples sent by a colleague, tagged them with P-32, and turned them over to Ken Luehrsen for the dicier next step: extracting and purifying radioactive RNA. Then from Luehrsen the stuff went to Linda Magrum—"our trusty Linda," Woese called her—for separation by electrophoresis and burning the films. Within a few months, they had their first catalog from a halophile. "It didn't disappoint," Woese wrote. It was another strange thing: not a bacterium after all, but a member of the archaea.

So much for the halophiles. He turned back to the thermophilic acidophiles. When his team finished fingerprinting the coal-refuse creature, Woese sent a manuscript to the journal *Nature*, presenting the new ribosomal RNA catalog and making a case that this creature too belonged among the archaea. *Nature* rejected the paper, with a return letter that essentially said: "Who cares?"

27

The three Germans cared—not just Otto Kandler, who became a great pal to Woese, but also Wolfram Zillig, an eminent biologist who directed the Max Planck Institute for Biochemistry, in Munich, and his younger associate, Karl Stetter, formerly a student of Kandler's. After meeting Woese and hearing firsthand about his evidence and his radical idea, Kandler carried the news back to Munich, where he shared it with Stetter, then still a junior researcher. Stetter was straddling two roles—teaching in Kandler's institute at the University of Munich, running a lab within Zillig's operation, commuting between them daily—and he brought Kandler's news from America across town. When he delivered his thirdhand account in a Friday seminar at the Max Planck Institute, Wolfram Zillig's initial reaction was cold. Zillig, born in 1925, was just old enough to remember Nazism and the war from the perspective of a soldier-aged young man. As the story comes from Karl Stetter, recounted to Jan Sapp decades later, Zillig in 1977 reacted sourly to Kandler's scuttlebutt about Woese's third kingdom of life. "A Third Reich?" he snapped. "We had enough of the Third Reich!"

But Zillig's resistance fell and his interest rose when he heard, a few months later, that Woese possessed data on the uniqueness of halophiles that nicely paralleled his data on the uniqueness of the methanogens.

Zillig and Stetter then reset their own research efforts, which involved something called RNA polymerase (the enzyme that helps turn DNA code into messenger RNA), to see whether anomalies in that molecule among salt-loving "bacteria," among heat-loving and acid-loving "bacteria," and among methane-producing "bacteria"—anomalies that might set them apart from typical bacteria—matched the drastic anomalies Woese was finding by his own method. They did match. So maybe these microbes weren't bacteria after all.

Derided in the United States, controversial at best, Woese was becoming a scientific lion in Germany, at least in those erudite circles where researchers studied the molecular biology of microbes. In 1978 Kandler invited him to a major congress of microbiologists in Munich. Woese declined. In a polite but cranky letter, he groused that the National Science Foundation and NASA were being stingy with him on grant funds while enjoying the considerable publicity from his work, and also that, quite apart from the costs, travel interrupted his research. Interruptions he found annoying. He was a driven man—toward results, not companionship. But the following year, Kandler tried again, and this time Woese accepted. His hosts paid the way. They treated him well. They asked only that he deliver a keynote lecture at another microbiology conference and then a seminar at Zillig's institute. On the night of a festive dinner, in a great hall at the University of Munich, Kandler laid on a brass section from a local choir. They gave Woese a fanfare of trumpets. Not many molecular phylogeneticists ever get that level of jazzy appreciation. It melted his frosty rime.

Two years later, his German friends organized another meeting in Munich, this time an international conference—though they called it a workshop, suggesting informality and collaboration—devoted entirely to the archaea. It was the first such conference ever, giving the third kingdom a new measure of recognition. The attendance was relatively small, about sixty people, but included researchers from Japan, the United States, Canada, Great Britain, the Netherlands, and Switzerland, as well as the Federal Republic of Germany (West Germany), where the archaea were now big; and its program encompassed a wide range of topics and approaches. Ralph Wolfe came. So did Ford Doolittle, George Fox, and Bill Balch. Woese not only traveled to Munich again but also delivered the welcoming address—and he made that a substantive lecture, rich with ideas and provocations, not just a ceremonial greeting.

"We are about to embark on a scientific meeting of historic significance," he told the group (as reported later in the proceedings, edited
by Otto Kandler). What they shared, this assemblage of scientists, was
their concept of the archaea, which "did not exist four years ago." They
had been working, in their respective labs, with "organisms that intuitively felt peculiar": methanogens, halophiles, thermoacidophiles. These
things had seemed idiosyncratic and unrelated. We had been slow to recognize their connectedness, their unity, Woese said, because the existing
framework of bacterial taxonomy was so misleading in its overview and
so wrong in its details.

"Generations of failure had discouraged the microbiologist about ever
uncovering the natural relationships among the bacteria." Here he was
talking about the generations that had included Ferdinand Cohn, C. B. van
Niel, and Roger Stanier. "With a few important exceptions, microbiologists were content to classify bacteria determinatively," he added, alluding
pointedly to *Bergey's Manual of Determinative Bacteriology*, the authoritative
handbook, and the cautious experts who had produced it for sixty years.
The problem with that approach, Woese complained, was that it tried to
understand bacteria only as static entities—items to be placed into categories of convenience. "Matters of their evolution became reserved for en-
joyable but idle after-dinner speculation." That's what was missing from
both microbiology and now molecular biology, he said: evolution.

Woese was casting down a gauntlet: telling some of the most brilliant
and influential figures of late-twentieth-century biology—his friend Francis Crick, Crick's colleague James Watson, the Nobel winners François
Jacob and Jacques Monod and Max Delbrück and Salvador Luria, who had
counseled Ralph Wolfe to stay away from Woese for the sake of his good
reputation—that they were shallow, mechanistic thinkers with no curiosity about life's history. That they were nothing but code breakers, riddle
solvers, and engineers. The questions and answers offered now by the
recognition of the archaea, he said, should go far to revivify evolutionary
thinking, and "hopefully divert biology to some extent from its present
course of technological adventurism." By that odd phrase, "technological adventurism," he seems to have meant not just high-tech molecular
biology for its own sake, without regard for evolutionary questions, but
also perhaps gambits in genetic manipulation. It was a condemnation so
damning and prescient, this whole 1981 rant, that you might imagine he

had foreseen gene patenting, the growth of the biotech industry, gene-editing therapies, preimplantation screening of human embryos, and full-on human germline engineering. He set this "technological adventurism" against "molecular evolutionary biology," his ideal, but an unspoken phrase, which at that time would have seemed oxymoronic.

That's the notable takeaway from his 1981 Munich talk: it reflects Carl Woese's compulsion to dig ever deeper into the narrative of life. He was a man possessed by the most deep-diving curiosity. This work he was doing, this door he had opened, this journey he was on—it wasn't just about the Archaea, a third kingdom. It was about the origins and history of the other two kingdoms also. How did they arise? How did they diverge from one another? How were each of the three related to the two others? Which came first? Why did just one of the three lineages lead onward to all visible, multicellular organisms—all animals, all plants, all fungi, ourselves—while the other two remained unicellular and microscopic, though still vastly abundant, diverse, and consequential? And what kind of creature, or process, or circumstance preceded them all? Where was the tree of life rooted?

Woese wasn't interested just in this separate form of life he had chanced upon. He was interested in the whole story.

Immediately after the workshop, which had gone well and given its participants a sense of momentum for the archaea concept, Kandler and his wife took Woese and Wolfe on a larkish field trip. They drove south from Munich into the Bavarian Alps and climbed a modest but picturesque mountain, the Hohe Hiss, along a graded path. "Woese and especially Wolfe were not in top physical shape, but with some huffing and puffing, they reached the top," according to Ralph Wolfe's own self-mocking account. At the summit, Kandler's wife took a photo of the three men, all of them sunlit and contented on a clear day. Wolfe and Kandler appear as what they are: middle-aged scientists, balding, amiable, savoring a day outdoors. To their right sits Woese, with a full beard, leonine hair, a sweater tied jauntily over his neck, a cup of champagne in his left hand, smiling an easy, full smile of triumph. He was fifty-two years old, at the height of his powers and fame, and looked like a man on his way to a Nobel Prize.

PART ✦ III

Mergers and Acquisitions

28

T he entrance of Lynn Margulis into this story occurred abruptly, with some fanfare, at a time when Carl Woese still labored in obscurity. Margulis was a forceful young woman from Chicago. Her role proved important because it brought new attention and credibility to a very strange old idea: the idea that living ghosts of other life-forms exist and perform functions inside our very own cells. Margulis, adopting an earlier term, called that idea *endosymbiosis*. It was the first recognized version of horizontal gene transfer. In these cases, rare but consequential, whole genomes of living organisms—not just individual genes or small clusters—had gone sideways and been captured within other organisms.

Margulis made her debut in March 1967 with a long paper in the *Journal of Theoretical Biology*, the same journal that had carried Zuckerkandl and Pauling's influential 1965 article on the molecular clock. This paper was much different. Its author was no canonized scientist like Pauling, and its assertions were peculiar, to say the least. Put more bluntly: it was radical, startling, and ambitious, proposing to rewrite two billion years of evolutionary history. It included some cartoonish illustrative figures, funny little pencil-line drawings of cellular shapes, and virtually no quantitative data. According to one account, it had been rejected by "fifteen or so" other journals before a daring editor at *JTB* accepted it. Once published, though, the Margulis paper provoked a robust response. Requests for reprints (a measure of interest, back in those slow-moving days before online access to journals, when scientists *mailed* one another their articles) poured in. It was titled "On the Origin of Mitosing Cells."

That was a quiet phrase for a huge subject, though the title's echoes of Darwin's *On the Origin of Species* suggest the loud aspirations of the paper's author. Never short of confidence, she was twenty-nine years old at that time, an adjunct assistant professor at Boston University, and a single mother raising two boys. She had been married as a teenager to a flashy young astronomer and, for the moment, was still keeping his surname. Her authorship on the paper read: Lynn Sagan. Later, she would be famous—venerated by some, dismissed and disparaged by others, including Carl Woese—under the surname of her second husband, Thomas N. Margulis. But to many of those who knew her, she was always and informally: Lynn.

The phrase "mitosing cells" is another way of saying eukaryotic cells, the ones with nuclei and other complex internal structures, the ones that compose all animals and plants and fungi (as well as some other intricate life-forms, less familiar because they're microscopic). "Mitosing" refers to mitosis, of course, the phase in eukaryotic cell replication at which the chromosomes of the nucleus duplicate, then split apart into two bundles within two new nuclei, as a prelude to the cell fissioning into two complete new cells, each with an identical set of chromosomes. You learned about it in high school biology, not long before you dissected the poor frog. Mitosis is taught along with meiosis, the yang to its yin. Mitosis occurs during ordinary cell division, whereas meiosis constitutes "reduction division," yielding the specialized sex cells known as gametes (eggs and sperm in an animal, eggs and pollen in a flowering plant). Meiosis in an animal yields four new cells, not two, after two divisions, not one, each resulting cell reduced to a half share of chromosomes. Later, sperm will meet egg, and, bingo, the full measure will be restored. It's a little hard to remember which of those terms is which, I concede, but here's my mnemonic: meiosis is reduction division because its spelling is reduced by the loss of the *t* in *mitosis*. Helpful? Granted, that leaves the inconvenient fact of meiosis containing the addition, not reduction, of an *e*. So, okay, never mind. But it works for me.

Mitosis defines all the cell divisions by which a single fertilized egg grows into a multicellular embryo and then an adult, and also by which worn-out cells are replaced with new cells. Your skin cells, for instance. The cells of a scar when a wound heals. The cells that replace your worn-out colon lining. Mitosis occurs everywhere in a body. Meiosis, by contrast, occurs only in the gonads. Lynn Sagan's paper, though, wasn't focused on mitosis as an ongoing process. The key word in her title was *origin*.

Her interest was the deep history, to the beginning, of eukaryotic cells. She quoted the statement from Roger Stanier and his textbook coauthors, declaring that the prokaryote-eukaryote distinction "probably represents the greatest single evolutionary discontinuity to be found in the present-day living world." It was the biggest leap in the history of life—an Olympic long jump, a high jump, a backward slam dunk—forever reflected in the differences between bacteria and more complex organisms. She proposed to explain how that leap happened.

"This paper presents a theory," Sagan wrote—a theory proposing that "the eukaryotic cell is the result of the evolution of ancient symbioses." Symbiosis: the living together of two dissimilar organisms. She gave her theory the more specific name endosymbiosis, connoting one organism resident *inside the cells* of another and having become, over generations, a requisite part of the larger whole. Single-celled creatures had entered into other single-celled creatures, like food within stomachs, or like infections within hosts, and by happenstance and overlapping interests, at least a few such pairings had achieved lasting compatibility. So she proposed, anyway. The nested partners had grown to be mutually dependent, staying together as compound individuals and supplying each other with certain necessities. They had replicated—independently but still conjoined—passing that compoundment down as a hereditary condition. Eventually they were more than partners. They were a single new being. A new kind of cell.

No one could say, not in 1967, how many times such a fateful combining had occurred during the early eras of life, but it must have been very rare that the resultant amalgams survived for the long term. Later, there would be ways of addressing that question. Sagan left it open. Microscopy, which was her primary observational mode of research, couldn't answer it.

The little entities on the inside of such cells had begun as bacteria, she argued. They had become organelles—working components of a new, composite whole, like the liver or spleen inside a human—with fancy names and distinct functions: mitochondria, chloroplasts, centrioles. Mitochondria are tiny bodies, of various shapes and sizes but found in all complex cells, that use oxygen and nutrients to produce the energy packets (molecules known as adenosine triphosphate, or ATP) for fueling metabolism. ATP molecules are carriers of usable energy, like rechargeable AA batteries; when the ATP breaks into smaller pieces, that energy is released for use. Mitochondria are factories that build (or recharge) ATP molecules. To drive the production, mitochondria respire, like aerobic bacteria. Chloroplasts are little particles—green, brown, or red—found in plant cells and some algae, that absorb solar energy and package it as sugars. They photosynthesize, like cyanobacteria. Centrioles are crucial too, but for now, I'll skip the matter of how. All these components, Sagan wrote, resemble bacteria by no coincidence but rather for a very good reason: because they evolved from bacteria.

The bigger cells, within which the littler cells were subsumed, had been bacteria too (or possibly archaea, though that distinction didn't exist at the time). They were the hosts for these endosymbioses. They had done the swallowing, the getting infected, the encompassing, and had offered their innards as habitat. The littler cells, instead of being digested or disgorged, took up residence and made themselves useful. The resulting compound individuals were eukaryotic cells.

Never mind that "compound individuals" is oxymoronic. The whole process, as Sagan described it, was oxymoron brought to life—paradoxical and counterintuitive, though supported throughout the paper by her detailed arguments.

Paradox is enticing, but was it real? Was it right? Had this adjunct assistant professor presented not just an astonishing cluster of possibilities but also a persuasive new vision of the origins of all complex life? The scientific consensus at first, and for some years afterward, was no. The early read on Lynn Sagan, soon to be Lynn Margulis, held that she was smart, knowledgeable, insistent, charming, and in thrall of a loony idea.

29

She was born and raised in Chicago, the eldest daughter of Morris and Leone Alexander, her father an attorney who also owned a paint company, her mother a homemaker who also ran a travel agency—enterprising and versatile people. Lynn Alexander was precocious but also a "bad student," by her own account, at least bad in behavior, enough so that she stood in the corner a lot. (Hard to know whether to take that literally. Later in life, she stood in remote corners of the scientific community often, proudly, and by choice.) Brilliant and impatient, she switched schools, worked through a little adolescent revolt, and was good enough to get early admission to the University of Chicago as a young teenager. She loved her experience there—more particularly her experience at the College, as the undergrad school was known, a haven of broad learning within the university, its system of pedagogy shaped by the educational visionary Robert Maynard Hutchins. She thrived in an introductory course, Natural Science 2, that had her reading not textbooks but the seminal writings of great scientists themselves: Darwin, Weismann, Gregor Mendel, and J. B. S. Haldane, among others. One day in her freshman year, as she bounded up the steps of the mathematics building, she literally ran into Carl Sagan, then a nineteen-year-old graduate student in physics. He was tall, handsome, articulate, and polished,

already something of a public figure on campus. "I was a scientific ignoramus," she would recall. "Carl, and especially his gift of gab, fascinated me." Three years later, one week after her graduation, she married
him and became Lynn Sagan. Photos from the event show her as a small,
pretty young woman, bare shouldered in a white gown and pearls, with
a dangerous smile.

She accompanied Sagan to Wisconsin, where he continued his grad
work at an observatory while she started a master's degree at the University of Wisconsin. That's where she met Hans Ris, a professor in the
Zoology Department, who taught her microscopy.

Ris was "a fine teacher—the best of my whole career," she wrote later.
She took his class on cell biology, probably in 1959, while she was pregnant with her first son (who would become the writer Dorion Sagan). In
addition to microscopy, Ris seems to have given her more: a pathway,
from obscure earlier sources and his own research and thinking, to her
theory of endosymbiosis. Some testimony to his influence remains, like
fossil fragments among the rocks of a dusty gorge, in the references at
the end of her 1967 paper. There she cited a paper coauthored by Ris a
few years before, and among many other citations, she included work by
two unconventional scientists from the early twentieth century, a Russian
and an American: Constantin Merezhkowsky (whom Ris had also cited)
and Ivan E. Wallin. These men anticipated some of the ideas that, pulled
together later by Lynn Margulis and affirmed still later with molecular
evidence, would radically change the understanding of how complex organisms arose.

Hans Ris was a Swiss-born cell biologist and biochemist who, coming to the University of Wisconsin in 1949, had reinvented himself as an
electron microscopist. In the early 1960s, with his colleague Walter Plaut
from the Botany Department, he used microscopy and biochemical methods to investigate chloroplasts, the tiny cell organelles found in plant cells
and some algae, enabling them to harvest solar energy by photosynthesis.
What *are* these chloroplast things, Ris and Plaut wondered, and what's
their origin? The two men looked closely at the chloroplasts of a certain
green alga. With biochemical staining, they found evidence of DNA.
They could see it with their electron microscope.

This was important because it suggested the possibility that genes can
exist in the cytoplasm (the stew of liquids and solids inside a eukaryotic

cell that includes everything but the nucleus) and not just in the nucleus itself. Genes in the cytoplasm were previously considered improbable if not impossible—except by a few earlier researchers. Chromosomes resided in the nucleus, protected there within the nuclear membrane, and that was thought to be that. Cytoplasmic inheritance, if it was real, represented an exception to the reliable rules of Mendelian inheritance, as articulated by Mendel, the Moravian monk who discovered those rules by crossbreeding peas. In Mendelian inheritance, determined by a sex cell from each parent, merging at fertilization, both parents contribute equally to the genetic makeup of the offspring. Cytoplasmic inheritance (also called maternal inheritance) was a very different proposition. If it existed, it wouldn't be so neatly Mendelian. It wouldn't be so binary. If genes were afloat in cytoplasm, that would tilt inherited genetic identity toward the female parent, in any sexual reproduction, because eggs carry a lot of cytoplasm, and sperm or pollen carry little.

But that wasn't the half of it. Ris and Plaut's detection of DNA in chloroplasts, within the cytoplasm of green algae, had even larger implications than challenging Mendel. Those implications pointed toward endosymbiosis, a wholly unorthodox vision of the origins of complex life.

By electron microscopy, Ris and Plaut detected aspects of their algal chloroplasts that closely resembled what was seen in certain bacteria: DNA fibrils, a double membrane, and other structural features. Specifically, these chloroplast traits seemed to match with the microbe group soon to be renamed cyanobacteria. This suggested that the chloroplasts, in a sense, *were* bacteria; or, at least, that they once had been. It suggested that cyanobacteria had been swallowed or otherwise internalized somehow in the deep past, that some or at least *one* of those captures had resisted digestion or ejection, that it had replicated inside a host cell, that those replicates were inherited through the lineage of algal cells, and that they gradually transmogrified from undigested prey, infectious bug, or neutral passengers to internal organelles. They survived and proliferated (by Darwinian selection) because they served a function: allowing the algae to derive energy from sunlight. Their role, as organelles, was photosynthesis. All this agreed with an old hypothesis, Ris and Plaut noted, that Constantin Merezhkowsky had proposed back in 1905. Merezhkowsky had been considered half crazy in his time. (In fact, half crazy and worse, as you'll see.) But now, Ris and Plaut wrote, "endosymbiosis must again be considered seriously as a possible evolutionary step in the origin of complex cell systems."

The paper by Ris and Plaut appeared in 1962, soon after young Lynn Sagan left Wisconsin to continue her life and studies elsewhere. She not only saw the paper and read it but also, from her personal contact with Ris, had presumably known it was coming. According to one account, she was introduced to these radical notions even earlier, and more directly, by Hans Ris himself. A classmate of hers in the Ris course on cell biology, back in 1959, remembers Ris laying out a very full treatment of endosymbiosis, along with supporting literature on its various aspects from obscure German and Russian sources, which Ris had pulled together. This classmate, Jonathan Gressel, now an emeritus professor of plant genetics at the Weizmann Institute of Science in Israel, says that "the theory was completely Ris's idea, well set out in his course. She did a great job of promulgating it." Gressel was friendly with Lynn Sagan at the time, and he recalls her "struggling to reach the microscope," game but slightly impeded, when she was heavily pregnant with Dorion. Later, he was "aghast" that she hadn't given Ris fuller acknowledgment for assembling the theory.

From the University of Wisconsin, with a master's degree, she followed Carl Sagan to the University of California at Berkeley, where he had a postdoctoral fellowship. Their second son, Jeremy, was born in 1960. While her husband studied the possibilities of extraterrestrial biology (soon to be known as exobiology, the same discipline that helped get Carl Woese funded by NASA), she started on a doctorate in genetics, working around or somehow beyond the demands of mothering. "I was interested in evolution," she wrote later, "and I always thought genetics was the way to study evolution in a deep way." Lynn wanted to explore the phenomenon of non-Mendelian genetics, aka cytoplasmic inheritance, as she had learned of it from Ris and others at Wisconsin. She felt enticed by its implication: that genes are afloat in the cytoplasm of complex cells, not just bound up in the chromosomes within their nuclei, and that those genes could be very different from the nuclear genes. If so, where had they come from?

But her advising professor disapproved of the topic. His disapproval was grounded within a broader problem that rankled her: the intellectual disjunction between scientific disciplines. "At Berkeley," she recalled, "there was absolutely no relationship between members of the department of paleontology, where evolution was studied, and those of the department of genetics, where evolution was barely mentioned." She called it "academic apartheid." The geneticists on her side of campus had mostly begun as chemists, progressed into biology only so far as studying bacteria and viruses, and knew little or nothing about her great interest, cytoplasmic inheritance in eukaryotic creatures. They were so arrogant and ignorant, these Berkeley geneticists, that "they did not even know they did not know." There she was, a twenty-three-year-old woman, with two young sons, and only a master's degree plus her boundless self-confidence to set against them. Lynn knew she needed the doctorate so, in lieu of delving into evolutionary genetics, she produced a safer and smaller dissertation on the pond-water microbe *Euglena gracilis*. If I quoted you its title, your eyes would glaze like a Krispy Kreme doughnut.

That was only a temporary setback for such a determined young woman. She was still busy at Berkeley in 1962 when the Ris and Plaut paper appeared, offering evidence of DNA in the chloroplasts of green algae. DNA in the chloroplasts meant cytoplasmic inheritance—genes passing down through a lineage of eukaryotic cells, complex organisms, independent of any role for fatherhood or the chromosomes. Her fascination with cytoplasmic inheritance grew, despite the discouragement of her dissertation advisor. She read into it more deeply, including a classic old book by the cell biologist E. B. Wilson, *The Cell in Development and Heredity*, in its 1925 edition, which mentioned still earlier works by Merezhkowsky and Wallin, both of whom had suggested that cell organelles such as mitochondria and chloroplasts were the evolved remnants of captured bacteria. Wilson called Merezhkowsky's proposal "an entertaining fantasy," dismissed Wallin somewhat more gently, and remained cautious on the subject himself, while admitting: "To many, no doubt, such speculations may appear too fantastic for present mention in polite biological society; nevertheless it is within the range of possibility that they may some day call for more serious consideration." Lynn Sagan figured that day had come.

Meanwhile, her marriage to Carl Sagan was foundering. Later, she would call Sagan "unbelievably self-centered," a neglectful father, and a husband more needy of adoration than she could bear. The marriage, she would say, was like "a torture chamber shared with children." She took the boys and left, moving into a place north of Berkeley with another young mother. But in 1963, when Carl Sagan accepted an assistant professorship at Harvard, she agreed to bring the boys and join him in an apartment west of Cambridge. Although he seems to have hoped they could save the marriage, to her it was "a move of convenience" with some ulterior motivation. Her doctorate hadn't yet been granted, but she could settle that at a distance. She was through with California, disinclined to return to Chicago, and she thought Massachusetts might work out. It did, but not with Sagan.

She divorced him in 1964. During this difficult period, working one job as a staff member with an educational services company, another as a lecturer at Brandeis University, raising two kids, and with modest financial help from her father, she found time to gather a vast array of facts, ideas, and references on endosymbiosis. She wrote her long paper on the subject, submitted it to a series of journals, and saw it rejected by those "fifteen or so" before the *Journal of Theoretical Biology* finally accepted it. She remarried, this time to Thomas N. (Nick) Margulis, a crystallographer, and took his surname. She got hired as an assistant professor at Boston University.

Then, in 1969, pregnant with her third child (another son, Zachary) and obliged to stay home for extended periods, Lynn returned to endosymbiosis. "Enforced home leave permitted uninterrupted thought," she wrote later. Her elder boys had started school. Her story of complex cells—that they had originated from the fusion of other life-forms, as encapsulated in the 1967 paper—now "sprouted, expanded, and eventually was pruned into a book-length manuscript." Her recollection of "pruning" it to book length (329 pages, as published, not counting the index) reflects the fact that she was never shy, slow, or laconic about putting words on paper. She had a contract with Academic Press in New York. "I typed late into many nights, determined to make the deadline." Finally, she boxed up the manuscript and the many illustrations she had commissioned for it, and mailed it all off to the publisher. This would represent a moment of triumph, and the beginning of tense expectation, for any writer. She waited. Five months later, the box came back, by the cheap rate for printed matter, without explanation. The peer reviews from other scientists had been bad, but at first Academic Press didn't give her even the courtesy of hearing that. Eventually she got a form letter of rejection.

Lynn went back to work, revising the manuscript and offering it else-where. This time it found appreciation and good editing at Yale University Press, by which it was published in 1970 under the title *Origin of Eukaryotic Cells*. She had dropped that bit about *"On the* Origin . . . ,*"* as she had used it for the 1967 paper, which was a wise and modest move, making the book title less closely an echo of Darwin's. Still, her book too was a landmark, if not a classic. For many scientists interested in cell biology and deep evo-lutionary history, *Origin of Eukaryotic Cells* was the work that introduced the ideas of endosymbiosis and the name Lynn Margulis. Some of them thought she was nuts and some didn't.

30

Lynn Margulis wasn't the first to propose these unorthodox ideas, nor was Hans Ris. As she had learned at least passingly from reading E. B. Wilson's old book, speculations about swallowed bacteria as the ancestors of cell organelles, about all complex cells having originated from combinations of simpler organisms, had been voiced for almost a century, though not always in "polite biological society." One of those voices, not the first, not the steadiest, but with a timbre all its own, belonged to that Russian mentioned by Wilson and cited by Ris: Constantin Sergeevich Merezhkowsky.

Merezhkowsky was born in 1855, in Warsaw, which was then part of the Romanov Empire. His father was a court official and a hidebound conservative, and Constantin, as eldest son among nine siblings, probably faced the first, fullest barricades of parental expectations. He went through a period of student radicalism, sympathetic with Russian revolutionaries opposing the tsar (whom they would assassinate in 1881), and defied his father's wishes about a pragmatic career track, choosing to study natural sciences rather than law. During his student years at the University of St. Petersburg, Merezhkowsky joined a summer expedition to the White Sea and got interested in marine invertebrate animals, including amorphous creatures such as polyps and sponges. At age twenty-two, he published a paper on protozoans (a loose group of single-celled eukaryotes, including amoebae). More field trips, ranging as far as the Bay of Naples, gave him more opportunities to study protoplasmic gobs of life that had gotten little attention, and at one point he identified what he thought was a new species of sponge. It's hard to imagine the crusty father back in Warsaw sharing much excitement at that discovery. In any case, Merezhkowsky was wrong. His sponge turned out to be another protozoan, but bigger.

He graduated in 1880, traveled through Germany and France for a few years, then returned to St. Petersburg and qualified as a privatdozent, a sort of freelance lecturer within the ambit of the university. He married a woman named Olga, and, after three years, they decamped to Crimea, which was then also part of the empire. On the southern coast of the Crimean peninsula, beyond the mountains, he worked as a pomologist—a fruit wrangler—supervising orchards. These facts come from a biographical study by a team of three scholars, including the excellent Jan Sapp; but where Merezhkowsky had picked up pomology skills, even Sapp and his colleagues don't say. They do note: "Merezhkowsky's career was unsettled between 1880 and 1902." From age twenty-five to forty-seven, in other words, he knocked around, investigating this and that, here and there, and paying the bills God knows how. He did research on grapes. He studied the physical development of children, an odd digression for an invertebrate zoologist, and he described a method of measuring their bodies. That could have been innocent but looks creepy in light of later events. In 1898 he suddenly left Crimea, without Olga, without their son, Boris, and perhaps incognito, moving one jump ahead of local outrage. He stood accused of child molestation.

The next place Merezhkowsky landed was California, traveling under a false passport as "William Adler." He must have spoken with a strong Russian accent (though he wrote English well, as judged by the sponge-discovery paper, which had been published in London), and presumably the name William Adler was a poor fit—but this was California in the Gilded Age, and he wouldn't have been the only pilgrim reinventing himself. While on the lam, he wrote a fantastical novel, *The Earthly Paradise*, ramblingly subtitled *A Winter Night's Dream. A Fairytale of the Twenty-seventh Century: A Utopia*, which was later published in German for lighthearted souls to enjoy. I suppose it bears saying that no, I'm not making this stuff up. Nor did Jan Sapp concoct or hallucinate it; one of his coauthors, Mikhail Zolotonosov, researched Merezhkowsky's later pedophilia case in seventy Russian newspapers and secret files of the police.

As the century turned, Merezhkowsky spent time at a research station on the coast south of Los Angeles, then at Berkeley, continuing his work on marine life when he wasn't writing fantasy fiction. He focused now on diatoms, a kind of single-celled alga with each cell enclosed in a shell-like silica wall. Diatoms are extraordinary; minuscule beings aren't supposed to seem so intricately and geometrically neat. Many of them contain chloroplasts, meaning they live by photosynthesis, like plants. Although their external walls show a great range of diversity, decorative and useful for classification, Merezhkowsky began reclassifying them by their internal anatomy, including the chloroplasts. Dwelling on those chloroplasts, which through a microscope looked very much like bacteria, may have led him toward the big idea of his life: that internalized bacteria had *become* chloroplasts, not just in photosynthesizing algae but also in plants. He eventually named this phenomenon *symbiogenesis*, defined by him as "the origin of organisms by the combination or by the association of two or several beings which enter into symbiosis."

The simpler word *symbiosis* had an earlier history of its own, originally as applied to a community of people. Its biological usage dates back to a German biologist, Anton de Bary, in 1879, for whom it meant any sort of merger or close cohabitation by two or more different forms of life. The term encompassed relationships ranging from parasitism, to temporary partnership (with one or both partners deriving benefit), to intimate and heritable integration of the sort that Merezhkowsky would later propose. De Bary himself recognized that a lichen, for instance, represents not one creature but a symbiotic association between at least two kinds: algae or cyanobacteria living amid fungi. A clown fish, feeding blithely on parasites amid the stinging tentacles of a sea anemone, is engaged in symbiosis. But the idea that one organism could exist permanently *inside the cells* of another, replicating itself as the cells replicate themselves and becoming part of a new, composite, and heritable identity, was taking the idea a step further: symbiogenesis.

Several researchers of the late nineteenth century had considered that possibility in connection with chloroplasts, including de Bary's student Andreas Schimper, an adventurous German botanist from a family of eminent scientists, who traveled widely during his early manhood for fieldwork in the West Indies, South America, Africa, and the Indian Ocean. Schimper grew up in that French borderland city, Strasbourg, where the Rhine River divides Germany from Alsace. One photo shows him looking youthful, earnest, and wide-eyed, weighted down with a large handlebar mustache as though he had pasted it on for a school play. During the mid-1880s, not yet thirty years old, he published two memorable papers, in one of which he coined the term *chloroplast* and mused that, if these things reproduced themselves within plant cells rather than arising anew from plant cytoplasm, the compound entity "would be somewhat reminiscent of a symbiosis." Just an analogy, just an aside, which Schimper didn't pursue further, partly because it seemed too bizarre to be taken literally, and maybe also because he died early, age forty-five, from ruined health after an expedition to malarial regions of Cameroon.

Merezhkowsky read one of Schimper's papers, and, lo, the idea of symbiogenesis came to him in what he called "a completely spontaneous way." In 1902 he returned to Russia, but not to Crimea, where memories of his alleged or real turpitude might have made him unwelcome or, worse, wanted. Instead he found a position at Kazan University, five hundred miles east of Moscow on the Volga River. He became a privatdozent again. Three years later, he published a paper, in German, like Schimper, on the origin of chloroplasts in plants, enunciating the symbiotic theory. That was his famous work of 1905, in which he identified cyanobacteria (he called them *cyanophytes*) as the strangers that came to stay. In a series of further papers over the following fifteen years, he expanded on the theory, gave it that name (symbiogenesis), and claimed it as uniquely his. Never mind poor Schimper, long gone. In truth, according to Jan Sapp and others, Merezhkowsky did more than anyone else to promote this partial version of the bigger idea—that chloroplasts, at least, are captured bacteria within composite organisms called plants—until the rise of Lynn Margulis.

The accepted view of how plants came to be plants, Merezhkowsky wrote in the 1905 paper, with a tinge of derision, was that chloroplasts were simply innate "organs" of each cell, which had "gradually differentiated" out of the otherwise colorless cytoplasm. That was the endogenous theory: chloroplasts had taken shape on the inside of plant cells, formed from internal materials. Not so, he argued. Rather than being homegrown organs, they are "foreign bodies, foreign organisms" that invaded the cytoplasm of animal cells sometime in the distant past and entered into a symbiotic coexistence. According to this theory, a plant cell is nothing but an animal cell with photosynthetic bacteria added. The plant kingdom derived from the animal kingdom by symbiogenesis. It had happened several times, in several independent events of conjunction, according to him—maybe as many as fifteen, giving the plant kingdom that many separate origins. How the animal kingdom originated was another question, which he mostly ignored.

After ten pages of close argument, Merezhkowsky ended with an evocative passage that has become famous to those who read the literature on cell origins, and that has remained unknown, incomprehensible, to everyone else:

> Let us imagine a palm tree, growing peacefully near a spring, and a lion, hiding in the brush nearby, all of its muscles taut, with bloodthirsty eyes, prepared to jump upon an antelope and to strangle it. The symbiotic theory, and it alone, lays bare the deepest mysteries of this scene, unravels and illuminates the fundamental principle that could bring forth two such utterly different entities as a palm tree and a lion.

How does symbiogenesis explain the palm tree and the lion? Well, the tree behaves peacefully because it contains all those pacific little workers—those docile "green slaves," the chloroplasts—nourishing it from sunshine. The lion needs meat. So it kills. But wait: imagine each cell of the lion filled with chloroplasts, Merezhkowsky suggested. They would generate sustenance for the lion from solar energy. Thus equipped, "I have no doubt that it would immediately lie down peacefully next to the palm, feeling full, or needing at most some water with mineral salts."

That would be sufficient diet for his green lion: sunbathing and Gatorade.

A nice idea, but he was wrong. As several alert biochemists from our own time have noted, a lion has vastly less surface area than a plant of equivalent mass (think about all those spreading palm fronds or the canopy of an oak), so that spackling its skin with chloroplasts, like a chartreuse sequined suit on Liberace, wouldn't catch nearly enough sunlight to power a robustly leonine life. The energy input would be insufficient.

The lion would wind down to a torpid, immobile groan, weak as an Energizer Bunny with bad batteries.

These papers on symbiogenesis didn't bring Merezhkowsky either scientific eminence or peaceful repose. He was a lion without meat, without even chlorophyll, hungry and mean. He shifted his politics from left to right, and during the Kazan years, he became an informer to the tsar's oppressive, anti-Semitic secret police. He denounced a Jewish colleague who was up for promotion. He seems to have continued "measuring" children. He absconded from Russia the second time in 1914, with charges swirling behind him that he had raped twenty-six little girls, including at least one he had tutored. Criminal cases were opened in Kazan and St. Petersburg. The charges never came to trial, so they are only accusations, not damnations, but presumably there was some basis. He went to France and continued writing, not just papers on symbiogenesis but also another dizzy philosophical book with science-fiction trappings, featuring a "seven-dimension oscillating universe," spiritualism, atheism, and eugenics, as well as cosmic evolutionary theory; then, in his final year, a short manuscript of "Instructions for My Disciples," offering himself as savior of the world. He was the L. Ron Hubbard of his day, crazed and megalomaniacal, but without the successful marketing or the celebrity followers. Not even Mikhail Zolotonosov, Jan Sapp's deep-burrowing Russian coresearcher, seems to know what became of Olga and Boris. By the time Merezhkowsky produced his last scientific article, "The Plant as a Symbiotic Complex," published in French in 1920, he was burning down to the end of the wick.

By then, he had moved to Geneva and holed up in a picturesque hotel. He tried to arrange some lecturing on symbiogenesis but was thwarted by a professor at the university, a botanist, who presumably viewed him as a blackguard or a flake, very possibly both. It was a difficult time for many people, just after the Great War, and not least difficult for expatriate Russian biologists with unsavory backstories, unorthodox theories, and delusions of grandeur. As he went broke, Merezhkowsky blamed that condition on the war. The only things he seems to have seen clearly, at this point, were chloroplasts and his own approaching end.

Sapp's group found an obituary note, with unusual detail, published in a Geneva newspaper, *La Suisse*, on January 11, 1921. Two days earlier, police had been called, after a porter at the hotel noticed a letter shoved out under the door of room 58, Merezhkowsky's. It warned: "Do not enter my room. The air in it is poisoned. It will be dangerous to enter for several hours." The police duly waited two hours. Then they went in, finding Merezhkowsky elaborately arranged and dead.

He had mixed up a brew of chloroform and several kinds of acid. He had poured that into a container rigged to the wall above his bed, like an IV drip. Instead of a needle for his arm, there was a mask for his face. But first he had sealed off the room, lain down, and tied himself to the bed, leaving one arm free. How do you tie yourself to a bed? Merezhkowsky was enterprising. How do you concoct such a recipe for death? Merezhkowsky was a scientist. Sapp and colleagues think it was a ritual suicide of some sort, related to his delusional metaphysics. Maybe so. The magistrate on the scene, according to *La Suisse*, found an inscription in Latin pinned to a cord. It might have been more esoteric ravings, or just a final yelp of despair, pretentiously expressed. But the Latin note disappeared when Merezhkowsky's Geneva police file was destroyed. Ritual or not, he had his plan. He put the mask on his face and opened a valve.

This made for a lurid ending to a strange life, but probably the strangest point in the whole story of Constantin Merezhkowsky is that, on the subject of chloroplast origins in plants, the pillar of his symbiogenesis theory, he was right. Fifty-four years later, that idea would be confirmed, with molecular data, using the methodology invented by Carl Woese.

31

The other pioneering theorist in this field who rose to a certain threshold of notice, the American, was Ivan E. Wallin. He was a son of Swedish immigrants but a corn-fed midwesterner, born in Iowa, eventually an anatomist at the University of Colorado School of Medicine. Wallin's publications in the 1920s, like Merezhkowsky's before him, became just well enough known to be mentioned, but not discussed, in the early works of Lynn Margulis.

Wallin offered a version of endosymbiosis that differed from, but complemented, Merezhkowsky's. He argued that mitochondria in all complex organisms, not just chloroplasts in plants and algae, are descended from captured bacteria. Mitochondria, you'll recall, are the little particles within cells that burn food and oxygen, packaging that energy into units of ATP (the energy-carrier molecule) for fueling the life of the cell. They perform other services too, but neither Wallin nor anyone else had yet figured out their functions. Wallin cared more about their origin. He wasn't the first biologist who, viewing mitochondria through a microscope, found their resemblance to bacteria striking. But for him, it became the crux of a research program. He set out to prove, with an exhaustive campaign of experiments, that the resemblance was more than coincidental.

Beginning around 1920, he became fascinated with the idea, mentioned by at least one earlier researcher but never persuasively developed, that mitochondria are the descendants of internalized bacteria. He launched a course of experimental investigations. His tools were simple, the basics of microscopy and microbial culturing in his day. He had no grant funding, just a bit of money given "from time to time" by a wealthy patron. He had no collaborators and no graduate students, just a pair of technical assistants, one also named Ivan. He was geographically isolated, in Colorado, from the leading centers of cell research on the East Coast, and he evidently didn't relieve his isolation (as many scientists did, including Darwin) by establishing close relations with colleagues through the mail. Wallin's day job was anatomy professor. His lab was a shed behind the medical school's classrooms. He set to work and, between 1922 and 1927, produced nine papers and then a book, making his case not just that mitochondria derive from bacterial symbionts but also that such partnerships had repeatedly changed the course of the history of life.

He coined a fancy term for this broad phenomenon, of which the bacterial origin of mitochondria was a cardinal instance: *symbionticism*. He defined symbionticism as an intimate and "absolute" symbiosis, one creature taking residence within the cells of another, in which the inner partner is always a bacterium. It was essentially the same as what Merezhkowsky had called symbiogenesis. Wallin wanted his own label. His penchant for coining new jargon, along with his vast claims about the implications of his idea, combined to earn him dismissal in his own time and footnote status in the longer view. In 1927, just after the series of articles, he published a compendium of his experimental findings and his theorizing as a book, *Symbionticism and the Origin of Species*. By now, you certainly recognize the echo in the title. He was implying, as Margulis would imply with her 1967 title, *I walk in the footsteps of Darwin*. He was suggesting, beyond that: *My idea explains, even better than Darwin's, the origin of diversity, complexity, and adaptation on Earth.*

Symbionticism, Wallin declared, was "the fundamental principle controlling the origin of species." Darwin's 1859 idea, natural selection, was secondary, determining only the retention or destruction of species once they have arisen. And there was a third force, an "unknown principle," accounting for evolutionary progress toward better and more complex forms. Symbionticism brought the emergence of new species by creating radical points of divergence. Natural selection eliminated the worst, least-promising of those innovations. And what was that third, unknown principle? Wallin didn't say.

He was devoted to empirical evidence, except when he wasn't. For that reason and others, *Symbionticism and the Origin of Species*, delivering its grandiose assertions, landed in January 1927 with a dull thud.

"Dr. Wallin's writings stirred up much interest, but little enthusiasm," according to one friendly account. A conspicuous review, in the journal *Nature*, brought London disdain against Colorado noodling. Ivan E. Wallin "asks us to believe" that mitochondria are bacteria, and he claims that "the origin of species has taken place largely owing to the activity of these bacterial symbiotes." That process, the reviewer sneered, "is called 'symbionticism,' a new and horrid word." These pounding rejections seem to have flattened Wallin's zeal for research, and for the next twenty-four years, until retirement, he contented himself with teaching anatomy.

By the mid-1960s, Wallin's ideas and those of Merezhkowsky and the other early seers of endosymbiosis had fallen beyond disrepute into forgetting. If you were a young biologist being educated at that time, you might never have heard those names or been exposed to those wild notions unless you happened to be at Wisconsin, taking a course from Hans Ris. One scientist that year, invoking the bygone theories of endosymbiosis for purposes of a broader argument, called them "certainly defunct." A year later, the paper by Lynn Margulis (under her Sagan name) announced their comeback.

The idea of mitochondria as captured bacteria traced to Wallin, and she knew it. Chloroplasts as captured bacteria traced to Merezhkowsky, and she knew it. In this 1967 paper, she added something more: another aspect of eukaryotic cells, possibly also originating from endosymbiosis. Her addition comprised three features she saw as related: the flagella of tiny swimming eukaryotes such as *Euglena gracilis* (on which she had done her dissertation); the cilia, little hairs that project from virtually every eukaryotic cell, including the cells of your body; and the centrioles, tiny structures in a cell, which I mentioned earlier. Flagella are threads that wiggle back and forth, powering single-celled organisms through liquid like the tail of a fish. Cilia (from the Latin for "eyelashes") serve various important purposes in larger eukaryotes, including mammals: moving mucus and unwelcome debris along the windpipe, for instance. Centrioles are cylindrical bodies that help organize and distribute the chromosomes during cell division.

Flagella, cilia, and centrioles share certain similarities, not just with one another but with bacteria of the group called spirochetes, which tend to be long, spiral, or corkscrewy in shape, capable of twisting motions that allow them to move. Can you see where this is going? Many spirochetes are parasitic, invading other creatures and, in humans, causing afflictions such as syphilis, yaws, leptospirosis, and Lyme disease. Margulis's innovative idea was that these three other crucial mechanisms in eukaryotic cells—the flagella, the cilia, and the centrioles—are also descended from captured bacteria. Maybe something wiggly and mobile, she suggested, like a spirochete.

She hypothesized that an ancient amoeboid creature, an early eukaryote, had acquired the wiggly thing by eating it. Or perhaps the wiggly thing had attached itself to the outside of the eukaryotic cell. Instead of being digested (if it was inside) or causing harm as an internal parasite, or being sloughed off (if it was externally attached), in at least one fateful case, it had become domesticated. It stuck, it stayed, it assimilated. Some of its genes, including those that coded for a particular structural feature Margulis noticed, were incorporated somehow into the coding genome of the host. Those genes were put to three purposes: building flagella, cilia, and centrioles, all of which helped lead the way to glorious new possibilities for the eukaryotic lineage.

Spirochetes, with their bad reputation, commonly viewed as nasty pathogens, were a counterintuitive pick for partners in the rise of complex life. That wouldn't have discouraged Lynn Margulis. Apart from the structural evidence, which she found so compelling, this idea had the merit of being outrageous.

If it was true, it was vastly consequential. Some bacteria have simple flagella, by which they propel themselves through a liquid environment, moving clumsily toward attractions, clumsily away from repellants. But the flagella (and cilia) of eukaryotic cells are entirely different from those bacterial versions, using a different kind of motive power to produce a different kind of motion—potentially faster and more efficiently directed. Adding spirochetes as flagella and cilia to the outside of eukaryotic cells, if that's what happened, might have been the first big step toward greater mobility and complexity. Cilia also facilitated, among other things, the movement of fluids along the internal surfaces of multicellular creatures. And the addition of centrioles, somehow derived from those spirochetes, Margulis thought, would have enabled the development of two new capacities: mitosis and meiosis. Systematic duplication and division of chromosomes. The words "enabling meiosis" may sound dull, so let me rephrase that: what we're talking about is the invention of sex.

The structural feature I alluded to, which persuaded Margulis that these three improvements are related to one another and to some spirochete-like bacterial symbiont, is simple. Imagine a thick utility cable, a main line. You cut that cable using an industrial hacksaw (turn off the cable power first) and view it in cross section. What you see inside are the cut ends of nine smaller cables arranged in a neat ring. That's what Margulis saw by electron microscopy: flagella, cilia, and centrioles all shared that arrangement of nine tiny tubules, distinct in cross section and ordered radially like numbers on a clock. She deduced that eukaryotic cells had inherited that feature, commonly, after acquiring some ancestral spirochete as a symbiont, which then became cilia, flagella, and centrioles. There seemed no other plausible reason—happenstance? no—that each of those three anatomical features would have nine inner tubules, just nine, arranged in a circle. It was their mark of Cain, ineradicable. Furthermore, the flagella and cilia of eukaryotic cells both showed two additional tubules inside the neat ring of nine. She called that the (9+2) structure. The centrioles of cells, with no such central pair, had a (9+0) structure. Still close enough to be persuasive, Margulis reckoned. The whole nine-tubules arrangement may have come straight from the spirochete, she suggested, or else evolved early from that common ancestor. And the difference between flagella in bacteria and flagella in eukaryotes was so basic, so telling, she decided later, that it was necessary to give the latter a different name. She revived an old one. From 1980 onward, she referred to eukaryote flagella, and cilia along with them, as *undulipodia*. From the Latin *undula* and the Greek *podos*, it meant: "little waving feet."

You can almost picture them, those teeny undulipodious eukaryotes, waving their little feet at us to indicate kinship.

There are two points worth noting about this bit of the story, which may otherwise seem rather arcane. The first is that Lynn Margulis was a cell biologist and a microbiologist of the old school, of the classical methods, meaning that she worked primarily from visual evidence with whole organisms: growing bugs in her lab, collecting new forms from the wild, peering at them through a light microscope or inspecting the electron micrographs produced by her colleagues. The recent improvements in electron microscopy, she said herself, had made some of these insights possible. She was also deeply knowledgeable about paleobiology, bio-chemistry, and geochemistry. Her books for a scientific audience, such as *Origin of Eukaryotic Cells* in 1970 and *Symbiosis in Cell Evolution* in 1981, are filled with graphic illustrations presenting her evidence in schematic form, as well as photos revealing the microscopic structures of all manner of living things, from purple sulfur bacteria to chloroplasts in a tobacco leaf, and from spirochetes in the hind gut of a termite to a centriole tak-ing shape within a human cell. Browsing those images, looking deep into cell structure and the primordial beginnings of complex life, is enough to make you dizzy. To a person who's no microbiologist—to you or to me—it's like abstract art painted in protoplasm. But at least you won't be dizzied by any long strings of letters such as AAUUUUCAUUCG. Mo-lecular sequencing was not her métier. RNA catalogs were not her kind of data. Some of her signature work preceded the Woesean revolution in molecular phylogenetics, but even amid the revolution and afterward, she showed little interest in that sort of evidence.

The second notable point is that Margulis claimed originality for only a single element of her three-element theory of endosymbiosis. "Every major concept presented in this book," she wrote in her 1981 tome, "has been developed by others"—except one. She acknowledged Merezhkowsky and Wallin and various other early speculators; she cred-ited E. B. Wilson and her old teacher Hans Ris with having alerted her to such precursors; she listed the many aspects of her own thinking that had previously been articulated by others—from mitochondria as captured bacteria to the role of endosymbiosis in triggering major evolutionary transitions. Having granted all that, she asserted her proprietary pride in just one idea: the bacterial origin of undulipodia, those little waving feet. It was she who had first imagined their origin—that they were spi-rochetes, serpentine parasites, which had come aboard for no benign pur-pose and, having gotten stuck, stayed to help.

All these ideas can now be tested by the newer methods, she added.

And so they were. Science marched forward, into the age of Woese, bringing new forms of supporting evidence and broadened credibility to some of her most radical ideas—but not to the one part that was hers alone.

32

〜〜 〜〜

From the cool distance of Halifax, Nova Scotia, Ford Doolittle regarded the wild ideas of Margulis and got interested. He decided they were worth testing.

In the early 1970s, Doolittle was an assistant professor, barely thirty years old and lately arrived in the Department of Biochemistry at Dalhousie University, his position funded by a scholarship from the Medical Research Council, Canada's version of the British agency that had supported Fred Sanger and Francis Crick. Doolittle wasn't doing medical research, not in any applied sense, but that didn't matter. He studied ribosomal RNA and its transcription from DNA within cells—especially an aspect called RNA maturation, the cutting of long, raw RNA molecules into those 16S and 5S and other sections for assembling the ribosome, which he had explored earlier during a postdoctoral fellowship in Denver. That postdoc was in the lab of a smart, young scientist roughly his own age, Norman R. Pace, who will figure later in this story. Biochemistry was Doolittle's field, but now his curiosity was shifting toward evolution. Big questions: What were the major events in the history of life, how did complexity arise, how did the eukaryotic cell originate? Three factors converged to reroute his path, one of which was the Margulis book. The other two were cyanobacteria (still sometimes called blue-green algae, in those days) and a skilled assistant.

Serious thinker though he is, Ford Doolittle often shows a certain whimsical detachment from the scientific enterprise and his own career within it. He spoke about all this in our several conversations. "When I came to Dalhousie, I was supposed to keep working on ribosomal RNA maturation," he told me one day as we sat in his office. "Which is what I had done with Norm." It was technical biochemistry, involving the isolation of certain enzymes. "And I'm not a biochemist." Not by disposition, he meant. "I hate doing that kind of stuff." In the department at Dalhousie, he met another fellow, a true biochemist, who was working with the cyanobacteria. "I thought, 'My, what lovely color.'" They grew in nice varied shades of blue and green, he recalled, and they were "fun." Wouldn't it be pleasant to work on them?

"Where do cyanobacteria live in the wild?" I asked, trying to get a grip on these invisibilities as living creatures.

"Everywhere. I mean, in water—a lot of pond scum is cyanobacteria. Greenish tinge on the side of an old building in Oxford would be cyanobacteria, probably." He knew I had attended Oxford and seen plenty of greenish tinge on old buildings. "They're common. They have a lot of interesting, different colors. So I had gotten into cyanobacterial molecular biology as a kind of . . ." He paused to consider: why exactly? Not for medicine; not just toward evolutionary questions. Doolittle fancied the subject, and he was a scientist. "For its own sake," he said bravely.

Furthermore, there happened to be a good international community of scientists doing similar work—on the molecular biology of cyanobacteria, these things that until recently had been classified with algae—of whom the leader in Doolittle's time was still Roger Stanier. Leaving Berkeley in 1971 with some distaste for American politics, Stanier had accepted a position at the Pasteur Institute, in Paris, on the condition that he be allowed to work exclusively on cyanobacteria. But none of those cyanobacteria experts, not even Stanier, had studied RNA maturation (that process of cutting longer molecules to fit function) for the production of ribosomes, Doolittle's own little specialty since his time in Norman Pace's lab. "So I thought I could just kind of repeat what I did with Norm"—repeat it but now on cyanobacteria—"and get at some low-hanging fruit."

Doolittle began working with various kinds of cyanobacteria. He published a couple papers on the RNA maturation topic and other aspects of their biology, and in response to one of those papers he received a congratulatory note from Stanier himself. Nice piece of work, it has given me some ideas, Stanier told him; would you please send a few reprints for my group? That was heady praise from a grand man in the field, gratifying for an assistant professor. (Carl Woese would tell him some time later that Woese was jealous of the note, having never gotten any praise from Stanier.) But still Doolittle knew that his line of research was unambitious, that he hadn't reached very high. Around that time, he read the Margulis book.

He found her endosymbiosis theory enticing and was struck also by the book's illustrations—of single-celled organisms, symbiosis events, even a tree of life—which had been drawn and labeled freehand by an illustrator named Laszlo Meszoly, in a style that Doolittle saw as "funky." That was a compliment. The 1960s had ended, but its cultural vapors remained and, if R. Crumb had sketched an amoeba swallowing a bacterium for some hallucinogenic report in *Rolling Stone*, it might have resembled Meszoly's work. Doolittle, with an itch for drawing himself, liked that cartoonish touch applied to serious science. "I think it was part of the inspiration for me to do figures," he said—a consequential inspiration, given that his own drawn figures would later help communicate a radical new vision of the tree of life.

The Margulis book also caught his attention because it was evolutionary, not just biochemical. And it tracked evolution far into the past, deep to the base of the tree. "You know, for somebody like me," Doolittle said, "I don't really care whether elephants are related to hippopotamuses or whales. Those kinds of things didn't concern us." The details of mammal phylogeny seemed like small beans, he meant, for microbiologists who were wondering how one vast kingdom of life had diverged from another. "But the relationship between eukaryotes and prokaryotes seemed like a pretty big question." The endosymbiosis theory, heterodox but concrete, went to that question. Doolittle began thinking: Hmm, we could test this.

Then one day, in about 1973, Linda Bonen walked into his lab. It was the same Linda Bonen who had worked as a technical assistant for Woese in Urbana, helping him use electrophoresis and X-ray films to sequence ribosomal RNA. She was now living in Halifax, having come with her husband, an exercise physiologist who had accepted a job at Dalhousie within the Phys. Ed. Department. Bonen wanted some interesting work. She was well qualified for certain onerous, difficult tasks along the borderline of biochemistry and molecular biology, and some of those skills might be useful to other researchers. It was Woese himself who had said to her, when he heard she was moving to Halifax, "I know who you want to work with." He meant Doolittle. Woese and Doolittle were well acquainted from the time period, just a few years earlier, when Doolittle had done his first postdoc in Urbana. They kept in touch intermittently by letter and phone. Woese may even have written to him on Bonen's behalf, or Doolittle got wind of her skills somehow; he doesn't recall. "But when she came to me," he said, "I knew who she was and what she could do."

So the little group constituting Doolittle's lab, now with Linda Bonen as technical assistant, took up a new question: the origin of chloroplasts in complex cells. Among their first efforts was a comparison of ribosomal RNA samples from five sources: the cytoplasm of a red alga (a eukaryote), the chloroplasts of that alga, and several kinds of bacteria, including the familiar *E. coli*. If the Margulis theory was correct, Doolittle knew, chloroplasts contained in the complex cell should resemble the bacteria because they had originated from bacteria themselves.

Doolittle and Bonen reenacted the same messy, dangerous preparations of material that Bonen had learned in Urbana from Woese. They grew their organisms in nutrient lacking phosphorus, then added P-32 (the radioactive isotope) so that the bugs would build it into their molecules, including their ribosomal RNA. Then they broke open the cells, extracted the rRNA, selected out the subunits they wanted (16S from the bacteria and 18S, its eukaryotic equivalent, from the alga), cut those into short fragments with enzymes, and ran the fragments in races against one another by electrophoresis. From the electrophoresis runs, they printed images on X-ray films. Like Woese, they called each image of spread fragments a fingerprint, although it looked more like a herd of stampeding amoebae. From the fingerprints, they deduced the base sequences of the fragments and compiled them into catalogs.

Bonen did most of the wet work, with Doolittle helping. He was the boss, but she knew the techniques. When they raced the fragments in a second dimension, pulling them sideways to learn more about their composition, the electrophoresis was powered by 5,000 volts and sizable amperage. To cool the paper racetrack, they kept each end submerged in a tank of Varsol, the same flammable solvent that Mitch Sogin had used. "We had a special room built here," Doolittle told me, "and the room had CO_2 tanks, huge CO_2 tanks." What was the need for carbon dioxide? "To put out the fire in the Varsol that might have been started," he

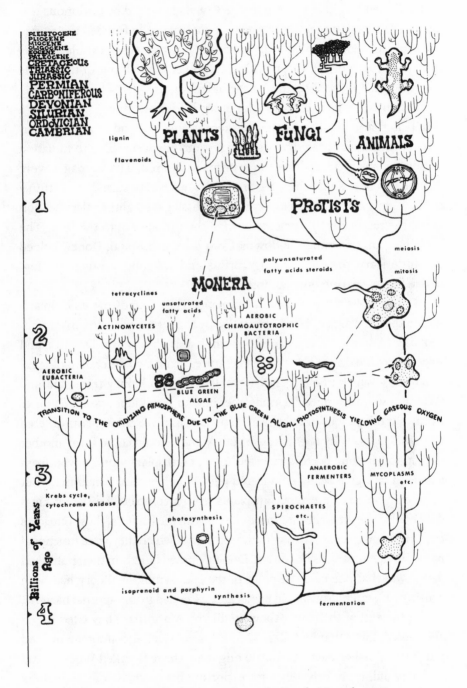

Margulis's tree of eukaryotes, drawn by Laszlo Meszoly, 1970.

said and then laughed. Of course, the CO_2 itself would be poisonous to a human if, in response to a fire signal, it automatically flooded the room's atmosphere. "You had, like, thirty seconds, between when the alarm rang and when the room filled up with CO_2, to get the hell out of there." He laughed again at the absurdity of the old methods.

Bonen, with less relish for the absurd, told me simply: "I was in a small lab that had all these special precautions." Instead of blowing herself up, getting suffocated by a safety system, or succumbing to radiation poisoning, she produced the fingerprints as desired. Those images were printed on chest X-ray films, big rectangles, each exposure done within a shallow plastic box known as a cassette, which gave lightproof protection while the radioactive fragments burned their images onto the film. The cassettes came as hand-me-downs from a local hospital. Bonen helped Doolittle learn to read the fingerprints and assemble catalogs of fragments for comparing one to another.

Their results were clear and dramatic. They found that chloroplasts within their red alga differed drastically, by this rRNA measure, from ribosomal RNA in the alga's own cytoplasm. It was almost as if Doolittle and Bonen were looking at two distinct creatures, from two different biological kingdoms—which in effect they were. If you had a kidney transplant and the donor was a stranger, the ribosomes in your new kidney wouldn't differ from your other ribosomes by nearly so much as these chloroplasts differed from their alga. Why not? Because your kidney would come from another human (or, if your doctors resorted to xenotransplantation, maybe from a baboon or a genetically engineered pig—in either case, a mammal). But these chloroplasts were xenotransplants from a completely different kingdom of life. They matched far more closely with the bacteria, those outsiders chosen for comparison, than with the red alga of which they were functional parts. What it meant, as Bonen and Doolittle noted quietly in their published paper, was that one cardinal point of the endosymbiosis theory had been confirmed: yes, chloroplasts in plants are descended from captured bacteria.

At the end of the paper, they cited Lynn Margulis. They cited Merezhkowsky. They thanked Carl Woese for "advice, encouragement, and much unpublished data." Doolittle might also have thanked Woese—and probably did, more privately—for exporting his methodology to Halifax in the person of Linda Bonen.

33

Ford Doolittle and Carl Woese had a longstanding friendship, founded in Urbana during Doolittle's postdoctoral fellowship of the late 1960s. It was fueled by common interests, strengthened occasionally by collaboration, stressed occasionally by competition, and became really complicated only in Woese's later years, when they disagreed about the tree of life.

Urbana was their crossroads by double happenstance. Doolittle, fourteen years younger than Woese, grew up in that little Illinois town long before Carl Woese arrived. His father was on the art faculty of the university, as I've mentioned. Doolittle had left for Harvard College and then gone on to graduate school at Stanford by the time Woese came in 1964. Four years later, now with his doctorate, Doolittle returned to the University of Illinois as a postdoc in the lab of Sol Spiegelman, who had gained fame from some intriguing, spooky experiments on RNA replication in vitro. These experiments involved creating what Spiegelman himself called "the little monster": a self-replicating molecule of synthetic RNA that produced unlimited generations of itself in a beaker. Sol Spiegelman was also the man who had recruited Woese to Illinois, and who had brought Sanger sequencing in the person of Dave Bishop, who had trained

Mitch Sogin, who had preceded Linda Bonen as Woese's key technician. It's a smallish world these scientist live in, much interconnected.

Doolittle's relationship with his postdoc boss, upon his return to Urbana in 1968, was complicated by overfamiliarity: Spiegelman's son Will had been Doolittle's closest pal in high school, and young Ford had worked summers as a junior assistant in Dr. Spiegelman's lab, washing glassware and doing other lowly chores. Spiegelman had given him his first brush with science, to that degree, and later advised him about graduate school. But Spiegelman was an intimidating figure, not a genial, avuncular mentor. "He had this nasty habit of wearing crepe-soled shoes and sneaking up behind you when you were pipetting P-32," Doolittle told me. Laboratory pipetting, in those days, involved using your breath to siphon liquid into a glass straw. Having snuck up behind his lab tech, Spiegelman would start to hum. "So, half the time, you'd swallow the P-32." Doolittle laughed still again, and, at that bizarre memory, I did too.

"People were afraid of him," Doolittle added. "I wasn't afraid of him." The word *afraid* didn't capture it. After all, this was just Will Spiegelman's brilliant, perverse dad. But when Doolittle returned as a twenty-six-year-old postdoc in the Spiegelman lab—probably reflecting a bad decision by each of them—the disparity of the old relationship never gave way to a more equal collaboration between two scientists. It lingered in stunted form, more of a barrier than a bond.

Scientific relationships are always shaped by personal chemistry as well as by work and ideas. Sol Spiegelman wasn't a fellow with whom you could enjoy a relaxed collegiality away from the lab, Doolittle told me. Not if you were Ford Doolittle, anyway, and Spiegelman had known you since you were a scrawny adolescent. But there was this younger professor in the same department, Carl Woese, less formidable, less remote. He wasn't lofty and gruff—not in those days, anyway. "You could go out and have a beer with Woese," Doolittle said. So by default and inclination, Woese became a kind of "social and emotional mentor" to Spiegelman's students and postdocs, including Doolittle.

That friendship endured through the 1970s, with the Woese and the Doolittle labs working sometimes on parallel projects, in a competitive spirit but also, at times, sharing ideas and unpublished data. In the very same issue of the *Proceedings of the National Academy of Sciences* in which Doolittle and Bonen published their chloroplast study, for instance,

Woese's group published a similar piece of work, likewise confirming that Lynn Margulis had been right about the bacterial origin of chloroplasts. Just as Bonen and Doolittle had thanked Woese for the courtesy of sharing unpublished data, so he thanked them. Overlapping and reciprocal collegiality; science as it should be.

One year later, Doolittle and Bonen published another paper, this one in *Nature*, offering evidence for two significant claims: that blue-green algae are not in fact algae but bacteria (hence they became known as cyanobacteria); and that chloroplasts, at least in some complex organisms, had originated not just as bacteria but specifically as this sort, cyanobacteria. The evidence had been gathered again with Woesean methods, and the paper was filled with respectful nods to Woese's work.

Then, in 1978, a new paper of high interest to both men appeared in a relatively obscure European journal. This paper, from a French team in Strasbourg, offered something never yet seen: the *complete* sequence of bases (not just a sampling of fragments) for Woese's molecule of record, 16S rRNA, from a single bacterium. The bacterium was that familiar bug *E. coli*. The method of sequencing was essentially Fred Sanger's, modified by use of a new ingredient: extract of cobra venom, to help cut the molecule. The information value of the full sequence, to researchers such as Woese and Doolittle, was huge. Until this point, they had been comparing fragments—those short blurts of letters—without knowing how the fragments might fit together. The French team, with their cobra venom, had revealed the fit.

But that European journal was slow in reaching the University of Illinois. Woese, having caught wind of the paper, was impatient. So he called Doolittle in Halifax, where the journal had arrived, and asked a favor: "Would you read me the sequence?" Doolittle obliged. Sitting in his office, with the October issue of the journal, he recited all 1,542 letters to Woese by telephone. He found that reading them in triplets gave a natural rhythm, making it easier to avoid omissions or duplications.

That is, he said: "AAA, UUG, AAG, AGU, UUG, AUC," and so on. He said, "AUG, GCU, CAG, AUU, GAA, CGU," and so on. He said, "UGG, GAU, UAG." UAG is a stop codon, a signal to terminate, when it appears within messenger RNA; but this wasn't messenger RNA, it was structural, and when Doolittle came to a UAG, he didn't stop. He said: "CUA, GUA, GGU, GGG, GUA," and furthermore, "ACG." He read the whole damn thing, the entire eye-crossing stream of letters, while Woese in Ur-bana copied them down carefully. Finally Doolittle said, "GGU, UGG, AUC, ACC, UCC, UUA," and they were done.

Ladies and gentlemen, this wasn't the Stone Age. It wasn't the era of campfire incantation by Celtic druids. It was 1978, and a former biochem-ist was helping a former biophysicist, both of them keen for using molecular methods to plumb evolutionary mysteries, by sharing the latest and hottest scientific data.

34

Linda Bonen and her Woesean skills also helped confirm the second major postulate of the endosymbiosis theory: that mitochondria, as well as chloroplasts, are descended from captured bacteria.

The big point of debate on mitochondria in the 1970s was whether these crucial organelles had arisen from increasing complexity within the eukaryotic cell or, alternatively, had come originally from outside the cell in the form of a captured bacterium. The first view embraced conventional wisdom: somehow the cell innards had evolved toward greater complexity by gradual differentiation, assembling new structures, including a nucleus, chloroplasts in plants, and mitochondria for energy packaging. Maybe these organelles coalesced from ambient materials, like stardust forming planets. Or maybe they pinched off from some other internal organ, like an appendix, then floated free. Nobody could say. The second view, proposing external origin, echoed Lynn Margulis and her assertions of endosymbiosis. By comparing catalogs of DNA fragments, Bonen and her colleagues in Halifax gave new support to what Margulis and Ivan Wallin had proposed: that mitochondria originated when some sort of single-celled organism (the pre-eukaryote host cell, whatever it was) swallowed a bacterium and then failed to digest it, or became infected by a bacterium and then failed to cure itself, or was otherwise entered by a bacterium and allowed the thing to stay. This signal event happened only once. Descendants of that internalized bacterium became the first mitochondria.

Among the biochemists at Dalhousie University was another young assistant professor, Michael W. Gray, lately arrived from western Canada by way of a postdoc at Stanford. Gray had grown up in Medicine Hat, Alberta, a "prairie kid," by his own later description, and then gone to the big city, Edmonton, for undergrad work and a PhD. He trained as an RNA biochemist at a time when that subject seemed entirely detached from evolution. He had never taken a course in evolutionary biology, and his work concerned cellular systems as they function in the present, not their deepest origins. His dissertation topic involved transfer RNA (tRNA), which carries amino acids to the ribosomes for building proteins. He had never heard of Carl Woese or Lynn Margulis. Then he happened to read a journal paper about transfer RNA in a certain fungus, and the paper noted that this tRNA seemed to be traceable to the fungus's mitochondria. Hmm, he wondered, what's transfer RNA doing in mitochondria? Since when does protein building have any role in their known functions? Ribosomes make proteins. Mitochondria make ATP. The presence of transfer RNA suggested something else about mitochondria, something that Gray and other RNA chemists hadn't known: mitochondria contain ribosomes of their own, almost as though they were (or had once been) independent cells. Why?

That little riddle led Gray to focus his next major research effort on mitochondria—more specifically, on the DNA and RNA found in mitochondria of plants. This seemed important because it might illuminate mitochondrial origins. He looked in particular at wheat. During his dissertation research, he had turned to commercial wheat germ as a source of the tRNA he needed, and now it occurred to him that wheat germ again might be a good source of mitochondria and the genetic material they contain. As Gray snooped through the literature, having set off on that project, he came upon the work of Margulis. He had missed her 1967 paper at the time, but now he read her 1970 book, *Origin of Eukaryotic Cells*, offering a fully elaborated endosymbiotic theory, including mitochondria origins. Conventional wisdom be damned.

"Was she perceived back then as being radical or flakey?" I asked Mike Gray on an afternoon forty years later.

"Uh-huh," he said. "Right from the beginning, I think."

We had shared lunch at a Turkish restaurant around the corner from Dalhousie, then returned to a small office at the back of Gray's laboratory. He was retired now, and just closing down lab operations after a quiet but distinguished career devoted largely to the study of mitochondria. He recalled the negative reception to Margulis's theory, the scientists who "pooh-poohed the idea and had their own scenarios" of nonendosymbiotic origin. There were debates. She was marginalized. But to Gray, her theory suggested that "maybe this was a remnant of a bacterial system that did protein synthesis inside mitochondria. Which I thought was pretty cool." Soon he too, like his colleague Ford Doolittle, found himself coming to her support.

It began as a spin-off from the work Linda Bonen did with Doolittle. Gray himself had a grad student—Scott Cunningham, the very first grad student he supervised—who happened to be talking with Bonen one day, and they hit upon the idea of applying the Woesean methods to mitochondria from wheat. They would test that pillar of the endosymbiotic theory—the bacterial origin of mitochondria in all eukaryotes—by looking at ribosomal RNA in the mitochondria of this particular eukaryote, an agricultural plant, familiar to Gray since the PhD work in Edmonton. Now he and his partners would extract rRNA from wheat mitochondria, chop the molecules into fragments, sequence the fragments, and see how their catalogs of short sequences compared with catalogs from other organisms, including bacteria. One problem: working with any extract of mitochondria was notoriously difficult, because mitochondria themselves are not very abundant, compared with chloroplasts within plant cells, and harder to isolate. But that was the beauty of wheat, as Gray understood. The germ of the wheat kernel—that little nubbin inside the grain, constituting the embryo from which a new plant would germinate—contains enough mitochondrial rRNA for such an experiment.

"Where would you get the raw wheat?"

"From western Canada," Gray said. A kid from Medicine Hat knows wheat. "I had it shipped down here." He had previously gotten his wheat germ from a local flour mill in Halifax, finding it very useful for some experimental purposes. But that wheat germ, having been processed, wouldn't germinate. You couldn't grow new plants from it, and if you couldn't grow new plants, you couldn't feed in the radioactive phosphate for labeling molecules. Gray and Cunningham wanted wheat seed, as it would come fresh from the field, or at least fresh from the grain elevator standing above a railroad spur on the western plains.

So Gray ordered it by the bag from a supplier back in Alberta. Cunningham worked out the laborious mechanics of extracting the viable wheat embryos from the seed wheat, using a kitchen blender and sieves, until they had a small, precious quantity of the stuff: 16 grams (about a half ounce). They sprouted those wholesome Canadian wheat embryos on filter paper, adding water and, oh yes, "a horrendous amount of radioactivity," for labeling that would show on the films. From the wheat embryos, they isolated mitochondria and extracted mitochondrial rRNA. Then, with Linda Bonen leading this part of the process, they cut the rRNA into fragments, exposed the films, read the fingerprints, and made catalogs.

Gray had chosen well when he picked wheat, because rRNA within plant mitochondria tends to mutate more slowly than the equivalent rRNA within animals. The similarity to its bacterial counterparts and their possible shared ancestors is therefore more pronounced. The work yielded clear findings, as reported in another journal paper, with Gray and Cunningham now added as coauthors alongside Bonen and Doolittle. The Margulis hypothesis was confirmed. Based on their ribosomal RNA, the mitochondria of wheat do not resemble wheat. They are alien little beings, co-opted into wheatly service. They came from elsewhere. They resemble bacteria.

35

For almost ten years after that publication on wheat mitochondria, a sizable (or at least, vocal) minority of biologists in the field continued to reject the endosymbiosis theory. Part of their resistance was to swallowing it whole, in the three-part version proposed and promoted by Margulis: chloroplasts derived from cyanobacteria, mitochondria derived from another sort of bacteria, flagella derived from spirochetes or some similar bacteria. That was too much.

There may also have been subliminal distaste for its implications, including one that still seems almost too personal for comfort: that all animal cells, all human cells, all your cells and my cells and the cells of the doubters, are powered by captured bacteria functioning as organelles. These are not the many teeming bacteria that live in your stomach or your armpits, remember; your gut microbes and other little passengers have been internalized at the body level but generally not *within* your *cells*. I'm talking about all those captured-and-transmogrified bacteria that are more fully integrated into your being. These captives have, over the course of maybe two billion years, become part of the machinery *inside* the cells from which your cells are descended. Their DNA is part of your DNA—the part received exclusively from your mother. Why mother only? Because mitochondrial DNA is passed along via eggs and not via sperm.

In plainer words: the endosymbiosis theory stipulates that we are all composite creatures, not purely and unambiguously individuals. Small wonder that it was slow to take hold. Nobody likes that Jason Bourne feeling of being told that you aren't who you thought you were.

Another challenge to the theory arose when researchers, using electron microscopes, began looking closely at the physical structure, and the amount, of the genomic material in mitochondria and chloroplasts. Both mitochondria and chloroplasts are organelles (that is, organ-like subsidiary units within a eukaryotic cell), yet they have their own genomes, distinct from the genome tucked away on chromosomes within a eukaryotic nucleus. Under microscopic inspection, the DNA in at least some mitochondria and chloroplasts appeared as circular chromosomes, ring shaped, rather than linear chromosomes, as found in the nuclei. Spinach, for instance, held chloroplasts containing tiny rings of DNA. That much supported the theory, because bacteria too carry their genomes on circular chromosomes. If mitochondria and chloroplasts *were* bacteria, or derived from bacteria, it stood to reason that their chromosomes would have the same shape.

But these circular chromosomes seemed much too small; far smaller than the chromosomes of any known bacteria. The amount of information they could carry—the number of base pairs, the number of genes—was minuscule compared with a bacterial genome. Again the example of spinach: scientists who scrutinized its chloroplasts, at the University of Düsseldorf in Germany, found teeny little rings of DNA, only one-thirtieth the size of a typical bacterial chromosome. So a new mystery had to be solved: if mitochondria and chloroplasts had indeed originated as bacteria, where were all their missing genes? Gone entirely, shucked off, withered away? Or had those genes somehow been transferred elsewhere within the host cell—maybe into the nucleus itself—from which they could still contribute their gene products to the life of the cell? This little conundrum hinted toward one of the most unexpected revelations (I'll explore it later) in the saga of the tangled tree: the importance, throughout life's history, of sideways gene transfer from one organism to another.

Still another reason for resistance to the endosymbiosis theory, back in the early 1980s, was an absence of evidence linking mitochondria with one specific kind of bacteria. If all mitochondria had originated from a single captured bacterium, okay, but *which one?* There was no prime candidate. That sort of match had been made for chloroplasts—aha, they were cyanobacteria—but not for mitochondria, no. The bacterial precursor hadn't been identified. Mike Gray and Ford Doolittle published a long review paper on the whole subject in 1982, titled "Has the Endosymbiont Hypothesis Been Proven?" Somewhat surprisingly, given their own 1977 paper on wheat embryos, they conceded that—with regard to mitochondria, at least—it had not.

Gray and Doolittle were fastidious about this matter of proof. Their paper began with a two-page introduction stipulating what sort of empirical data would potentially confirm the theory, and distinguishing such data from circumstantial evidence. Some forms of "proof" were more proovy than others. Mere correlations could have alternate explanations. This section was a display of thinking and writing that bore the philosophical style of Ford Doolittle. Then followed a section on chloroplasts and cyanobacteria, drawing on Doolittle's research focus. After that came a disquisition on RNA and mitochondria, played from the strengths of Mike Gray.

The gist of the whole paper was that, yes, the chloroplast hypothesis stands as proven but, no, the mitochondria hypothesis doesn't—not quite, not yet. Near the end, Gray and Doolittle mentioned their own work on wheat mitochondria as suggestive but inconclusive. Those wheat mitochondria did resemble (based on rRNA sequences) the bacterium *E. coli*. Anyway, no better match had been found among other bacteria, by other researchers, in the five years since their earlier paper. But that was just wheat, one plant. And when it came to the mitochondria of animals and fungi—two other big branches of the eukaryotic limb of life—there was no good match at all. There was no persuasive proof of endosymbiosis. There was no favored candidate for the role of the Bacterium That Became Mitochondria. (Conclusive evidence wouldn't be offered until fifteen years later, when another paper coauthored by Mike Gray matched a full mitochondrial genome with a particular bacterial group.) For the meantime, that question was still in play.

Carl Woese reenters the story here. In 1985 his lab published a paper titled "Mitochondrial Origins," announcing that the link had been found. They could place the progenitor of all mitochondria, Woese's team claimed, within a group of bacteria that still thrive on planet Earth. Turns out the progenitor's modern relatives are all around us, quietly parasitic, causing galls in walnut trees, in grape vines, and in other plants. They belong to the alpha subdivision of the purple bacteria, an unusual group now known as proteobacteria.

Woese and his colleagues had again used 16S rRNA, his preferred molecule, as their standard of evolutionary relatedness. Mitochondria have an rRNA molecule closely equivalent to 16S in bacteria, but how close, and to which bacteria? By now, sequencing techniques had improved. Fred Sanger had made still another important contribution, devising a new method of sequencing DNA (not RNA) that was faster and more accurate than anything done earlier. Woese adopted it for his own effort, and thereby made a change that may seem subtle or obscure but was important: instead of extracting the rRNA molecule from ribosomes (in tiny quantities) and sequencing portions of it, his team extracted DNA from genomes and (after multiplying the quantities by DNA cloning) sequenced portions of that master molecule—the portions that coded for 16S rRNA in bacteria and its mitochondrial equivalent. They read the blueprint, that is, instead of measuring the house. It was a methodological adjustment yielding the same sort of information as the old way, but better and quicker. Carl Woese had earned the right to such a shortcut—sequencing DNA, which could be had in largish quantities, rather than rRNA, eked out in tiny smidgens—after all the mind-numbing efforts he had made for a dozen years. This was the beginning of the new information flow, and there again he was, an early adopter of novel methods, but asking deeper questions than most other researchers.

First author on the 1985 paper was Decheng Yang, a doctoral student who had come to Illinois from a university in northeastern China, eventually joining the Woese lab. He and Woese and their collaborators compared rRNA genes from seven different organisms to see which of six prokaryotes (five kinds of bacteria and one representative of the Archaea, Woese's new kingdom) best matched the mitochondria from a eukaryote. The archaeon was included as a sort of outlier, thrown into the lineup for breadth. Bacteria were the real suspects. For their eukaryote, the team chose wheat, using the 16S rRNA sequences produced from wheat mitochondria by Mike Gray's group. The five bacteria they picked for comparison included the ever-available *E. coli*, the cyanobacterium *Anacystis nidulans*, two others you can forget about, and a microbe known as *Agrobacterium tumefaciens*, named for its troublesome habit of causing tumors in agricultural plants. The last of those, *A. tumefaciens*, was an alpha-proteobacterium.

Many alpha-proteobacteria have evolved lifestyles that entail living inside eukaryotic cells. The pathogens that cause typhus and Rocky Mountain spotted fever in humans, for instance, are alpha-proteobacteria, doing their damage as parasites within their victims' cells. Woese's bug, *A. tumefaciens*, is innocent of causing human disease but plays hell on plants. And the comparative study showed that it's the one: the closest match to wheat mitochondria. Mike Gray was on sabbatical back at Stanford at the time. Woese called him to share the news.

This finding was more vast and dramatic than it might sound. It involved more than wheat. Because other research had already established that mitochondria originated just once in all the history of earthly life, from a single instance of a captured bacterium, Woese and his young team could claim that they had narrowed the search for the ancestor of all mitochondria in complex cells. That ancestor was an alpha-proteobacterium of some sort. Its descendants exist in all of us, powering our cells, making complexity possible.

On the fourth page of their paper, Woese and his group offered a simple figure: another tree. It was a tree of mitochondria and bacteria. It showed wheat mitochondria and mouse mitochondria and fungal mitochondria as twigs, all clustered closely on the same little branch with the twig representing that bacterium, the tumor maker, *A. tumefaciens*. If they had sequenced the equivalent rRNA gene in human mitochondria, that would have sprouted here as well. Everything else stood apart—other branches, other limbs, other parts of our genome, our being, our identity (whatever that is), all of them too distant to fit within the figure. Arboreal illustration was starting to get complicated.

36

By the time this vindication arrived, in 1985, Lynn Margulis was a full professor at Boston University (where she had begun as an adjunct assistant professor) and an elected member of the National Academy of Sciences, America's foremost scientific advisory body. Her marriage to Nick Margulis had ended, and her youngest of four children was sixteen. Three years later, she moved to Amherst, as distinguished university professor at the University of Massachusetts, and remained there to the end of her life. She taught. She advised graduate students. She made instructional films and videos. She welcomed houseguests and cooked meals. She seemed to love people, conversation, engagement with humans as much as she loved engagement with ideas.

She continued to publish journal papers, on an increasingly wide range of topics, some provocative, some technical, as well as popular articles and books, promoting her ideas about endosymbiosis and embracing certain other notions, theories, and viewpoints that seemed even more far-fetched than endosymbiosis once had. She became an AIDS skeptic, challenging the reality that an agent known as human immunodeficiency virus (HIV) is the cause of the syndrome. She became a 9/11 doubter, calling it a "false-flag operation" contrived by unknown parties for dark political purposes. She had long since espoused the Gaia hypothesis, developed by her and the English chemist James Lovelock, which views planet Earth as a self-maintaining system that regulates its own biochemistry, analogous to a single living organism. That brought her a lot of acclaim from people who took the living-organism part literally (she didn't) and viewed Gaia as an almost mystical insight. Margulis herself wasn't mystical. She favored evidence, argument, and the material world of nature (especially as observed through microscopes), even when she followed those leads into strange terrain.

She embraced the medically controversial idea of chronic Lyme disease: that the Lyme pathogen, a spirochete, can hide in the human body as a chronic infection, unconquerable by normal doses of antibiotic. She endorsed for publication, in the *Proceedings of the National Academy of Sciences*, an extraordinarily odd and unpersuasive paper by a retired British zoologist named Donald I. Williamson, who argued that butterflies and their larval form, caterpillars, evolved as separate species, which had later become linked into one life-history form by some ineffable process of hybridization. Caterpillars and butterflies as different creatures conjoined, yes: like a tadpole hybridized with bird, to yield a single new beast with two stages of life. It was analogous, Williamson argued, to the symbiogenesis of eukaryotic cells.

Margulis defended the butterfly paper's publication if not the hypothesis, saying, "We don't ask anyone to accept Williamson's ideas—only to evaluate them on the basis of science and scholarship, not knee-jerk prejudice." Other scientists pointed out that genetic data already existed, before Williamson even published his monster-butterfly notion, to refute it.

Margulis thrived in this context: the challenges to authority, the dustups among scientists, the tension between caution and daring. She loved going out on a limb and bouncing there while others warned that the limb was weak and might break. Her attitude: *If it breaks, fine, that's science too!* She became famous and much admired among nonscientists, over a lifetime of such provocations, and infuriating to some of her colleagues. She gave interviews, organized meetings, traveled to lecture, and contended robustly in controversies. A profile in one serious journal called her "science's unruly Earth Mother." Throughout all this, she remained cheerily energetic, open to discussion, confident to a fault, generous with her time, and likable. "I quit my job as a wife twice," she once said. "It's not humanly possible to be a good wife, a good mother, and a first-class scientist." Something had to go. She preferred to be a scientist and a mother, unruly or otherwise. And she mothered, to one degree or another, a lot more people than four.

In 1986 she published *Origins of Sex: Three Billion Years of Genetic Recombination*, the first in a series of books coauthored with her eldest son, Dorion. It was characteristic of her reach, to write a book covering three billion years. A year later, she and Dorion produced *Micro-Cosmos: Four Billion Years of Evolution from Our Microbial Ancestors*. Their 2002 book was *Acquiring Genomes: A Theory of the Origins of Species*, proposing at length what she had claimed elsewhere: that neo-Darwinism (the twentieth-century school of thought merging Darwin's theory with Mendel's genetics) is wrong about the main source of genetic variation that drives evolutionary innovation. That crucial element, variation—it *doesn't*, according to Margulis and Sagan, come mostly from the tiny random mutations that seem sufficient to neo-Darwinists. "Rather," they wrote, "the important transmitted variation that leads to evolutionary novelty comes from the acquisition of genomes." It comes from symbiosis, the real origin of species.

Symbiosis in this sense includes endosymbiosis—those bacteria you've been reading about, captured and transformed into the first mitochondria, the first chloroplasts, within eukaryotic cells. But it also includes, as I've mentioned, a broader variety of cases—less drastic, less epochal—in which two organisms, two genomes, amalgamate into one living partnership. For instance, Margulis and Sagan described sea slugs of the species *Elysia viridis*, which feed on green algae during their immaturity and then, instead of digesting the algae completely, retain algal chloroplasts within their own cells. The acquired chloroplasts allow the slugs to photosynthesize like plants, gathering their energy from sunlight in the tidal shallows where they live. As adults, they become in effect "plant-animal hybrids." That sort of dramatic combining, Margulis and Sagan claimed, and not the incremental mutations of the neo-Darwinists, is the main way (so far as presently known) new species originate.

Margulis invoked the green sea slugs years later during an interview. "The evolutionary biologists believe the evolutionary pattern is a tree," she told a writer from the magazine *Discover*. "It's not. The evolutionary pattern is a web—the branches fuse, like when algae and slugs come together." She was right: the tree of life is not a tree.

Despite the peculiarity of her views and her reputation as a rebel,

Margulis got showered with awards and honors—enough to belie the popular image of her (and her self-image) as a rejected outsider. Election to the National Academy of Sciences in 1983 was only a start. Then followed election to the Russian Academy of Natural Sciences, far more rare for an American, and to the American Academy of Arts and Sciences, another august body. Election to the World Academy of Art & Science, whatever and wherever that is. She accepted the Alexander von Humboldt Prize in Berlin, shared the Darwin-Wallace Medal in London, and accumulated sixteen honorary doctorates. President Bill Clinton, in 2000, hung the National Medal of Science around her neck. The list is long. In 2010 she flew to the small city of Bozeman, Montana, to receive an award (from a wonderful institution called the American Computer Museum) named in honor of, and presented by, the biologist Edward O. Wilson, another scientific pioneer. Wilson had also flown in for the event. There was a banquet. That's where I met her.

Next day, very early on a blustery October morning, she and I and twenty other people, including Ed Wilson, boarded a bus for a field trip through Yellowstone National Park. Lynn and I sat together for much of eight hours while the vehicle rolled amid lodgepole pine forest, steaming hydrothermal vents, geyser basins, multicolored mineral springs, trout-filled rivers winding through meadows grazed by bison and elk, and other Yellowstone scenery. We talked about endosymbiosis, she and I, and the origin of species, the events of 9/11, the etiology of AIDS, and Lyme disease, while the others talked about—I'm just guessing—bison and bears and elk. Lyme disease especially engaged me, because I was writing a book on infectious diseases. We talked about much else too, probably including the local wildlife; she was interested in everything. Come to my seminar in Amherst, she said. Although I made no notes of our conversation during or after the bus ride—not foreseeing that I would ever write about her—there is a photo in my files to serve as a memento. It shows Lynn, myself, one other scientist, and Ed Wilson, posing with linked arms at an overlook into the Grand Canyon of the Yellowstone River. In the background is a great waterfall. Lynn wears a bulky gray sweater. Ed sports a park ranger hat—a Smokey Bear, with a flat brim—borrowed for this larky shot, and a smirk. He had suffered some of the punishments of voicing heterodox theories himself. He liked being out in the woods with Lynn Margulis, what the hell.

Two weeks later, I did go to her seminar at the University of Massachusetts. Driving out to Amherst after other business in Boston, I kibitzed the class and then returned to her house for dinner, a hearty and simple stew she had cooked herself. I met her dog. We found no time to sit down, just Lynn and I quietly, so that I could interview her about Lyme disease, as I had hoped; but it didn't seem to matter. There were other guests, there was far-ranging conversation. As always, Lynn stirred the pot. I drove back to Boston and never saw her again. She died a year later after a massive stroke. She was seventy-three.

"There's a role in science for iconoclasts," Ford Doolittle said of Lynn, not in eulogy after her death but years earlier, when she was so tempestuously alive. "It would be a great mistake to jump on her with both feet." Iconoclasts such as Lynn Margulis, he added, "raise questions even when they're wrong. And, of course, they're occasionally right, as she was." Right about endosymbiosis, he meant, on two of the three cardinal assertions she had made: mitochondria, yes; chloroplasts, yes; undulipodia (those little tails), apparently no. While the bacterial origin of mitochondria and chloroplasts had been confirmed by molecular evidence—the genes were still there, linking them to bacteria—no molecular evidence matching undulipodia with spirochetes had ever been found. The only evidence was from microscopy, such as that notable similarity of the nine-tubule cross sections. But in the molecular age, microscopical synonymy wasn't enough.

Doolittle himself had gotten to know her in Boston, during a sabbatical he spent at Harvard through the 1977–78 academic year, while she was still at Boston University. He couldn't recall, when I asked, just how he met Lynn, but they would have had a congenial starting point: his 1975 paper with Bonen, confirming the chloroplast postulate of her endosymbiosis theory. Lynn was still married to Nick Margulis then, living just west of the city. One thing Doolittle did remember: "She had a lot of good parties at her place." For which, read: she lived life.

Her openness to ideas, her personal confidence, and her love of intellectual stew pots gave her a knack for maintaining friendships, or at least amicability, across deep divides of scientific disagreement, as reflected in her relations with some of the more strongly opinionated biologists of her era. She disagreed on important points with Ernst Mayr, one of the founding neo-Darwinists, but he was happy to write a foreword to one of her books, and she was happy to let him say in that foreword just where he felt the book went wrong. Stephen Jay Gould performed a similar service, vouching for another of her books; so did Joshua Lederberg, Lewis Thomas, and G. Evelyn Hutchinson. These are big names. Although she differed strongly with Richard Dawkins on neo-Darwinism and once debated him at Oxford, he said: "I greatly admire Lynn Margulis's sheer courage and stamina in sticking by the endosymbiosis theory, and carrying it through from being an unorthodoxy to an orthodoxy." That was nicely, and carefully, put. Ed Wilson handed her an award and, as mentioned, so did Bill Clinton.

But with Carl Woese, it was different. Woese had little patience for her more venturesome ideas, and she was scorned also by some of the scientists closest and most loyal to him, whom she called "Woese's Army," as though they were arrayed in battle against her. One of those loyalists, when I asked his opinion of Margulis, answered silently by miming expectoration. These feelings went deep. Woese disliked her, according to Jan Sapp, who knew them both well, and he resented the army metaphor. He told Sapp, "If I hear her say it again, I'm going to sue her." Woese himself had moved away from militaristic and sovereign words such as *empire*, even *kingdom*, in choosing a categorical term for what he preferred to call the three great domains of life. Although he doesn't seem to have been an especially pacific person, this point was important to him. Meanwhile, she coauthored a book titled *Five Kingdoms*, published in 1982, delineating five major divisions of life on Earth, which ignored and contradicted Woese's own great discovery of a third form of life in 1977. Was it three domains? Was it five kingdoms? They couldn't both be right.

Woese expressed his considered view of Margulis, caustically, in a private communication written in 1991. It was his response to a request from the dean of the College at the University of Chicago, her alma mater, which was considering her for an honorary degree. (She never got this degree, for one reason or another, and had to content herself with the other sixteen.) Carl Woese may have seemed, to the Chicago dean, a logical choice for supportive words: another world-renowned biologist concerned with cell evolution and based there in Illinois. Woese preferred to be blunt. "If you wish merely a complimentary letter to support the case for Prof. Margulis' receiving an honorary degree, you have come to the wrong individual," Woese wrote.

"I have fairly deep scientific disagreements with her," he added. He might have left it there, but he didn't.

Granted, she was a good teacher, Woese told the dean. Her reputation in that regard was deserved. By his lights, she was "primarily a teacher," only secondarily a contributor of new science. She had done "more than anyone to promulgate the idea" of endosymbiosis in the origin of eukaryotic organelles, "and for this she deserves great credit." But, of course, that idea wasn't uniquely hers. In fact, the correct parts of the theory weren't original with her, he noted, and the parts original with her weren't correct. She was wrong about flagella (her undulipodia). She was wrong about the primordial host cell—within which captured bacteria became embedded—having been a bacterium itself. Woese acknowledged that her efforts at "spreading the word" about cellular evolution, through teaching and popular writing, had been very effective. "Unfortunately," he added, "that word itself has been somewhat defective." She was sowing confusion.

He disapproved in particular of her book *Five Kingdoms*. The first edition was bad enough, but its mistakes were "excusable," he told the dean. The revised edition, published in 1988, annoyed him still more. This one was inexcusable because, in the six years between printings, she and her publisher had been advised (presumably by him) about its deficiencies regarding "the newer findings in microbial evolution," and they had done almost nothing to fix the problem. The problem, for Woese, was that Margulis and her coauthor had persisted in treating his great discovery of 1977, his separate form of life, as something not separate at all. The problem was five kingdoms versus three domains, and none of those five labeled Archaea. The problem, which Carl Woese couldn't forgive or ignore, was that her tree of life differed so fundamentally from his.

PART ✳ IV
Big Tree

37

In late February 1864 Charles Darwin received an unusual package in the mail—unusual even for him, to whom the faithful postman in his little village routinely lugged correspondence and natural-history specimens (dead pigeons, French peas, pickled barnacles, and so forth) from contacts all over the world. This package weighed more than seven pounds. It contained two big folio volumes: one filled with stunning copperplate engravings, one of text, comprising a work titled *Die Radiolarien*. It was a monograph on the radiolaria, a group of single-celled planktonic marine creatures that produce elaborate silica skeletons, varying species by species like a gallery of crystal chandeliers. Darwin had seen the book—so erudite, yet so decorative—sometime a year or so earlier during a visit to London, probably at the home of his friend Thomas H. Huxley. Now he had his own copy, sent with courtesies by its author, a young German zoologist and artist named Ernst Haeckel.

Darwin had never met Haeckel and knew him only through the exchange of a few polite letters. It was a difficult time for Darwin; in addition to the new crush of fame and controversy brought by *On the Origin of Species*, he had suffered a recurrence of the mysterious illness that punished him for much of his adult life. Haeckel knew Darwin from his *Beagle* journal, a best-selling travel book with no overt evolutionary message, published decades earlier, and more importantly from *The Origin*, recently made available in German (a bad translation, but good enough to be inspirational) promptly after its second edition in English. Haeckel's life, vision, and sense of scientific purpose had been transformed by reading *The Origin*, and he wanted his hero to know that. Hence the gift, which also implied an expectation: *Please read me.*

Darwin replied graciously just a week later, telling Haeckel that *Die Radiolarien* was "one of the most magnificent works which I have ever seen, & I am proud to possess a copy from the author." This was the careful language of a polite but busy man, and while seeing and possessing could be reported easily, reading the thing was another matter. Darwin read German only slowly and painfully, so he may not yet have gotten far into the 570-page text, nor come across the footnote on page 232 in which Haeckel gushingly saluted him and his theory. But he had at least browsed the illustrations. "It is very interesting & instructive to study your admirably executed drawings," Darwin told him, "for I had no idea that animals of such low organization could develope such extremely beautiful structures." He had never studied the radiolaria himself, not even aboard the *Beagle*.

Another thing of which Darwin had no idea—not in 1864, when he couldn't have foreseen it—was the huge role that this effusive young German artist-scientist would play over the next fifty-five years in promoting his theory and depicting a Darwinian concept of the tree of life. By the early twentieth century, in fact, long after Darwin's death, Haeckel would be the world's most famous living Darwinian, and his trees would be well rooted in the popular understanding of life's history.

38

Ernst Haeckel was born on February 16, 1834, making him almost precisely twenty-five years younger than Darwin, perfectly spaced to reach an age of fervid impressionability just as Darwin exploded into midcareer fame. Haeckel's father was a jurist and a counselor to the Prussian court at Potsdam until the family moved to Merseburg, a smaller town in Saxony, where Ernst grew up. His parents exposed him to great literature and serious ideas—the poetry of Friedrich Schiller, the nature philosophy of Johann Wolfgang von Goethe—and as a boy, he read Alexander von Humboldt, and Matthias Schleiden's vivid botanical book *The Plant and Its Life*, and Darwin's *Beagle* journal, all whetting an appetite for adventurous scientific travel. From Humboldt and Schleiden, he learned that "proper evaluation of nature required aesthetic as well as theoretic judgment," according to his biographer Robert J. Richards, a scholar of German Romantic thought. Haeckel leaned toward botany, but when he reached age eighteen, his father, concerned for the practicalities, pressured him to study medicine. He enrolled at the University of Würzburg and hated the medical curriculum but stayed with it, stealing time to read more Humboldt and Goethe amid his studies. Then came his clinical training, which by Richards's account was devoted largely to "horrible worms, rickets, scrofula, and eye diseases" among the poor of Würzburg.

Haeckel detested it. He wasn't born to be Albert Schweitzer. The part of medicine that he did enjoy was autopsies—a grim exercise fitting his interest in anatomy. He also liked histology, studying the microscopic anatomy of cells and tissues, and he discovered an aptitude for drawing tiny structures in fine detail, one eye on the microscope eyepiece, the other on the page. That would later come in handy for the radiolaria.

He passed his state medical examinations in March 1858 but never practiced, renouncing medicine at that point to do research in zoology. Haeckel was fascinated by morphology, which encompasses anatomy but goes beyond it—anatomy being the study of bodily structures, morphology being the study of *relationships among* and *comparisons between* bodily structures. Morphology led toward questions of whence and why, evolutionary questions, while anatomy remained descriptive. Haeckel followed that lead fairly soon, but at first, for him it was just nature study. In particular, he came to love marine creatures, including those radiolarians, which offered a universe of weird bodily structures.

Haeckel was a fervent young man, prone to deep intellectual, aesthetic, and emotional excitements, a German Romantic in the tradition of Schiller and Goethe. His turn to marine biology began during a summer getaway from his medical studies in 1854, when he and a friend caught a boat from Hamburg to one of the islands of Heligoland, a small archipelago fortysome miles offshore in the North Sea. It started as a lark but became a formative experience for Haeckel when they fell in with Johannes Müller, a renowned zoologist from Berlin, who was there to study starfish, sea urchins, and other echinoderms. Scooping up invertebrate animals from the sea and examining them with this scientist was revelatory. Müller's friendship and guidance swung Haeckel from botany to marine zoology, with special attention to invertebrates. Back in Berlin, Müller's journal published Haeckel's first zoological research article while he was still a medical student. The mentorship might have continued, but in April 1858 Müller died of an opium overdose—probably a suicide, provoked by depression, Haeckel suspected.

That spring was a dark and confusing time for Haeckel. He was twenty-four, quitting medicine for a scientific career, and now his best teacher had jumped into an early grave. The turmoil gives some context for what happened next: two days after Müller's funeral, Haeckel became engaged to a young woman named Anna Sethe. He had known her almost six years, and she was his first cousin, daughter of his mother's brother. Haeckel's own brother had married Anna's sister; it was at their wedding that he had met Anna, whom he saw as a dancing seventeen-year-old "elf." This sort of tangled intermarriage within families wasn't rare or titillating in nineteenth-century middle-class circles: Charles Darwin himself married his first cousin, Emma Wedgwood, and Darwin's sister Caroline married Emma's brother Josiah Wedgwood III. But despite the cozy family linkage, Haeckel (unlike Darwin, who could never be mistaken for a German Romantic) embraced his chosen partner with a great passion. Anna to him was a "true German child of the forest, with blue eyes and blond hair and a lively natural intelligence," and a soulmate with whom he looked forward to sharing "every thought and every action." In a letter that summer, he told her: "When I press through from this gloomy, hopeless realm of reason to the light of hope and belief—which remains yet a puzzle to me—it will only be through your love, my best, only Anna." Unfortunately, he couldn't marry her yet because he didn't have a job.

Haeckel's other great passion was science. Rather than turning back toward medicine for a livelihood that might have supported marriage, he launched himself toward marine zoology as a sort of freelancer, again with more heart than practicality. He gathered up some equipment and, in early 1859, went to Italy. After seeing Florence and Rome as a tourist, and disliking the tedious religious flavor of the art, he tried Naples for its access to the sea. That didn't work out so well either. He was living in crummy digs, annoyed by the Neapolitans, losing his religious faith, and frustrated at the work, trying to do zoology from the buckets of bycatch he got from fishermen on the waterfront. After almost six months there, Haeckel felt he was getting nowhere. So he crossed the Gulf of Naples to Ischia—another island but now a Mediterranean one, warmer and more picturesque than Heligoland—taking a palette and easel to do some painting. His luck in travel encounters was good, and this time he met Hermann Allmers, a German poet who was also a painter. Allmers was a small man, older, with a formidably hooked nose and a thrust chin—a strange match to Haeckel, who was tall and handsome, long-faced, with curly blond hair, a beard, and presumably by now bronzed from the Mediterranean sun. But their interests, talents, and dispositions meshed well, and after a week of knocking around the island, they were fast friends.

From Ischia they went to Capri, where they swam, painted, and danced (Haeckel did, anyway) the tarantella. It was just what a troubled twenty-five-year-old should be doing, that sort of spirit-cleansing interlude, but it wasn't getting Haeckel any closer to marrying Anna. After a month on Capri, he and Allmers went to Messina, on Sicily—which was not just another island but a turn back toward science for Haeckel, because in this very place, Johannes Müller had done fieldwork. After five more weeks of cavorting and painting, crisscrossing Sicily, climbing Mount Etna, Allmers had to leave, and Haeckel got serious about zoology. The landscape of Sicily he found boring and threadbare, but the sea was full of abundance and variety. Based on what came from the water, Haeckel called Messina "the Eldorado of zoology."

He was still torn between art and science, and between his romantic vision of an artist's rich experience, on the one hand, and an academic scientist's salary, which would allow him to marry. He was overwhelmed by the diversity of sea creatures he was seeing and trying to study. At the end of November 1859, according to Robert J. Richards, "with just a few months left for his research in Italy, Haeckel finally decided to focus on just one group of animals, the almost unknown radiolaria." They aren't actually animals, the radiolarians, they're something else, but never mind; that fact was unknown and unimportant to Haeckel at the time.

39

Ernst Haeckel found a boggling diversity of these tiny, glassy creatures in his sea-water samplings at Messina, and, within a few months, he'd shipped home specimens representing more than a hundred species previously unknown to science. His revered teacher Müller had led him to the topic by way of a short monograph on the radiolaria that Müller published shortly before killing himself, and which Haeckel carried with him to Italy. He depended heavily on that guide as he got started. But he quickly went beyond Müller, not just discovering new species but also drawing the creatures in microscopic detail, beginning to scrutinize their soft internal anatomy as well as their external silica skeletons, and trying to classify them in some orderly way. Back in Germany after the Italian rambles, Haeckel got permission to work on his harvest at the Berlin Zoological Museum, and he started writing a report. Around the same time, in the summer of 1860, one other factor added to his focus and momentum: he read the German edition of *On the Origin of Species* and fell in love with Darwin's theory.

His report on the radiolaria grew into a dissertation, and the dissertation would soon grow into Haeckel's large, two-volume, lavishly illustrated monograph. But he still couldn't afford to marry Anna, or felt that he couldn't, so he interrupted the radiolarian effort to accept paid work.

A friend on the medical faculty at the University of Jena, southwest of Berlin, offered him an assistant's position, which might lead to something more. Jena was a special place, considered the cradle of German Romanticism, site of a distinguished university, rich in cultural ferment as well as intellection—suffused with the spirits of Schiller and other philosophers and poets, and just down the road from Weimar, where Goethe lived. About the offered job, Richards tell us: "He had no choice. Jena, that warmhearted and energetic pulse of Romantic élan, could not be refused, especially since its embrace might also bring him into the arms of that other love, Anna." Haeckel took the assistantship and qualified as a privatdozent at the university. Working hard, burning long hours, he did his teaching and also finished *Die Radiolarien*, both volumes, and had them printed in Berlin. On the strength of that book, extraordinary coming from an unknown junior scholar—extraordinary coming from anyone—he was offered a professorship at a good salary and directorship of Jena's Zoological Museum.

That summer, back in Berlin, Haeckel married Anna. She shared his love of nature and art, understood something of his intellectual passions, and called him "her German Darwin-man." Their happiness would last eighteen months.

His new status at Jena was gratifying, but it didn't make him famous. His transition from obscure young zoologist to prominent explicator of Charles Darwin's theory happened suddenly with a public triumph at the thirty-eighth meeting of the Society of German Natural Scientists and Physicians, a large scientific organization, on September 19, 1863. That year's gathering was in the Prussian town of Stettin. Haeckel was invited to give the big opening lecture.

Before an audience of two thousand people, society members and guests, to whom the topic had been advertised, he spoke for an hour about evolution by natural selection. Haeckel noted that earlier thinkers had floated evolutionary ideas but that no one before Darwin had offered a material theory, involving laws of inheritance and variation. He described the mechanisms of selection and adaptation. He cited three kinds of evidence supporting Darwin's big idea: the fossil record, the clues inherent in embryology, and the patterns of relatedness suggested in systematic classification. Relatedness implied divergence from common ancestors. "The whole natural system of plants and animals," Haeckel said, "appears from this perspective as a great stem tree, and so each genealogical table of relations can be represented intuitively in the form of a ramifying tree whose simple roots lie hidden in the past."

Stammbaum was his word, a stem tree, a family tree, upon which every branch and twig represented a form of life evolved by natural selection from some other form of life—every form except, possibly, the very first and simplest. (He was still undecided, or anyway ambiguous in public, about whether life began from an act of God or a chemical accident. Later, he would opt for the latter.) And he made explicit what Darwin had hinted in *The Origin*: that this evolutionary process, this tree, included humans.

Just how the tree should be drawn was another question, which Haeckel would address soon enough. For the moment, he enjoyed the great applause of his audience and the extensive coverage next day in a local newspaper, *Stettiner Zeitung*, a clipping of which he sent proudly to Darwin just a few days after Darwin had thanked him for the gift of the books.

40

This phase of Haeckel's life, beginning from the splash at Stettin, was glorious but brief. Back in Jena with Anna, he taught a very popular lecture series on Darwin's theory, illustrated with large drawings by his own hand. One student remembered decades later how Haeckel would stride into the auditorium with "the victorious rush of an Apollonian youth." He was slender, looked thoughtful but dashing, with "great golden locks flowing from his large head, which itself evinced a great brain." His large blue eyes were "blazing yet friendly—he was probably the most handsome man I had ever seen up to that time," the dazzled student testified. The lectures he poured out were as brilliant as his appearance. Off the podium, Haeckel continued his study of Darwin's theory and its application to real groups of organisms, including the radiolarians. In late January of the new year, 1864, his life turned dark.

Anna fell sick with what Richards reports as pleurisy, a lung-related inflammation. She improved, then she relapsed, or suffered something new, this time with abdominal pains suggesting appendicitis. Haeckel himself later called it typhoid fever. By an alternate account, she had a miscarriage with fatal complications, and the details were concealed from her husband. Whatever her trouble, on the night of February 15, the agonies came to a peak, and she lost consciousness the next morning.

February 16 was a bizarre day for Haeckel, bringing a gruesome mix of events: he got word that he would receive a prestigious award, the Cothenius Medal, for his scientific work; he turned thirty years old; and that afternoon, Anna died.

The central theme of the Richards biography is that Anna's death was the defining moment in Haeckel's life, killing whatever remained of his religious faith, his sense that there existed a spiritual dimension apart from the material dimension, and turning him to Darwinian theory as a kind of substitute theology. Going beyond even Darwin (who underwent his own loss of faith, catalyzed partly by the death of his favorite daughter), Haeckel replaced God with natural selection, as the central force in what he called his "religion of monism." What he meant by *monism* was a bit paradoxical and woozy: God is nature, nature is God, mind and matter are two manifestations of some single underlying reality, neither can exist without the other, and therefore (by implication) immortal souls and eternal rewards don't exist. Haeckel called this "the purest kind of monotheism," but Judeo-Christian theologians wouldn't agree. For orthodox believers of Haeckel's day, Richards writes, monistic metaphysics "could only be viewed as transparently shrouded atheism." Whatever Haeckel's monism was, ineffable or just dreamy, it guided his version of Darwinian theory—as he promulgated that theory (or anyway, that version) in his writings and lectures over the next fifty-five years.

More important to us, though, is how he promulgated that theory in his art: by drawing evolutionary trees of life that portrayed actual creatures and their actual patterns of descent and divergence, not just letters and dots in a hypothetical figure. He made phylogenetics concrete.

41

Haeckel's first response to Anna's death was prostrate grief. He spent eight days in bed, half delirious. When that passed, his parents sent him to Nice, France, for a convalescent getaway. While walking along the seashore, he caught sight of a beautiful jellyfish in a tide pool and watched it for hours, its long tentacles reminding him of Anna's golden hair flowing out below a headband. Seeing his dead wife in the shape of a jellyfish sounds a little ghoulish, but it helped reconnect Haeckel with his passion for nature and science, his efforts now rededicated to her memory. The jellyfish represented a new species, and he named it *Mitrocoma annae*, meaning "Anna's headband."

His second response to her death was work. He buried himself in a new writing project: an ambitious book synthesizing his views of Darwinian theory, monism, and a whole edifice of "natural laws" discovered by him, framed in new terminology invented by him. Darwin's natural selection was his primary law, but he had no shyness about adding Haeckelian corollaries. He offered "laws" of inheritance and "laws" of adaptation. There was a *law of uninterrupted or continuous transmission* and a *law of interrupted or latent transmission*. There was a *law of correlative adaptation*. There were many others in that vein—more than 140, according to

Richards, who describes this new book as "stuffed with as many lawlike proposals as the municipal code of a small city." The neologisms with which Haeckel salted it have been more useful and durable than his laws. He coined the term *ecology*. He coined the term *phylogeny*. He coined the term *ontogeny* and propounded what he called the *biogenetic law*, asserting that the embryological development of an individual retraces the course of its evolutionary descent. A human embryo, by this argument, passes through stages at which it looks like the embryo of a fish, then of a salamander, then of a rabbit. Put in three words, as it would later be famously known: ontogeny recapitulates phylogeny.

Haeckel wrote the opus, two volumes again, a thousand pages, with illustrations, in a year of ferocious effort. "I lived then quite like a hermit, allowed myself barely 3–4 hours sleep daily, and worked all day and half the night," he said afterward. The work appeared in 1866 as *Generelle Morphologie der Organismen*, a study of the shapes of living creatures—including the origins of those shapes. According to Richards: "It contains the foundation for all of Haeckel's later thought." It also contains an arresting series of trees. He drew them in ways no one else had: as an artist-scientist, illuminating evolution with flair.

The idea of drawing evolutionary trees may have been planted in Haeckel's imagination by Darwin's diagram in *The Origin*, which was artless but important. Or it might have come from two other sources, one or both: Heinrich Georg Bronn, a paleontologist who translated Darwin's book into German, or Haeckel's friend August Schleicher, a linguist. Bronn had published an essay in 1858 on the laws of development of living creatures, and that essay included a sort of stick-figure rendition of a tree. It was another bare abstraction, a naked sketch, suggesting the idea of progressive lineages but without committing to specifics. Bronn himself was no evolutionist at that point, one year before Darwin's book appeared in English, and Bronn attributed his progression of forms to a "creative force," not natural transmutation. So his tree was an old-school tree. Still, Haeckel may have seen it and envisioned a new meaning. Haeckel's other possible inspiration was a book by Schleicher presenting what its author called a "Darwinian" theory of linguistic evolution, in which the divergence of languages, ancient to modern, was depicted in a branching diagram. Influenced or not, Haeckel seized on the tree shape and went far beyond Darwin, far beyond Bronn or anyone else, in drawing evolutionary trees. He bundled eight of them into his *Generelle Morphologie*, covering eight major categories of living things.

And his were different, as I've said: specific, not hypothetical. The
limbs and branches of these trees fruited with the names of actual creatures and groups of creatures, not vague letters. Each tree offered a concrete proposal about which animals, plants, or other life-forms share
ancestors with which others. Some of them also presented phylogenies
(histories of ancestral lineages), putting that new word of his to good
use. Not least important, his trees were visually rich and deftly executed,
displaying Haeckel's graphic talent and his maniacal attention to detail.
The most intricate of them, his *Stammbaum der Wirbelthiere*, a Family Tree
of Vertebrates, has so many long, slender branches, rising vertically but
pulled a bit sideways, that it looks more like a great clump of seaweed
swaying gently in the current than like a rigid maple or elm. Nobody had
ever drawn a tree of life like Haeckel's trees.

His vertebrates tree included mammals, reptiles, amphibians, fish,
and birds, of course, with a vertical axis along the left border showing
increments of geological time, throughout which these classes had arisen.
There was a Family Tree of Coelenterates, within which grew the jellyfishes, now his sentimental favorites, among many other forms. There
was a Family Tree of Mollusks, a Family Tree of Plant Kingdoms, and
a Family Tree of Articulate Animals, which contained arthropods and
worms. Mammalian forms, besides being placed on the vertebrates tree,
got another tree of their own, his *Stammbaum der Säugethiere* (Family Tree
of Mammals), and in its upper-right-hand corner, you can see a small
branch denoted "Homo sapiens," just beside an equally small branch for
gorillas. That branching implied: we are another ape.

Haeckel threw everything together in his great *Stammbaum der Organismen*, a tree of all organisms, portrayed in a complicated figure rising
vertically but sliced horizontally with three baselines across the bottom
parts of the image, reflecting three alternate hypotheses about the origins
of all life. Take each baseline as a starting point, and you get a different
hypothesis. The tree has nineteen secondary limbs, all rising vertically,
all cut across by the topmost baseline: suggesting that life began independently nineteen times. The tree has three major limbs from which

those nineteen diverge: suggesting three independent origins of life—one leading to plants, one to animals, one to creatures of all other kinds, such as Haeckel's beloved radiolarians. (He called that third group Protista, and it represented an important departure from orthodox thinking, to which I'll return in a moment.) At the base of the tree is its single big trunk: suggesting all life as one tree, derived from one origin, one primordial ancestor. He labeled the trunk Moneres, by which he seems to have meant the simplest of single-celled organisms, resembling bacteria. (For technical reasons, their name, meaning "single," was later corrected to Monera.) This hypothesis, among his three, was the boldest application of Darwinian theory that could be made at the time. It asserted that all living creatures, including humans, have descended from some common ancestor resembling a bacterium. But in 1866 and for some years after, Haeckel himself was still undecided about which of his three hypotheses was correct.

The other innovation presented in Haeckel's big tree was that group he named Protista—his kingdom of nonplant, nonanimal organisms. From the time of Aristotle, through the work of Linnaeus and until Haeckel's era, naturalists had viewed all life as divided into two kingdoms: plants and animals. That was simple. It conformed to common sense. If the large living forms we see all around us—the trees and the grasses, the birds and the fishes, the flowers and the elephants—fall neatly into two categories (as they seemed to do, before any microscopic investigation of fungi), then all life must be binary. A creature moved, or it didn't. It ate, or it greened. Even microbial organisms—the bacteria and the amoebae, the radiolarians and the ciliates, the diatoms and all the others—were considered either animals or plants. That's why Leeuwenhoek wrote of the animalcules he saw through his lenses, and why the misleading term *protozoa* (earliest animals) was coined in 1818. But Haeckel said no, wait: life *isn't* binary. It's trinary. And he placed those puzzling little creatures in a kingdom all their own. It was a visionary act of almost Copernican daring, which went largely unnoticed and unfussed-about at the time.

Generelle Morphologie was a dense book, written stubbornly amid sorrow, and though it influenced scientists, it didn't reach a broad audience. He rectified that two years later, producing a work for the general public under a title that best translates as *Natural History of Creation*. This one

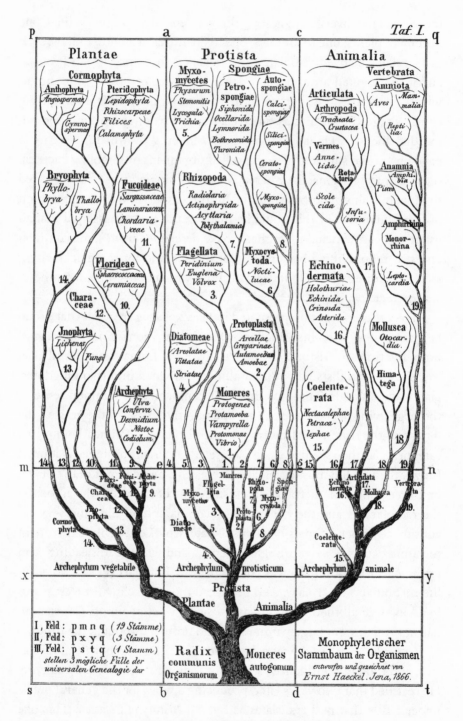

Haeckel's tree of organisms, 1866.

emphasized the human part of the big story, always a good tactic when you're writing for a human readership, and it celebrated Darwin, Darwin's theory, and some of Darwin's precursors (Lamarck, Lyell, Darwin's grandfather Erasmus, Wallace, and, because Haeckel was German, Goethe) in language somewhat more accessible and fluid than Darwin's. It presented the biogenetic law, about ontogeny and phylogeny, with help from an illustration showing similarities among the embryos of a turtle, a chicken, a dog, and a human. It sold well, blazing through twelve German editions and being widely translated. In English, it appeared as *The History of Creation* (the word *natural* dropped evidently because it sounded too materialistic), and that version went through a number of printings too. By the early twentieth century, one historian called this book "the chief source of the world's knowledge of Darwinism." It also contained trees, mostly depicting relationships among the higher animals. There was a distinctly racist tree of human ethnicities, purporting to show which were primitive and which advanced. By strange coincidence, Germans were on the uppermost branch.

42

Busy and prolific, Haeckel continued his popularizing as well as his research, and in 1874 he published a book devoted entirely to human evolution—or, as put in his subtitle, *The Developmental History of Man*. His tree making came to a crescendo here, in an image known later as Haeckel's "great oak." This is probably his most widely recognized work of art. You may have seen it as a poster. You could get it on a T-shirt. It's his version, in graphic form, of the lineal ancestry of humans, from Monera, through worms and amphibians and reptiles, straight up to us. This tree has no canopy and not much branching. It's thick at the bottom and tapers skyward, looking less like a great oak than like an enormous rutabaga, slightly hairy, pulled from the ground and turned point up. At its very topmost branch sits the word Menschen, "people," flanked below (not side by side) with gorillas and orangutans and chimps.

Does this drawing reveal Ernst Haeckel as a throwback anthropocentrist in the guise of a Darwinian, as some scholars have argued? Was his great oak just another ladder of nature, of the sort Aristotle proposed and Charles Bonnet had been designing back in 1745? Does it show that Haeckel took humans as the crown of creation, the end point toward which a teleological evolutionary process was directed? Not necessarily. It might look that way at first glance, yes, but his thinking and his illustrating were more complicated.

PEDIGREE OF MAN.

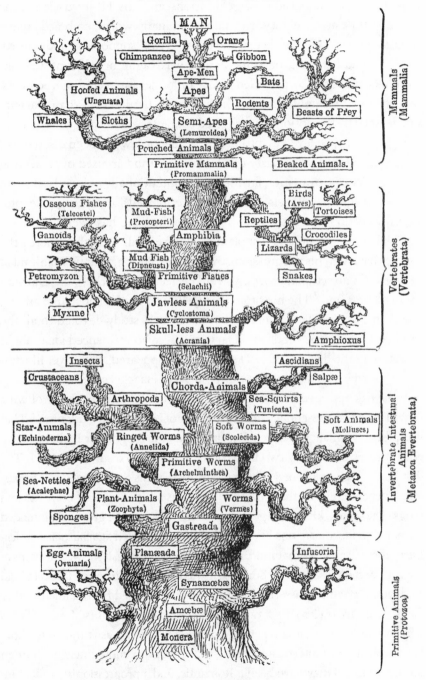

Haeckel's great oak, English version, 1879.

Haeckel is controversial among historians of biology, who disagree about his real or supposed views, confusions, and sins. He lived a long time (dying in 1919, at age eighty-five), published many works, embraced many variations on Darwinian theory and many "laws" supposedly under its rubric, changed his mind on some points, was vague or self-contradictory on others, and thereby provided plenty of fodder for such disagreements. One school of thought dismisses him as more Lamarckian than Darwinian, deluded by the notion that acquired characteristics can be inherited (although Darwin believed that too). His biogenetic law has been discredited. During his lifetime, in a nasty crisis, Haeckel stood accused of presenting bogus evidence—drawings, falsified either intentionally or by mistake—in support of that hypothesis. He has been criticized for making paleontological assertions without much fossil evidence. He has been called "superficial, inconsistent, and just plain muddleheaded." And then there's the charge that, with his progressionist view of evolution, placing mankind at the apex of a directional process toward perfection, Haeckel was "Darwinian in name alone." The historian Peter J. Bowler made that charge among others in *The Non-Darwinian Revolution*, his 1988 study devoted to all the "pseudo-Darwinians" and "anti-Darwinians" who influenced ideas about evolution in the decades after *The Origin* first appeared. The cover illustration of Bowler's book, pressing this point, was Haeckel's great oak.

The oak appeared inside Bowler's book too, offered in evidence of what Bowler called "the essentially linear character of Haeckel's evolutionism." That was probably unfair. Other scholars, including the biographer Richards, have noted that Haeckel drew two kinds of trees, with two distinct purposes. The first kind were phylogenetic illustrations, such as his *Stammbaum der Organismen*, intended to show the breadth of all life as a full canopy of branches and twigs. The second kind were genealogical trees such as the great oak, intended primarily to illuminate one lineage. The oak was labeled *Stammbaum der Menschen*, after all: a tree of mankind, not a tree of everything. It showed humans at the apex because it was *about* humans. It was drawn to illustrate Haeckel's bold (and, yes, Darwinian) assertion that we have descended from a line of other forms, traceable back to the simplest single-celled creatures: his Monera. That's why its trunk is thick and tapering, with few limbs. It wasn't meant to show life's diversity and interrelatedness. It was meant to show lineage—human lineage. Haeckel may have been a Romantic, and a progressionist, and an inconsistent Darwinian, but even he knew that the tree of life is not a rutabaga.

43

Haeckel had started something that didn't stop. Well into the middle of the twentieth century, paleontologists and biologists created phylogenetic trees, graphically executed by their own hands or with help from illustrators, portraying inferred evolutionary relationships and histories of descent. They based these trees mainly on similarities and differences among the shapes of living creatures (comparative morphology), on the fossil record, and on embryology. That was the evidence they had. Molecular phylogenetics didn't yet exist.

Some of these images were trees of all life and some were trees of particular groups. Henry Fairfield Osborn, a paleontologist at the American Museum of Natural History in New York, published a tree of proboscidean mammals, in 1936, festooned with mammoths and mastodons and elephants. The tree limbs were just simple arrows, but the pachyderms were nicely sketched. A British entomologist named William Edward China, a specialist in the taxonomy of the Hemiptera (true bugs, which live by sucking fluids from plants or other animals), produced a tree of hemipteran families in 1933. It resembled a geisha fan as much as a tree, but it showed a stinkbug's degree of kinship with a bedbug, for those who wanted to know. Herbert F. Copeland, a biology teacher tucked away at Sacramento Junior College in the late 1930s, published a memorable paper on the kingdoms of life that, besides reprinting Haeckel's tree of all organisms, included a graphic figure of Copeland's own. Copeland's was not exactly tree shaped, but it expressed the same basic idea: that lineages have risen through geological time and, as they diverged, expanded through ecological space. Instead of branches, the Copeland illustration shows smoothly tapered cones ascending vertically in a cluster, like organ pipes ready to belch out a fugue.

Alfred S. Romer, a paleontologist at Harvard, was one of the great tree makers of the midcentury, and his reached a broad audience of young scientists-in-training through his influential textbooks, most notably *Vertebrate Paleontology*. That book was first published in 1933, revised in 1945, revised again, and still a standard source when I bought my copy in 1982. Its tree branches rise and thicken like plumes of black ink on the page, then in some cases grow thinner again, their breadth representing the comparative abundance and diversity of different groups throughout the vicissitudes of time. Romer shaped a tree of all vertebrates, trees of the vertebrate classes, and trees of some of their subdivisions: fishes, amphibians, reptiles, mammals, even-toed ungulates, whales, rodents. Glancing at his figures, you can see that dinosaurs crashed at the end of the Cretaceous period but, for some reason, crocodiles didn't, and that rats thrived in the Pliocene Epoch. Romer's trees speak of the shifting fortunes, the pulses and fades, that are so basic to evolution's story.

Then in 1969 came an unusual tree from Robert H. Whittaker, a plant ecologist at Cornell University for whom "broad classification," as he called it—numbering and delineating the kingdoms of life—was a sidelight. Whittaker's tree had little to recommend it by way of artistic flair but delivered some provocative content. It looked like an annotated balloon animal. Or maybe, to be more polite and botanical: a prickly pear cactus. It consisted of five roughly oval lobes: three, stacked upon one, stacked upon another. The lobes represented what Whittaker proposed as the five kingdoms of all life.

Five kingdoms? That was a new count at the time, a radical proposal, and Whittaker had come to it in stages. At the start of his career, studying insects and plant communities, he gave no hint that he would ever be interested in, let alone drastically revise, how life-forms are classified at the highest level.

Robert Whittaker grew up in eastern Kansas during the Dust Bowl and the Depression, collecting butterflies as a boy, wandering through meadows and woods, becoming a man of sternly traditional values, with a personality that some colleagues would later call "stoic" and "intense." But he was capable of recognizing ambiguity where he saw it. He went to college in Topeka, did a hitch as a weatherman for the Army Air Forces during World War II, then went to grad school in a place that keeps reappearing in this story: Urbana, Illinois. He wanted to study ecology. Rejected by the Botany Department, he joined Zoology instead but became a plant ecologist anyway. He distinguished himself at Illinois by presenting a dissertation in zoology that had nothing to do with animals.

During that PhD research, on the vegetation of the Great Smoky Mountains, Whittaker began to challenge a canonical principle in his branch of science: the idea that plant communities are stable, highly integrated associations, with consistent species compositions, clear boundaries, and hard reality as units, almost as though they were living organisms. This idea had been enshrined in ecological thinking through the influence of an ecologist named Frederic Clements. Whittaker dismantled it in his dissertation. He showed that plant communities are loose associations, not integrated units, with blurry boundaries and a "low degree of reality." By way of this experience and his other work, he brought two forceful predispositions to his efforts, years later, on the tree of life: he saw things ecologically, and he appreciated that boundaries are often ambiguous.

Somewhere along the way, Whittaker got interested in delineating not just plant communities but also the most fundamental categories of life. In 1957 he published his first effort, a short paper titled "The Kingdoms of the Living World." How many, and what were they? The traditional view recognized just two kingdoms of life: plants and animals. Haeckel had said three: plants, animals, and his Protista, the everything-else category composed mainly of microbes. Several earlier naturalists, including Richard Owen and John Hogg in Britain, had also found a third kingdom of life, Owen calling his Protozoa and Hogg coining the name Protoctista ("first created beings"). But neither Owen nor Hogg was an evolutionist, and their influence on phylogenetics was far less than Haeckel's. Herbert F. Copeland, who published that 1938 paper with the organ-pipe illustration, came up with four kingdoms. He took Haeckel as his inspiration but, drawing on better microscopy and newer thinking about microbes, separated Monera (bacteria) from Protista (simple creatures with cell nuclei). So for Copeland, using the formal names, it was Monera, Protista,

Plantae, and Animalia. He argued his case more fully in a 1956 book, which is full of intricate stipple illustrations of microbial creatures but, oddly, omits any tree-of-life art whatsoever. What it does contain, as its reverential frontispiece, is a photo of Ernst Haeckel in his prime—with the beard, the wavy blond hair, the blazing eyes—suggesting how forcefully Haeckel haunted this field even into the middle twentieth century. Robert Whittaker seems to have been provoked to his own declarations on kingdoms in reaction to Copeland's book.

Whittaker's approach was unique in that he tried to answer the big question with ecology, not with morphology. "Ecologists are familiar with divisions of the living world which correspond neither to Copeland's nor to the two-kingdom conception," he wrote. Ecologists see distinctions that microscopists miss. Among the boldest of those distinctions, Whittaker noted, are three categories of organism: producers, consumers, and decomposers. Animals are consumers, swallowing other creatures for their sustenance. Plants are producers, gaining sustenance from sunlight and water, creating their bodily substance from nonliving materials. Bacteria and fungi are decomposers, taking their sustenance by gently dismantling other creatures, dead or alive, and putting the pieces to new use. Each of those three categories, in Whittaker's 1957 proposal, represents a kingdom. Another way of explaining it, he wrote, was that "kingdoms are, most essentially, major directions of evolution" and that those directions reflect three different means of gaining nutrition: eating, photosynthesis, and absorption.

"The kingdoms are man's classification," he added, and their meaning derives only from the fact that we humans choose to recognize them, purely for our convenience in organizing biological knowledge. This is similar to what he had said about plant communities: that the communities, as distinct from individual plants or populations of this species or that, have a "low degree of reality" in the tangible world. Plant communities are how we think about plant diversity, which *tends* to get arranged so that plants with similar needs share similar habitats. Kingdoms of life are how we think about life's diversity, so that we don't get nosebleeds and despair of the whole enterprise of biology. It's worth noting too that Robert Whittaker, like Herbert Copeland, was a professor as well as a scientist. He understood that teaching biological information, along with organizing and retrieving it, required putting its bounteous variousness into handy categories.

Two years later, after some further thought, Whittaker revised his schema, in a paper titled "On the Broad Classification of Organisms." Evidently he felt he had simplified too far. Three kingdoms, however broad, didn't cover the full breadth of nature's diversity. There were four—but not precisely the same four as Copeland's. For the Whittaker of 1959, it was Protista, Plantae, Animalia and . . . Fungi. Although fungi absorb their nutrients, he had become uncomfortable at lumping them with all those single-celled absorbers among the Protista. So the plants (producers), the animals (consumers), and the fungi (absorbers) were defined by ecology, he explained, and the protists (unicellular) were defined by morphology. "These themes are inconsistent," he admitted, but it was the best he could do.

This paper is also memorable for containing his first prickly-pear-of-life illustration. It showed four lobes: a basal lobe, labeled Protista, with three lobes for Plantae, Fungi and Animalia sticking up from it. The reason he had chosen such an unconventional form—a figure with lobes rather than limbs—wasn't obvious at the time. But it would become more clear in a later iteration.

Whittaker returned to the problem in 1969. His paper "New Concepts of Kingdoms of Organisms" appeared in *Science* that year and, somewhat surprisingly, given his previous waffling, influenced a generation's worth of biology textbooks. The new concepts it offered were primarily Whittaker's old concepts from 1957 and 1959—classification should be ecological, except for single-celled organisms—plus one important addition: another kingdom. Meanwhile, as I've described, Roger Stanier and C. B. van Niel had offered their own forceful idea on broad categories in 1962—dividing all life into prokaryotes and eukaryotes. Whittaker nodded to that dichotomy, used it for characterizing bacteria, then otherwise essentially ignored it. He split the bacteria away from Protista and gave them their own realm, going back to Haeckel's label: Monera. His new illustration was a prickly pear cactus with five lobes: Monera at base, Protista rising from that, then Plantae and Fungi and Animalia. Five kingdoms, not four, comprising all life on Earth.

Why had he changed his mind? "Recent work has made more evident the profound differences of organization between bacterial cells and those of other organisms." Whose recent work? He cited the 1962 paper by Stanier and van Niel. They had added those two words to his scientific vocabulary and those two categories to his view of life: prokaryote and eukaryote. It seemed undeniable now, to Whittaker, among others, that bacteria and single-celled protists are distinct from each other, belonging to utterly

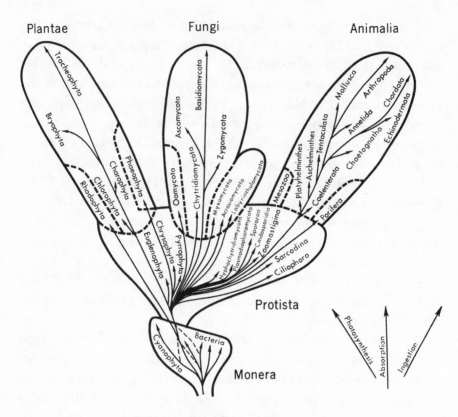

Whittaker's prickly pear of life, 1969.

different kingdoms, because the cells of protists contain nuclei and the cells of bacteria do not. There were other big differences too, such as the mitochondria and chloroplasts within eukaryotes, the capacity for mitosis, the flagella. Those contrasts, he now conceded, echoing Stanier and his two textbook coauthors, defined "the clearest, most effectively discontinuous separation of levels of organization in the living world." Whittaker accepted that discontinuity as convenient for defining his first kingdom, Monera, and then essentially ignored it for everything else. The prokaryote condition served a purpose within his "broad classification." The eukaryote condition didn't.

Still, he had implicitly recognized those two major divisions of life, even while explicitly defining five kingdoms. How had such a huge evolutionary leap, from prokaryote to eukaryote, occurred? One good hypothesis, he added, lay in the idea of "ancient cellular symbioses." At that point, in 1969, Whittaker's thinking converged with the work of Lynn Margulis.

44

Convergence is a theme that will be salient throughout the rest of this book. Convergence is something the limbs on a tree never do—not in the natural world, not in the normal course of growth among oaks, elms, maples, hickories, pines, larches, sycamores, beeches, banyans, baobabs, or other actual trees living in the wild. Limbs *diverge*, they don't converge. Branches diverge. Flannery O'Connor, the extraordinary Southern short-story writer and novelist famed for her dark humor and her pet peacocks, once published a story titled "Everything That Rises Must Converge." It was a grim tale of racism and rage, labeled ironically with those five words, which she had drawn from a hopeful quote in the writings of Pierre Teilhard de Chardin, a French Jesuit and paleontologist. Chardin's woozy philosophical writings were popular among liberal Catholics (but not at the Vatican) in the late 1950s and the 1960s. Flannery O'Connor was a liberal Catholic. The story title became the book title for a collection released after her early death. It's an interesting axiom, her *everything that rises must converge*—not always applicable to biology, but sometimes. The path of Robert Whittaker's efforts to classify life converged with Lynn Margulis's arguments on endosymbioses, for instance, and in 1978 the two published a paper together.

At the heart of their collaboration was a friendly disagreement about how to define Protista, the everything-else kingdom. They proposed a compromise. Their discussion of the particulars was long and technical, reflecting the fact that this paper was first presented (by Whittaker, speaking for both authors) as a keynote address to the 1977 meeting of the Society for Evolutionary Protistology, a group that Margulis had helped found. Margulis knew much more about protists than Whittaker did, having spent much of her career to that point gazing at them through microscopes. But she had liked Whittaker's 1969 paper on the five kingdoms, shared his frustrations at trying to teach new biological discoveries within old-fangled categories, and chose to throw in with him for this joint effort, which emphasized their agreements—not just about Protista but also about all five kingdoms and the entire enterprise of classification.

One point they agreed on was that kingdoms of life are hard to define. The lines dividing one kingdom from another are inescapably blurry. This had always been a problem, with some creatures left in the borderlands no matter how those borders were drawn, and the problem hadn't gone away. Fungi: Are they plants or something else? Blue-green algae: Are they, in fact, algae, or are they bacteria? Sponges: Can they be animals, despite the fact that they don't move around and have no nervous, digestive, or circulatory systems? Protozoans: They're not really "zoans," proto or otherwise. Placozoans: What the hell *are* those flat little things? Whittaker and Margulis noted that classification is a human endeavor, not an inherent reality of the natural world—it's a matter of discovery plus decision—and that simplicity is important. You can't have sixty-one or ninety-three major kingdoms of life, it just isn't practical, if you want to organize biological knowledge and teach it.

The second point on which Whittaker and Margulis agreed was that classifications of life should, as far as possible, reflect evolutionary events and relationships. They should be phylogenetic. The third point of agreement was that phylogenetic classification *isn't* always possible.

So they compromised, resigning themselves to the embarrassment of *polyphyletic taxa*. A polyphyletic taxon is any group, such as a kingdom, that encompasses creatures from more than one evolutionary lineage. The phrase itself seems a bit paradoxical, because *taxon* means a unit of classification—but it might be classification by convenience (as in the old days before Darwin) rather than classification by evolutionary ancestry. If you chose to call "marine vertebrates" a taxon, for instance, it would be polyphyletic because whales belong to one distinct lineage of descent, sharks belong to another, and saltwater crocodiles to still another. Each of those three, though indisputably a marine vertebrate, shares closer ancestry with some terrestrial animals than the three share with one another. Therefore a polyphyletic taxon can't be drawn as a tree limb, with all its branches diverging. It's something else: a human construct, an organizing principle, that allows the convergence of lineages. In this picture, branches come together.

Some lineages do converge, even in the real world—by endosymbiosis, for instance, Margulis's favorite process. At the time Whittaker and Margulis were coauthoring their paper, though, that sort of convergence was taken as a rare event. Most biologists considered it anomalous. In later years, and thanks in great part to Carl Woese as well as Lynn Margulis, convergence of genetic lineages would be seen differently. Classification itself would be seen differently. In 1978 the main issue was how to draw lines, between one kingdom and another, one class of creatures and another, that were "natural" insofar as possible yet still orderly and convenient. To the evolutionary classifier, Whittaker and Margulis admitted, polyphyletic taxa are "unwelcome"—but not quite so unwelcome as the prospect of having just too many kingdoms, too many classes, too many small but precisely defined groups.

This dilemma explains why Whittaker drew his systems—the four kingdoms of 1959, the five kingdoms of 1969—as prickly pear cactuses, not as trees. He had recognized all along, and represented candidly, the complication of polyphyletic groups. You can see this if you look closely at his five-kingdom prickly pear. There are multiple lines, like veins, running from one pad to another. Three lines cross from the Protista pad into the Plantae pad, implying three separate origins of the plant kingdom. Five lines cross from Protista into Fungi: five separate origins of the fungi. Two lines from Protista into Animalia, one of which leads only to the sponge group: indicating that sponges evolved from nonanimals along a lineage separate from that of all other animals. Animalness was invented twice. The two distinct lineages converged upon animal identity, at least as Whittaker (and Margulis with him in 1978) chose to define that identity.

Contrary to Flannery O'Connor's title, contrary to Tielhard de Chardin, not everything that rises must converge. But some limbs of the tree of life do. You'll see much more of this in what follows.

45

The years 1977 and 1978 marked a notable stage in the tree-building saga because of Whittaker and Margulis's paper and one other: the announcement, by Carl Woese and George Fox, in November 1977, of a third kingdom of life. Both these offerings were highly influential, and both imprinted themselves on biology textbooks. The striking thing about them in retrospect is that they have so little in common. They don't agree, and they don't disagree. Five kingdoms, according to Whittaker and Margulis; three kingdoms, according to Woese and Fox—and not just the numbers of kingdoms but also the kingdoms themselves were discrepant. The two pairs of authors were talking past each other. They were worlds apart.

Fox, the young postdoc who had served as Woese's main collaborator on the discovery of the archaea and the 1977 paper, had by then left the Woese lab in Urbana and accepted a job, his first independent academic position, at the University of Houston. He was an assistant professor, thirty-two years old, without tenure, and besides teaching and establishing a lab of his own, he needed to publish more work. The sooner he did that, the better his prospects of a career. Fortunately, he was still much engaged with Woese, by telephone and mail, on the effort to classify

life-forms, especially bacteria and archaea, and deduce their deep relationships, using the evidence of ribosomal RNA.

Up in Urbana, in Woese's lab, another young researcher was filling
the role Fox had played, and the data continued to pile up. Woese got
more catalogs of 16S rRNA, that very special molecule he had hit upon
as his Rosetta stone for the early history of evolution, and those catalogs
revealed things that microscopy and biochemistry couldn't. New patterns
emerged. There was more to be said than simply that the archaea are a
separate form of life. Woese felt it was urgent that he and his team publish
an overview. His informal name for that project, in correspondence with
Fox and others, was "big tree."

Late in November 1977, just weeks after his Warhol moment on the
front page of the *New York Times*, Woese wrote to Fox in Houston, enclos-
ing the catalog of still another organism, voicing concern about certain
aspects of their data-analysis method, and then raising a larger subject:
"Please give big tree the top priority. Unless we get that out soon, you
will find our credibility will be eroding." They had based their procla-
mation of the kingdom Archaea on just four organisms, a shockingly
small sample to represent such a large category of life, and he was eager
to present more of their data. They needed the overview paper. "Just
rough out the text and generate the tree," Woese told Fox. "That will get
the ball rolling. The situation I think is more critical from your point of
view than you may realize. If you become controversial to your peers,
you'll find that the scientific friends you've made at Houston will tend to
become enemies."

Soon afterward, they began drafting a paper.

It would be a highly collaborative publication, culminating a decade
of work, coauthored by Fox and Woese and a list of their partners. What
"coauthorship" means on a scientific publication is that these people have
contributed in their various ways. Some might have done painstaking,
dangerous laboratory chores. Some might have donated microbes, from
cultures in their own lab, or offered fruitful discussion and ideas. A senior researcher, the group's leader, most likely established the context,
mentored at least some of the team, and provided the funding. Another
scientist—call this the primary active researcher, who was possibly a grad
student or a postdoc—might have chosen the topic, conferred with the

senior researcher on conceptual details and experimental design, and done a good share of the hands-on work. Then somebody, usually just one person or a few, almost certainly including that primary active researcher, would do most of the writing. In the purely biological sciences, the primary active researcher would likely appear as first author; the senior researcher, the mentor and sponsor, would appear as final author, a position known to suggest overarching responsibility and ultimate credit. For the field of biophysics, though, in which Woese had been trained, the opposite was true—and that may have contributed to a small conflict between Woese and Fox.

In this case, the two men worked closely together on the text. Fox had functioned as primary active researcher in Urbana, handling some of the more difficult tasks, with support from others of Woese's lab team. Fox had labeled microbes with radioactive phosphorus and extracted their ribosomal RNA. Linda Magrum had cut those molecules into fragments and run the fragments through electrophoresis. Magrum had printed films of the stampeding fragments, from which Woese could deduce their sequences, creating catalogs. It was Fox again who had devised the similarity coefficient by which those catalogs could be analyzed, and he had written the computer code to do it. (That skill derived from his training as a chemical engineer: unlike any other biologist in Woese's lab, and almost any in America at the time, he knew the computer language Fortran.) He had entered the code and the data onto hundreds of IBM punch cards, state of the art in those days, and run them through a mainframe machine. He had sketched the resulting trees. Now he and Woese produced a series of drafts, which were converted to clean typescript by a departmental typist in Urbana. Some of those early typescripts, which have survived, show many scribbled revisions and handwritten additions by Fox.

The first page of the first version began with a sentence that was accurate but not catchy: "For at least a century microbiologists have attempted to determine the relationships among the myriad microorganisms that inhabit almost every conceivable niche on the planet." They could do better. Woese wanted more sense of drama from the get-go. A second version began: "Biology has reached an important turning point with regard to the study of evolution." Better, but with a pencil, he changed it to: "An important turning point has been reached in the study of evolution." That put some oomph into his first four words: *an important turning point.* At the top of the page, too, now appeared Woese's informal title: "Big Tree."

Fox's fourth version, as the effort progressed, contained twenty-five handwritten pages of new and reworked material. It went to the departmental secretary for typing, with a cover sheet that said: "Fox. Draft. Need ASAP. Put ahead of all my other work. Priority 1." The tinkering continued until, in a seventh version, the opening sentence was replaced by a new one, its initial four words setting the tone better still: "A revolution is occurring . . ."

There would be more versions, more back-and-forth of typescripts between Houston and Urbana during 1978 and well into 1979, more changes and tweaks and at least one caustic dispute. But from that point, version seven, the opening paragraph of the Big Tree paper was cast in a form that Carl Woese saw no need to improve:

> A revolution is occurring in bacterial taxonomy. What had been a dry, esoteric, and uncertain discipline—where the accepted relationships were no more than officially sanctioned speculation—is becoming a field fresh with the excitement of the experimental harvest. For the most part the transition reflects the realization that molecular sequencing techniques permit a direct measurement of genealogical relationships.

This was a sonorous understatement. Measuring genealogical relationships with molecular techniques was what Francis Crick had proposed back in 1958. It opened a new perspective on the evolutionary past, equivalent in revelatory scope to all the fossils in all the museums of the world. It was a revolution not just in bacterial taxonomy but also in something broader: the way scientists understand the shape of the history of life.

46

The typescript of version seven also carried a new title: "The Phylogeny of Procaryotes." It showed a roster of authors, including Ralph Wolfe, Linda Bonen, Linda Magrum, and fourteen others. George Fox's name led the list, and Carl Woese's came last, signaling his role as the senior researcher. That order of names would become an issue as vexed as any disagreement over content. Somewhere along the way, Woese had proposed that he should be first author, and that Fox should be satisfied with the honor of last.

George Fox talked candidly about this disagreement as we sat in the pizza parlor in Urbana. "I told him I wanted to be a first author on that paper. Because, I mean, I typed all those goddamn IBM cards. I made all those tree diagrams, right? Participated in all the discussion." Then he had gone to Houston and, as new junior faculty, faced some exigent expectations. Thirty-six years after the fact, he seemed to remember those pressures as though he were still in the moment. "Look, I'm in a different university now, right? I'm trying . . ." He paused. *Trying to establish myself*, were the words that went unspoken. "I'm an assistant professor. I need to get tenure, and I need to ultimately become an associate professor and full professor. You know," he said, "it's not doing me any good to collaborate with him if I don't get any serious credit for it."

On August 27, 1979, Woese wrote to Fox about several "potential points of conflict." The first involved Fox's forceful request for first authorship, an important distinction for a young scientist. "I agreed to that," Woese wrote, "and still do. However, you should know how I feel, as the matter is touchy." Then, mindful of all those long hours and days spent blurring his eyes before the light board, gazing at films of galloping fragments, taxing his brain to infer their sequences and recognize their patterns, he unloaded:

"Big Tree is the major statement from my lab. The work is my conception. And I have put far more hours directly into that work than anyone else, including yourself. By all these criteria, I should be first author."

Woese recognized that Fox would soon be up for tenure; that Fox faced a struggle, also, to win grants for his own lab. Since "you feel your funding is at stake," Woese added grudgingly, "I will allow this to be the overriding consideration, and so will defer to your wishes on authorship. But know that I do so with ambivalence. It would certainly not be proper if this became known as your work or that I needed you to analyze my data for me. That must not happen."

Back at the point when Woese had proposed taking first authorship himself, Fox told me, "I put my foot down, and resisted that. And that's when he terminated our . . ." Another pause, for care in phrasing: ". . . basically terminated our collaboration at that time."

Woese signaled the termination in his August 27 letter. He was dissatisfied with their method of data analysis, he explained—the method Fox had devised, generating a similarity coefficient. Woese hoped to find a better approach and apply it to some of the more peculiar bacteria. "I frankly wanted to do all of this with yourself, but your recalcitrance has made that impossible, so I now proceed alone." In fact, he was proceeding with two other colleagues and, just as Fox was jealous of his authorship, they would be of theirs, Woese claimed. So he was taking Fox's name off that other paper.

"These matters are a bit sticky," he ended the letter, "but if we are open about them they will be only differences of scientific opinion." It was a nice analgesic platitude meant to dull, slightly, the sting of the slap. His relations with George Fox, once the scrawny postdoc, smart enough to learn from him, smart enough to challenge him, were never the same. Their names would appear together on just a few further papers, in the early 1980s, as the two men came to closure on work done in the 1970s. But they didn't talk science anymore—not in the fresh, daring, mutually energized way that they had.

Meanwhile the Big Tree paper went to press. It was published by *Science*, in the issue of July 25, 1980, with its title corrected for spelling: "The Phylogeny of Prokaryotes." Instead of comparing just four strains of archaea and nine other creatures, it drew upon molecular sequences from more than 170 different organisms. First among its assertions was that the special molecule, 16S rRNA, could be and had been very useful in discerning evolutionary relationships. Its second main point was that, notwithstanding the title, "prokaryote" is a meaningless category. There are no prokaryotes. There are only bacteria, archaea, and eukaryotes. This was reflected, ever more clearly, in the new data. Its third point, somewhat unexpected and stated indirectly, was that Lynn Margulis had been right about endosymbiosis. The eukaryotic cell "is now recognized to be a genetic chimera," a compound creature, resulting from ancient convergence events among several lineages, including the bacteria that became mitochondria and chloroplasts. Beyond these three points, the paper focused on classification of groups within the first two of its three kingdoms, the Bacteria and the Archaea. And, of course, it offered, in graphic form, a big tree.

This one resembled a grand candelabra, with stems for thirty candles. Five of those stems represented groups of archaea, three led to all the eukaryotes (a vast kingdom, in our perceptions, but tangential to the paper's focus), and the rest stood for groups of bacteria. The particulars of relatedness among those five archaeal and twenty-two bacterial stems held high interest for microbiologists at the time, but I don't ask you to share that interest. It's microbe arcana. Never mind. The more intriguing aspect of the Big Tree is how Woese and Fox chose to root their three kingdoms: ambiguously. The three major limbs don't rise from a single trunk. Each one emerges separately from a rounded base—like a great mound of dirt—bearing the somewhat mysterious label "Common Ancestral State."

"Here's one of the things that's important to me," Fox said across his cheese and my pepperoni. I had brought out my annotated, underlined copy of the 1980 paper. "See that tree?"

"Yeah," I said. The article was folded open to page 459. "Yeah, that is the Big Tree, right?"

"That is the Big Tree. And you'll notice the root."

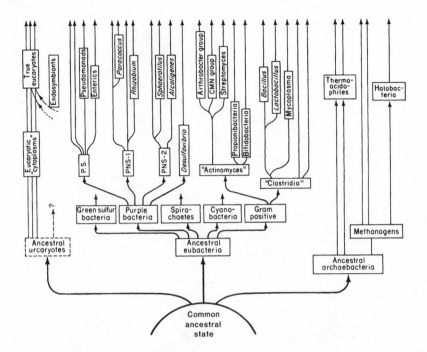

The "Big Tree" of Fox, Woese, and colleagues, 1980.

Yep. *"Common ancestral state,"* I read. "Not a single ancestor, but an 'ancestral state.' Right?"

Right, he said. This mound represented a world of precellular life; a world of primeval twitching almost four billion years ago; a world of naked molecules (maybe RNA in particular) that had acquired the capacity to replicate themselves; a world before living things could be sorted into species, let alone into kingdoms. "We refused to decide the triumvirate," he said. Sprouting three limbs independently from the mound was a way of saying: *Who knows how the three kingdoms are ultimately related? Who knows which came first, or how they diverged, or which is most closely akin to which other? Not us!* It was an act of candid agnosticism, the best that could be done at the time.

"Okay. Now the secret thing," Fox said. For a moment, he looked sly, insofar as a sixty-nine-year-old molecular evolutionist seated over a Coke and a plain pizza can look sly. "The secret thing to realize is . . . I don't believe in trees. And Carl didn't either."

"Okay," I echoed, wondering what would come next.

"Because that's not the way evolution works."

47

The way evolution works was still being discovered and contested. Charles Darwin had been deeply percipient in recognizing the mechanism of natural selection, and he had persuasively described its workings in *The Origin*; but he didn't explain everything about how lineages change and diverge, and he wasn't right about everything that he did explain. Darwin knew nothing about the mechanisms of inheritance, for instance, and very little about the lives of microbes. Ivan Wallin and Lynn Margulis, among others who did study microbes, argued that symbiotic combining of one life-form with another had done more, in generating evolutionary novelty, than the incremental variations, the random mutations, that supposedly fueled natural selection. By the time Fox and Woese and their colleagues published Big Tree, there were other new ideas and new questions too. Some of them focused on that mysterious mound, at the base of the Fox-and-Woese tree, labeled "Common ancestral state." What did the words mean? What was in there?

George Fox got his associate professorship at Houston in 1982 and continued working on early evolution, but not with Carl Woese. As for Woese, he found new collaborators and proceeded with his own long-term project of discerning the course of deep evolutionary events, the shape of the tree of life, from the evidence coded in ribosomal RNA.

His laboratory methods changed during the 1980s, as new technologies for molecular sequencing made that part of the work less arduous, less toxic, more precise, and much faster, but his interests and goals remained constant. By the mid-1980s, he was drawing trees based on complete sequences of ribosomal RNA, not just catalogs of short fragments.

Molecular sequencing, even then, wasn't the only approach to discerning a deep tree of life. Some researchers still embraced the idea of using morphology—comparison of shapes—rather than comparison of linear sequences of RNA or DNA. Among them was James Lake, at the University of California, Los Angeles (UCLA), who studied ribosome structure by way of electron microscopy. Lake was another of those scientists who gravitated to biology after training in physics. He had gotten interested in ribosome shapes back in the late 1960s during a postdoc fellowship at the Harvard Medical School, when he discovered that he could detect the three-dimensional structure of a ribosome from electron micrographs. That led to his recognition that ribosome shapes seem to differ consistently, between one group of microbes and another, and then to the notion that comparing such differences might be a valid way of defining phylogenetic units—maybe even kingdoms. In 1984 he and several colleagues published their analysis of ribosome shapes as seen in representative bacteria, archaea, and eukaryotes.

Some of those ribosomes looked like rubber duckies. Some looked like a fist with its thumb out to hitchhike. Some looked a bit more like popcorn. Lake's team carefully measured and quantified the differences. They found four basic shapes that, by their argument, served to characterize four fundamentally distinct kinds of life. This method split away the sulfur-dependent microbes that live in hot, acidic environments from the rest of Woese's archaea. Lake named that sulfur-breathing group the Eocyta (its members would be "eocytes") and declared it an independent kingdom. So now again, if you bought this, there were four kingdoms of life. Furthermore, by Lake's analysis, his eocytes seemed more closely related to eukaryotes, such as us, than to what remained of the archaea. Woese wasn't happy.

In his view, as he told Lake by letter, "all your proposal does is muddy the waters." They had been on genial terms, these two men, following an invitation by Lake and a visit to UCLA by Woese the previous year, but now that geniality ended. Woese's camp, including German colleagues such as Wolfram Zillig who shared his fondness for the archaea, sharply criticized Lake's methodology and conclusions; Lake criticized theirs in return. Woese told him directly: "Your apparent need to have there be a new kingdom detracts from your presentation of a solid contribution." Jan Sapp called this "the battle of the kingdom keepers," waged not just through personal letters but also from the podium at scientific meetings and on the news and letters pages of *Nature*. Zillig dismissed the whole episode as a "ridiculous intermezzo" and told Woese: "I must admit that I have greatly underestimated the stupidity of the scientific community and/or the persuasive power of Lake." The ferocity of the intermezzo, as documented by Sapp, left me surprised by the mildness of James Lake himself when I met him thirty years later at UCLA.

He was a tall man, slightly bent by the years if not the battles, with pale-blue eyes and thick gray hair, wearing a lavender cardigan and chinos, gently solicitous and hospitable as he led me to his office. It was Friday afternoon, and his manner and looks suggested a retired Presbyterian minister just after a round of golf. We talked across his desk, piled high with scientific papers, above which on a shelf sat a two-color plaster model of a ribosome, roughly the size of a human heart. He spoke of his soured relationship with Woese, and Woese's response to Lake's eocyte theory of four kingdoms, as though still baffled by the depth and rancor of negativity.

"Either you're with him or you're not with him," Lake said. At first, Lake had been with Woese on the integrity of Archaea as a kingdom. When that changed, when Lake's research led toward an alternate view, "I told him the reasons, and there was a sort of intransigence there." It was as though a switch had been flipped, terminating their cordiality, making Lake not just an intellectual competitor but also a perpetrator of falsehood and confusion. At that point, as Lake recalled, "It became, 'Pull out all the stops and try to ruin this theory.'" By the consensus of other expert voices, Woese won that battle of the kingdoms, if not the war.

In 1987 Woese published a thick review paper, himself as sole author, surveying the field of bacterial evolution and, beyond that, the whole enterprise of phylogenetics. He began with a historical recap of ideas in microbial classification, as it was done formerly on the evidence of morphology via microscopy, and amid that discussion, he inserted a full-page reprint of Haeckel's tree of all organisms, with its rooting in the Moneres and its three limbs. It was classic and it was orderly, Woese granted, but it was wrong. In contrast to all such earlier efforts, from Linnaeus with his two kingdoms of life, to Haeckel with his three kingdoms, to Whittaker and Margulis with their five kingdoms, to Lake with his four kingdoms, Woese touted his own method (phylogeny by sequences of ribosomal RNA) and his own three kingdoms. "The cell is basically an historical document," he wrote, "and gaining the capacity to read it (by the sequencing of genes) cannot but drastically alter the way we look at all of biology." The ribosomal RNAs were the best and most reliable texts within cells, ancient and revealing, because they occur in all living organisms, they contain much information, and that information has changed slowly over the vast reaches of time. They contain evidence for distinguishing one kingdom from another as those kingdoms diverged, so long ago, in the period of early life. Having made that argument, he offered another tree.

The most notable aspect of this one was that it had no trunk. It had no roots. The three limbs emerged from a point at the center of the page, like a starburst pattern from a lone fireworks rocket exploding against a dark sky. That rootlessness was another way of saying what Woese and Fox had suggested with the mystery mound in Big Tree: *Events here, unknown. Not even 16S rRNA can tell us.* But he put a label to that zone of ignorance. He gave a name to the kind of creature that had existed there, existed then, just before the Big Bang of life led to the Big Tree: *progenote.* Somewhere at the nexus of his three major limbs—but unmarked on his figure—resided the progenote.

The progenote was a theoretical construct, as he explained. It was a hypothetical entity that came before—that *had* to have come before— the more knowable evolutionary history of cells. It was something more simple and less organized than a cell. "The certainty that progenotes existed at some early stage in evolution," he wrote, "follows from the nature of the translation apparatus"—that is, from the universal utility of ribosomes, turning code into proteins to make life possible. This progenote thing wasn't a brand-new idea in 1987. Woese had coined the name and sketched the concept a decade earlier, in another paper with George Fox during that interesting year, 1977. Now he elaborated. A progenote was an organic unit capable of self-replication, but its genome probably consisted of RNA, not DNA. Its mechanism for translating that genome to produce

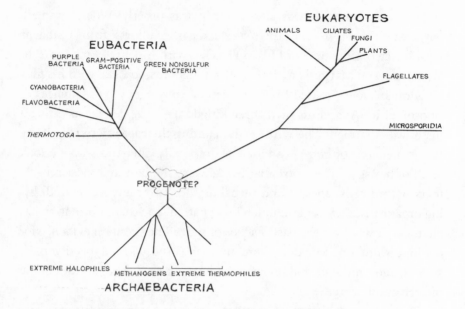

Woese's unrooted tree, 1987, and the "progenote."
Redrawn and modified, from Woese (1987), by Patricia J. Wynne.

proteins was rickety and imprecise—very prone to mistakes. Its proteins, therefore, were small and inefficient. It was still in the process of inventing—evolving, by trial and error—the ribosome. "Like the radio, the automobile, and similar devices, translation had to evolve through stages, from a much more rudimentary mechanism to the present precisely functioning one." What evidence did Woese have for the existence and nature of the progenote? Logic. Supposition. Informed guesswork. There were no data from the progenote era. There were no fossils, and there was no 16S rRNA. He was simply pondering the topic—what came just before life as we know it?—more carefully and yet more daringly than anyone else.

Whatever did occur during that phase of Earth's history, its most consequential result was a single lineage, some manner of creature, that became the universal ancestor of all three kingdoms of life. For this certitude, we do have evidence: the universality of the genetic code itself, a system that uses the same coding—these bases specify those amino acids—in bacteria, in archaea, and in eukaryotes. The genetic code is the ultimate shared character, uniting all forms of life within one ancestry. And that ancestry had its origin among progenotes.

"The progenote is today the end of an evolutionary trail," Woese wrote, "that starts with fact, progresses through inference, and fades into fancy." The end of a trail tracing backward in time, he meant. "However, in science endings tend to be beginnings." The flow of genomic data would soon become a deluge, he predicted correctly. The root of the universal tree of life would be identified. We were beginning, he promised, to understand the beginning.

48

Back in early 1980, around the time Woese and Fox finished their work on Big Tree, Woese wrote to his friend Otto Kandler in Munich. The letter's main purpose was to request help on a subject keenly interesting to them both: those unusual cell walls of archaea, Kandler's specialty. Woese needed electron micrographs, and some help with chemical structure, on which to base illustrations for an article he planned to write for *Scientific American*. Pivoting from that thought, he told Kandler: "Someday you and I must write a scientific article together. Our knowledge and approaches complement one another." What do you say, he asked, about teaming up to make a formal proposal that the archaea should be a full kingdom? He and Kandler and others had been talking that way, treating archaea as a kingdom, but informally, without trying to muster formal taxonomic consensus. "It's a sort of wild thought."

The wild thought seemed less wild with passing years and intervening events. In 1984 Woese received a MacArthur Fellowship for his efforts in phylogenetic analysis and his discovery of the archaea, and in 1988 he was elected to the National Academy of Sciences. Despite the MacArthur honor, and because the Academy had elected him relatively late (at age sixty, whereas Lynn Margulis had been welcomed at age forty-five), he still thought of himself as a neglected outsider. That gave him some latitude to continue being ambitious, bold, and ornery. And he wanted to revisit the status of his beloved archaea. It wasn't just to get them recognized as a kingdom. He also chafed at the fact that, for a dozen years, by his own doing, they had been burdened with a misleading name.

Now I'll clarify something that I've left streamlined for simplicity: the original name that Woese and Fox gave these creatures back in 1977, archaebacteria, had endured. Woese himself, his American and German colleagues, and everyone else in the field called them that throughout the period I've been describing. The big international meeting in Munich, after which Woese and Kandler and Ralph Wolfe celebrated with champagne in the Bavarian Alps, was billed as "The First Workshop on Archaebacteria." But that name increasingly rankled Woese, because it undermined the group's uniqueness, suggesting they were just another branch of bacteria.

In 1989 he sent an email to Wolfram Zillig, his other close German colleague, addressing the problem. "As time goes by it becomes more and more obvious," Woese wrote, "that I made a major mistake in naming the archaes." They weren't bacteria, of course. They weren't quasibacteria or ancient precursors of bacteria. Bacteria weren't even their nearest kin. Some evidence had emerged by that time, in fact, indicating that archaea are more closely related to eukaryotes—more closely related to *us*—than to bacteria. Woese repeated to Zillig the idea he had earlier suggested to Kandler: Let's write a paper, a group of us big dogs in the field, proposing the archaeans formally as a kingdom and giving them a new name.

This wild thought became reality a year later, in a paper by Woese with Otto Kandler and one other scientist (but not Zillig) as his coauthors. For publication, Woese turned to the *Proceedings of the National Academy of Sciences*, a journal in which—as a member now of the Academy—he could be a bit more speculative and face less rigorous peer review than he would at *Nature* or *Science*. The paper, published in June 1990 and titled "Towards a Natural System of Organisms," made several main assertions. First, any system of classification should be strictly "natural," as the title suggested—meaning phylogenetic, reflecting evolutionary relationships, and not compromised by convenience for memory and teaching (as Whittaker and Margulis preferred). Second, there should be three major divisions of life, higher in rank than Whittaker's kingdoms, higher also than Haeckel's kingdoms or Copeland's or Lake's, and those divisions should be known as *domains*. Three domains, recognized *above* the old kingdoms rather than replacing them: it was ingenious strategically, transcending rather than rejoining the battle of the kingdom keepers. But incidentally, it led to a small conflict between Woese and Kandler regarding the third coauthor, Mark L. Wheelis.

Wheelis was a younger microbiologist at the University of California, Davis, not previously known for work in phylogenetics. He had trained under Roger Stanier at Berkeley in the 1960s, encountered Woese passingly during a postdoc at Urbana, and become friendly with Mitch Sogin while teaching molecular evolution in Sogin's summer course at Woods Hole. How those contacts or other factors may have brought him back to Woese's attention, in the late 1980s, even Mark Wheelis doesn't know.

"It's always been sort of a mystery to me," Wheelis said when I reached him by telephone. He had been interested in what he called "the kingdom problem"—the glaring illogic of treating plants and animals each as a distinct kingdom coequal with all bacteria, a much more overarching group—and he may have mentioned in conversations that a new, higher category of classification was needed; but he hadn't published on that subject. And then, "out of the blue," he received a draft of what became the 1990 paper. "This manuscript showed up in my mailbox one day, with Carl asking if I had comments." Wheelis made suggestions, including the one about a higher category, and returned the draft. Woese incorporated many of Wheelis's notes, raised other questions, and the manuscript flew back and forth four or five times. At that point, as Wheelis recalled, he asked Woese to consider adding him as a coauthor.

Woese agreed, notifying Otto Kandler in Germany. Kandler was surprised at the addition of a third collaborator, but he put on a game face and acquiesced. "I have no objections if Mark becomes a coauthor," he wrote Woese, "although I am not convinced that it is fully justified." Kandler noted that the Wheelis suggestion about a higher category merely "reactivated" an idea that Woese had already entertained. Wheelis was nevertheless "brought on board," according to Jan Sapp, his contributions rewarded with coauthorial credit. This is the decorous version of Woese's decision process, anyway, as told by Sapp in his superb book *The New Foundations of Evolution*. In private, over a glass of wine, Sapp gave me a slightly different version.

Woese, fond as he was of Kandler, didn't want it to appear that his German colleague had codiscovered the archaea. That was Woese's great distinction, shared with George Fox and, to a lesser degree, Wolfe and Balch, but no one else. He wanted his full measure of glory. Sapp himself loved Woese like a brilliant uncle and worked closely with him while researching *New Foundations*, but he was clear eyed about Woesean foibles. "I know this man." If the three-domains paper was a milestone, with just two coauthors, "then he and Kandler would be seen as codiscoverers. But he puts Wheelis in, and it waters Kandler down."

"That's bizarre," I said. Yet it's how science sometimes works—and human nature, often.

"Yeah," said Sapp. "That's Carl."

Last of the paper's main points was that these three domains should henceforth be known as the Bacteria, the Eucarya, and . . . the Archaea. The word *archaebacteria* should now disappear, the authors argued. So should the word *prokaryote*. Prokaryotes didn't exist as a phylogenetic

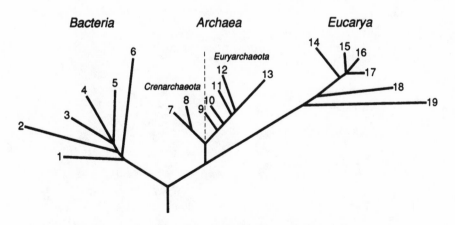

The "Natural System" tree of Woese, Kandler, and Wheelis, 1990.

category—it was a false unit—because Archaea and Bacteria stood so utterly distinct from each other.

And, of course, there was a tree. It was drawn in straight, simple lines, but it was rich and provocative nonetheless. Unlike his rootless 1987 tree, this one was rooted, using a complicated technique (involving duplicated genes traced back into deep time, but never mind those details) that Woese had come upon more recently in work by some Japanese researchers. Its trunk rose vertically from a single origin, then split into two big limbs, and then one limb split again. The left limb was Bacteria. The two limbs on the right were Archaea and Eucarya (that spelling later corrected to Eukarya, as I've mentioned, a better transliteration of the Greek root). This arrangement asserted what Woese's 16S rRNA data showed: that we humans, and all other animals, all plants, all fungi, all eukaryotes, have arisen from an ancestral lineage that was unknown to science before 1977. It was the last of the great classical trees: authoritative, profound, completely new to science, and correct to some degree. But it entirely missed what was coming next.

PART ✦ V

Infective Heredity

49

What came next was an exploding awareness of the role played by horizontal gene transfer in this whole story. That explosion occurred during the 1990s but had deep precedents. The long, bizarre history of the HGT phenomenon goes back about four billion years, in fact, and the first recognition by science that any such thing might be possible dates to 1928. It grew from work published that year by an Englishman named Fred Griffith, though no one at the time, not even Griffith himself, saw the implications of what he had found.

Griffith was born in a small town in North West England and educated in Liverpool, just down the Mersey estuary, with a degree in medicine and a fellowship in pathology. He worked for a while in a local laboratory and a hospital, picked up a diploma in public health at Oxford, and did research on tuberculosis before joining Britain's Ministry of Health in London during the Great War. He had a long, straight nose and a steady gaze. As a medical officer at the ministry's Pathological Laboratory on Endell Street, just north of Big Ben, he shared an upstairs lab space and kitchen with two technicians and his colleague William M. Scott; downstairs was the post office.

Griffith was a fastidious bench scientist and a consummate civil servant, narrow of focus, wary of speculation, described as "very shy and aloof and difficult to get to know," whose assigned topic was bacterial pneumonia. Scott's work is less known, but they became friends and remained side by side to the end, in 1941. By one account, in their little underfunded laboratory, Griffith and Scott "could do more with a kerosene tin and a primus stove than most men could do with a palace."

The bacterium *Pneumococcus pneumoniae* (now called *Streptococcus pneumoniae*), which Griffith studied, was a dangerous bug that could cause severe, often fatal, pneumonia. During the 1918–19 influenza pandemic, this kind of pneumonia took hold as a secondary infection in many patients and probably killed more millions of people than the flu virus itself. Antibiotics didn't exist then. The best treatment was antiserum therapy, using blood serum rich with antibodies drawn from inoculated horses; such serum could bolster a patient's immune response and help clear the bacterial infection. But there were at least four different types of pneumococcus—Types I, II, and III, plus a catchall type known with sublimely unnecessary confusingness as Group IV—and therefore several different sera, which were specific for type. When giving treatment, you wanted to know which pneumococcal type a patient was inflicted with, so as to pick the right serum. That's the sort of thing medical bacteriologists did. Griffith's work during the 1920s involved telling one type from another, investigating their properties, and tracking the prevalence of different types in different pneumonia outbreaks around the country. He studied almost three hundred cases between 1920 and 1927 and saw, among other things, that pneumococcus was rife in Smethwick district, just west of Birmingham, but that Type II was giving way there to Group IV. Such intelligence was useful for tracking outbreaks and judging which sera to prepare and ship. Griffith got his data by examining sputum coughed from the lungs of the ill. He had an ice chest full of hacked-up gobs.

In 1923 Griffith discovered something important: that in addition to different *types* of the pneumococcus bacterium, there existed two different *forms* within each type—one that was ferociously virulent, one that was mild. The virulent bugs clustered in colonies that looked smooth under microscopy, so he labeled that form S. The nonvirulent form clustered in rough colonies, so that was R. Sometimes the S form might transmogrify into the R form, he noticed. He didn't know why. Maybe it was mutation and natural selection, possibly in response to serum. Or not.

Then he made a second discovery, far more surprising even to him: under certain experimental circumstances, the R form of, say, Type II bacteria could change into the S form of, say, Type I. *What?* It seemed as though the pneumococcus had morphed into a different species. But Griffith was firmly grounded in the Ferdinand Cohn school of bacteriology, which held (as your keen memory will tell you) that bacterial species are fixed, stable, knowable—not protean and capable of shape-shifting magically from one into another. Yet here was a change. Griffith doubted his own lab technique and tested again. Same result. Having taken such care to exclude the possibility of contamination, he wrote later, admitting his skepticism, that "there seems to be no alternative to the hypothesis of transformation of type." And so that's what he called the phenomenon: *transformation.* One form of bacteria transformed into another. It was mystifying.

These transformations happened, for Griffith, in the bodies of his laboratory mice. His experimental procedure involved injecting mice with one type or another of pneumococcus (say, Type I or Type II), one form or another (S or R), and sometimes also a dose of serum, then watching to see whether the mouse died or survived. If the mouse died, he would extract a blood sample, culture the bacteria that had been raging within the animal, and by microscopy determine its type and its form. In the most revealing of his many experiments, he gave each mouse two forms of bacteria, injecting a dose of dead form S (virulent, if it had been alive) and living form R (mild). He had killed the dose of virulent bacteria (but not entirely destroyed its biochemical components) by heating it at a carefully chosen temperature. His first interesting result from this method was that a mixture of dead S and live R was capable of killing a mouse.

Even more surprising were the results when he mixed types as well as forms. In one experiment, for instance, he injected heat-killed S form (virulent) of Type I bacteria along with living R form (mild) of Type II bacteria into five mice. When all five keeled over within a few days, and Griffith drew blood, he found living Type I that was virulent. Note again this change: dead virulent I, plus living mild II, becomes . . . living virulent I. Something weird had happened. It sounded like zombie bacteria. Either the mixing had brought the virulent Type I back to life, or else the dead Type I had somehow transformed the living Type II into a version of itself. This wasn't a sci-fi movie, and neither of those options was supposed to be possible.

Griffith himself struggled to explain it when he published a long paper on his pneumococcus work. It seemed to him that the living bacteria "actually make use of the products of the dead culture" for generating their own capacity to be virulent. How could that work? Well, maybe some part of the wreckage of the killed, virulent bacteria served as a kind of "pabulum" from which living, nonvirulent bacteria built up their own virulence. *Pabulum*, British spelling for our *pablum*, simply means a pureed and easily absorbable food. The groping vagueness of that suggestion reflects the difficulty of interpreting Griffith's very peculiar experimental results, even for Griffith. Still, vague or not, it was roughly right.

In the summary of his pneumococcus paper, Griffith declined to speculate about the "pabulum" at all, merely listing his results like the steady, empirical government employee he was. At the top of his list: Type II pneumonia has declined in Smethwick, though the incidence of Type I has held steady. Near the end of his list: oh, and by the way, dead bacteria seem to be capable of transforming live bacteria from one type to another.

Griffith never pursued this topic further. By one account, he seemed uninterested in the phenomenon of transformation, possibly even irritated by it (because it seemed to contradict the fixity of species) and happy to leave it "up to the chemists" to explain. He shifted his research from pneumococcus to other bacteria. His paper, with its passing notice of transformation buried within forty-seven pages of precisely described experiments, became widely read and influential—one historian has called it "a bombshell which fell into a fused situation"—as other researchers (and not just "the chemists") picked up where he had stopped. He never knew, never lived to see other scientists discover, what the mysterious pabulum was.

As for himself, he had no taste for drama or scientific limelight. He rarely went to meetings or gave talks. When the International Congress of Microbiology met in London in 1936, Griffith "had to be practically forced into a taxi" to get him to the event, though he was committed to speak. He never married, and at one point, he shared his London flat with William M. Scott, his friend and colleague since the early days at the Pathological Lab. He died on a night in April 1941, during the Blitz, and so did Scott, when that apartment took a direct hit. His two-paragraph obituary, as it ran in the *British Medical Journal*, didn't mention transformation.

50

Griffith's long paper appeared in January 1928. His reputation for rigor was excellent, but his discovery seemed so bizarre that other bacteriologists remained unsure, at least for a year or two, whether to swallow it. A few worked to redo his experiments and confirm, or not, his findings. They confirmed them: yes, there was no mistake, transformation occurred. But what was the stuff, Griffith's pabulum, that came from dead bacteria and caused live bacteria to change identity? What might it reveal about the very nature of heredity?

These questions arose in a context—the context of early-twentieth-century genetics—in which the word *gene* was used freely, but no one really knew what it meant. A gene was an abstraction made concrete only by the word itself, that nice little unit of jargon coined in 1909 to stand for an entity of some sort that determines hereditary traits.

Geneticists suspected by then that genes reside on the chromosomes in a cell, and chromosomes could be seen through a microscope, but the genes themselves couldn't. Were they physical realities, discrete chemical units arranged on a chromosome like beads on a string? Or was each "gene" just the net effect of some measured-out quantity or fluctuating process, as Darwin had (wrongly) guessed? As late as 1934, the eminent American geneticist Thomas Hunt Morgan, who pioneered studies of mutation and heredity using fruit flies, said in his Nobel Prize acceptance lecture: "There is no consensus of opinion amongst geneticists as to what the genes are—whether they are real or purely fictitious." It didn't matter, Morgan added, because "at the level at which the genetic experiments lie," the results were the same either way. He meant genetic experiments *like his*, concerned with relative positions of genes on chromosomes, and how such positioning influences new combinations during sexual reproduction.

But as Oswald Avery, a medical researcher at the Rockefeller Institute, and others looked more closely at phenomena such as transformation in nonsexual creatures such as bacteria, it damn well did start to matter whether a gene was a material reality or a cloud of influences. If there was a pabulum that carried heritable change, what was the recipe? If the hereditary substance was a physical thing, a chemical entity, which molecule or molecules composed it?

Once this line of inquiry began, the leading hypothesis held that genes are made of protein. A protein, remember, is a long molecule consisting of different amino acids linked as a chain, their sequence highly variable from one protein to another, and such variation offers vast possibilities for encoding biological traits as linear information. (You heard about this back when I discussed Francis Crick and his 1958 paper, suggesting the use of proteins to chart phylogeny.) In fact, "vast possibilities" is an understatement. Start with the twenty amino acids of life; mix and arrange them in every possible way to make a molecule that's, say, three hundred amino acids long; do the math; and you get a gazillion possible sequences. That's enough for genetics.

An alternate hypothesis was that nucleic acids might be somehow involved. Nucleic acids, which we know as DNA and RNA, comprise one of the four major categories of molecule in living creatures—the other three being fats, carbohydrates, and proteins. But nucleic acids just didn't seem to have enough chemical complexity and variability to serve as an all-purpose alphabet.

DNA, for instance: its structure wasn't known in the early twentieth century, but its components were, and those components—the sugar ribose, a bit of phosphate, and the four bases designated by their initials, A, C, G, and T—didn't appear to offer a gazillion structural options. In fact, well into the century, the prevailing supposition about DNA's structure was radically simplistic. It presumed that all four bases were present in equal amounts and repeated themselves in a fixed sequence, such as ACTG-ACTG-ACTG-ACTG, on and on. That's why DNA was considered a "boring molecule" or a "stupid molecule" by some of the smartest whips in biology, even into the 1930s and 1940s. DNA was underrated, mistakenly assessed, like Albert Einstein as a sixteen-year-old high school dropout, when his father urged him to buck up and become an electrical engineer.

There was also a middle view, suggesting that the hereditary material might combine both protein and nucleic acid—the protein providing an alphabet of variation, the DNA serving some sort of supportive function. That might explain the abundance of DNA in chromosomes. But no one really knew. Opinion had fluctuated for decades. The tools didn't yet exist, the methods didn't yet exist—or, more accurately, the imagination didn't yet exist—to settle the question. Into this arena of uncertainty stepped Oswald Avery.

Avery was another of the eccentric, heroic paragons of early research in molecular biology. His story begins in Halifax, Nova Scotia, a city to which, like Urbana, this whole tale circles back and back. He was born there in 1877, second son of a zealous Englishman who had converted from Anglicanism to become an evangelical Baptist preacher and emigrated to Canada when a "strange impression" of holy duty seized him. Ten years later, another strange impression carried the family to New York's Lower East Side, where Reverend Avery became pastor of the Mariner's Temple, a mission in the Bowery. Oswald and his older brother took to music, got hold of cornets, and as young teenagers, helped the family enterprise, by standing outside the temple on Sundays and playing their brass horns to attract congregants. It sounds like a scene from *Guys and Dolls*: two kids on a street in the Bowery, blaring out "Follow, follow, before you take another swallow." The older brother got sick and died when Oswald was fifteen, Reverend Avery died too—a bad year for the family—but Oswald himself, the steady middle boy, went off to prep school (somehow) and Colgate University and eventually drew the youngest brother, Roy Avery, into bacteriology on his coattails.

Photos of Oswald, beginning from age six, show an angelic boy with big eyes and then a man with a great domed forehead, presumably full of brains, going gradually bald, stretching intelligence almost beyond the limits of the skull. Below that, always, a small, straight mouth. At Colgate, he played cornet in the band. From there he went to the College of Physicians and Surgeons at Columbia University, back in New York City, where he did well, except in bacteriology and pathology. His friends called him "Babe" because he still looked like a megacephalic little kid. With his medical degree, he went to a laboratory in Brooklyn, where he helped with administration and worked on topics such as the bacteriology of yogurt, but also a bit on influenza and tuberculosis. Six years later, Avery joined the Rockefeller Institute to do research on pneumonia, which by then had killed his mother.

The Rockefeller Institute Hospital, where Avery had his lab, was America's leading center of research on pneumococcal pneumonia, which Avery studied for much of the following two decades. But after the Griffith paper in 1928, and with some tugging by his junior colleagues, his focus changed gradually from the purely medical aspects to something broader. In summer 1934 a young man named Colin MacLeod, like Avery a Canadian, arrived. MacLeod had read Griffith's paper as a medical student and wanted to work on transformation. Avery was gone on sick leave at the time, for a thyroid disease, and by the time he returned, MacLeod had taught himself the methods and gotten started. Avery supported the effort. He was still recuperating and weighed barely a hundred pounds, but he and MacLeod worked long hours and weekends.

Avery seems to have sensed, with quietly increasing confidence over the next handful of years, that transformation in pneumococcus was more than just a medical issue—that it had huge implications for biology in general. He and his lab members began to speak of "the transforming principle," their name for the magical pabulum, and to suspect that it achieved its effect by transferring genetic information. If that suspicion proved correct, their transforming principle *was* the hereditary material—and not just of pneumococcus pneumonia, maybe, but of all life. In other words, as they tried to identify the transforming principle, they were looking for the physical reality of the gene.

Medical applications, though, were still their primary imperative at the Rockefeller Institute. The work on transformation flagged a bit in the late 1930s, as MacLeod struggled with the experimental challenges and as the first appearance of sulfa drugs—early antibiotics—promised the possibility of curing pneumococcus infections without need of distinguishing one type from another. If there was no need for typing, transformation of types might be moot—medically moot, anyway. MacLeod, needing some practical publications to advance his academic career, diverted his attention for a while to the sulfas. In the meantime, no one else in the scientific world seemed to share Avery's strong sense that transformation was a large and ripe scientific question. After the hiatus, in autumn 1940 he and MacLeod returned to it.

To identify the transforming principle, they first had to isolate it, whatever it was, in quantities that allowed for chemical analysis. So besides killing pneumococcus cells with heat treatment, they would break open the cells and derive an extract from the cell puree, then try to determine which component of the extract—a protein? a nucleic acid? some other kind of molecule?—carried the transforming power. MacLeod did most of the hands-on experimental work. In contrast to Avery, so precise and methodical, he was "much more impulsive and impatient," according to one colleague, and that may explain his effort to scale up their operation using a cream separator as a centrifuge.

They cultured their pneumococcus in beef broth and then had to centrifuge the culture—spin it, separating broth from cells—to get concentrated masses of bacteria. This took time, and the ordinary lab centrifuges could spin only a liter each, yielding just smidgens of bacteria. Scaling up would give them more cell puree, therefore greater quantities of extract, making the chemistry and the biological testing a bit easier. So MacLeod somehow got hold of an industrial cream separator, with a high-speed cylinder and separate outflow taps, that could process gallons of culture in a continuous cycle. The only problem was its tendency, when spinning at full throttle, to spray "an invisible aerosol laden with bacteria" throughout the room. This was more than inconvenient. A fine mist of low-fat milk might have been okay, even a fine mist of yogurt bacteria, but not a fine mist of virulent pneumococcus. To fix that, MacLeod found a technician from the Institute's machine shop who helped him design a housing for the separator, a sort of containment vessel, with a gasket-sealed door that bolted safely shut and could be opened with a tire iron. Before opening, the interior of the vessel would be sterilized with a blast of steam. Then they would lug-wrench it open, scoop out their masses of bacteria, and the work continued.

Colin MacLeod left in 1941 for a job elsewhere, and another young medical doctor with training in biochemistry, Maclyn McCarty, became Avery's next chief collaborator in the search for the transforming principle. By this time, Avery's nickname among his associates was no longer "Babe," far from it, but "the Professor." For short they called him "Fess." His team, at work on the transforming principle, had nearly convinced themselves that the mystery stuff wasn't a protein. McCarty devised a series of experiments aimed at narrowing the possibilities further, and by the summer of 1942, he and Fess had evidence suggesting it was probably DNA. That seemed counterintuitive, given DNA's reputation as boring, repetitive, a "stupid molecule," incapable of carrying hereditary information. "We were not unaware that this idea would be greeted with skepticism," McCarty wrote later. "We had already been told by more than one person"—among them, a crotchety scientist with a lab two floors above them at the Rockefeller Institute, who did DNA-related work himself—"that the transforming principle could not possibly be deoxyribonucleic acid because 'nucleic acids are all alike.'" Despite such discouragement, they trusted their evidence and wrote a paper, carefully limited but unambiguous in its assertions. They said: DNA causes transformation of pneumococcus. They didn't say: DNA is the substance of genes.

They didn't say that publicly, anyway. But in May 1943, around the time their work was reaching a crescendo, Oswald Avery wrote a letter to his brother, Roy, by that time a professor of microbiology at Vanderbilt University in Nashville. It was a long letter, discussing family matters and his impending retirement, but then he shifted into a detailed description of the work he had been doing with MacLeod and McCarty, and their discovery that the transforming principle was DNA. "Who could have guessed it?" He mentioned plans for one more batch of pneumococcus, one more phase of purification and testing to confirm their results, and then they would write up the work. "If we are right," he told Roy, "& of course that's not yet proven, then it means that nucleic acids are *not* merely structurally important but functionally active substances in determining the biochemical activities and specific characteristics of cells." It meant that DNA could carry, from one cell to another, changes that are predictable and hereditary. It meant that DNA could take hold somehow in the second cell, and be passed along through multiple generations, then recovered in quantities far greater than the amount originally introduced. "Sounds like a virus—may be a gene," Avery wrote. But he was judicious as ever, disinclined to overreach. "One step at a time—& the first is, what is the chemical nature of the transforming principle? Someone else can work out the rest."

He wrote the paper with McCarty in the following months, and they added MacLeod as a coauthor. It went to the *Journal of Experimental Medicine* in November and was published in February 1944. Maclyn McCarty, who was thirty-two at the time, sent a reprint to his mother, with a few scribbled words expressing his pride: "This is it, at long last." Mrs. McCarty was no bacteriologist, unlike Avery's brother, and what she made of that comment, or of the paper itself, is unrecorded by history. Does a mother back in Indiana sit down and read "Studies on the Chemical Nature of the Substance Inducing Transformation of Pneumococcal Types," or does she just put it on her coffee table and say, "That's my boy"? Maybe the latter. Anyway, the international community of biologists concluded, gradually, as more evidence accrued, that this *was* it—that Avery, MacLeod and McCarty had discovered the physical nature of the gene.

It's a good story, a true story. But the point of telling it here is that they had done that and also something more.

51

Transformation, in the sense of that word as used by Fred Griffith and Oswald Avery, is one of the three cardinal mechanisms of horizontal gene transfer, the most counterintuitive phenomenon discovered by biologists in the past century. Griffith had shown that some mysterious pabulum could transform a nonvirulent type of bacteria into a virulent type. Avery's group had shown that Griffith's pabulum is DNA, the physical carrier of genes. And the Avery team had demonstrated that in its naked form—floating loose in the environment after having been liberated from a busted bacterial cell—DNA is capable of getting into another bacterium and causing heritable change. Avery and his colleagues had no inkling at the time that this sort of sideways passage can carry DNA not just across minor boundaries, type to type among *Pneumococcus pneumoniae*, but also across huge gaps—from one bacterial species to another, from one genus to another, even from one domain of life to another. The transformations that result from such horizontal transfer can be far more consequential than merely changing a pneumonia bug from mild to virulent.

The two other primary mechanisms of such sideways genetics came to light in the decade after Avery's team published its paper. One involved a sort of "sex" between bacteria, and was dubbed *conjugation*. The other involved viruses carrying foreign DNA into the cells they infect, and that was called *transduction*. Both discoveries came from the ambit of a brilliant young scientist named Joshua Lederberg.

Lederberg was a twenty-one-year-old junior researcher in a laboratory at Yale University, with no doctorate, on temporary leave from medical school at Columbia, when he detected the phenomenon he named conjugation. He had requested bacterial cultures and guidance from Edward L. Tatum, a microbiologist whose specialty was bacterial genetics, in order to chase a question that interested them both: Did bacteria practice some sort of genetic exchange? If not, where did they get the diversity and plasticity that allowed them to evolve within changing environments? If they did exchange genes, *how*? Genetic exchange usually suggests sex, at least in multicellular creatures. Bacteria were thought to be asexual, reproducing by simple fission, when one cell splits into two. Where was the opportunity to get new genes, rearrange combinations of genes, and adapt to new circumstances? Lederberg was fascinated by Oswald Avery's discovery: the uptake of naked DNA from a dead bacterium into a live one. Did something like that occur between living bacteria too?

Within a year, working on the bacterium *Escherichia coli* under Tatum's mentoring, with an ingenious experimental design he had concocted himself, Lederberg made his own discovery: yes, living bacteria trade genes. He didn't see it happen, but he proved it by inference. Take a strain of *E. coli* with, say, a useful gene A and a disadvantageous gene B; put that strain together in culture with a strain carrying a disadvantageous version of gene A (call it a) and a useful version of gene B (call that b); then, as the bacteria make the best of their circumstances—some flourishing, some not—you would get a new strain, Lederberg found, carrying both the useful genes, A and b. He didn't need to do gene transplantation by some kind of fancy technique. The bacteria did it themselves. The genes were traded sideways into a more adaptive combination.

"In order that various genes may have the opportunity to recombine," he wrote in a short paper coauthored with Tatum, "a cell fusion would be required." Recombine: meaning, rearrange or swap genes. Cell fusion: meaning, a temporary clinch. It might be brief—a quickie—but prolonged enough for genes to be transferred. Although this was a rare kind of event, "only one cell in a million" getting the recombined genome, Lederberg reproduced the effect on numerous tries. "These experiments imply the occurrence of a sexual process in the bacterium *Escherichia coli*." He hadn't yet reached his twenty-second birthday when the paper appeared in *Nature*. So he was well and early launched; but it would take him until age thirty-three to win his Nobel Prize.

Lederberg grew up in New York City, son of a rabbi, oldest of three boys—a precocious kid who devoured books on science history and microbiology, received the textbook *Introduction to Physiological Chemistry* as a bar mitzvah present, and went off to Columbia College at age sixteen. After three years as an undergraduate, and despite some wartime work in clinical pathology at a US Navy hospital on Long Island, he was ready for med school. He started, and then came the interlude with Tatum in New Haven—a brief period but fruitful in that, on the strength of his discovery, Yale retroactively decided he was a graduate student and handed him, after modest additional work, a PhD. Before he could pack his bags to go back to Columbia and finish his MD, the University of Wisconsin offered him an assistant professorship in its Department of Genetics. Since boyhood, Lederberg had been pointed toward a career in medical research, solving urgent clinical mysteries in the tradition of Pasteur and Koch, but now he found himself a bacterial geneticist, paid to teach, to supervise graduate students, and to do basic research.

Among his first grad students was Norton Zinder, another prodigious teenager from New York City who had steamed through Columbia in three years and then come to the Midwest. Zinder began his work in Madison following up on what Lederberg had done with Tatum, which was natural for a new doctoral student in the new lab of a new assistant professor. His assignment was to look for conjugation in a different bacterium, not *Escherichia coli* but *Salmonella typhimurium*, a bug in the same genus as those that cause typhoid fever and food poisoning. Zinder used penicillin to distinguish one mutant strain from another—a process that worked because penicillin, when introduced to a cell culture, killed only the mutant strains that were growing, not the mutant strains that were resting dormant. Separating mutant strains, as Lederberg had already shown, was a crucial step toward learning how those strains might exchange genes. But in his *Salmonella* cultures, Zinder found no sign of conjugation. Instead, he detected a different mode of genetic exchange. In this new mode, as far as Zinder could tell, only a small section of DNA was transferred, enough to account for a single genetic trait. And the donor bacterium never came into contact—not for a quickie, not even for a smooch—with the recipient bacterium. They remained separated like lovers on opposite balconies. Whatever carried the DNA was so small it could pass through a fine ceramic filter (that was the gap between balconies), too fine to allow passage of bacteria. Zinder recognized that the filter-passing agent was a virus—it had to be, since no other biological entity was so small—which evidently picked up some genetic material from one bacterium and carried it into another. This was so different from conjugation that Zinder and Lederberg, when they published the work, gave the process its own name: transduction.

At almost the same time, Lederberg's wife, Esther, who was also a bacterial geneticist, made another key discovery about sideways gene transfer. Through experiments on the old standby *E. coli*, she detected a system of "sexual compatibility" and incompatibility between various individual bacteria, which did or did not allow them to "mate"—that is, to exchange genetic material by conjugation. The evidence again was inferential. At first, Esther Lederberg could say only that compatibility was determined by a mysterious particle or factor of some sort, which she called F, for fertility. If one bacterium had F (a condition designated as F+) and the other did not (F-), then those two could do the deed, yes, with a passage of genes from one to the other. If both carried it (each of them F+), then too they could conjugate. If neither had F (a pair of F- virgins, clueless and pure), then no, they were incompatible for mating. It was a fresh insight on bacterial dynamics and the flow of genes through the invisible world. But there was more.

Esther Lederberg found that this curious F factor, bestowing the capacity to initiate mating, could be acquired by a bacterium that didn't have it. How? Through a different mechanism, which I've just mentioned: transduction. That is, carried in by a virus. This was a dizzying compoundment of two kinds of horizontal gene transfer, functioning as a double-stroke process to move DNA between microbes. Take a deep breath and relax with your puzzlement as I say it again: transfer of the F factor—whatever it was—from one bacterium to another, by a virus, gave the second bacterium an ability to mate with still other bacteria. An F- bacterium became F+. A virgin became a player.

But wait. Let's pause here and remember that "sex" is only a metaphor for what these bacteria were doing. It applies nicely in some ways but in others does not. The Lederbergs tended to treat the term literally, speaking often in their work about bacterial sex, but other biologists disagree, noting the most important distinctions. Bacterial "sex" doesn't involve the fusion of two gametes, egg and sperm, each bearing a half share of an entire genome. And it doesn't result in reproduction. A bacterium generates offspring by splitting, not by mating. The net result of conjugation is genetic recombination—mixing—which often proves helpful in the evolutionary struggle. But conjugation itself doesn't produce babies.

It all sounds weird because it *is* weird. Esther Lederberg published her peculiar discovery, in collaboration with her husband and one other coauthor, and near the end of their paper, they noted passingly that the ability to conjugate among *E. coli* was conferred by some other sort of "infective hereditary factor." Her husband, as sole author of another paper published just a month earlier, had alluded likewise to "infective heredity." Transduction was infective, just as conjugation was sexy. And the phrase had an enduring ring.

Norton Zinder got his PhD in Wisconsin, and by 1952, he was back in New York as an assistant professor at the Rockefeller Institute, where Oswald Avery had worked. One year later, he published an overview of this whole perplexing business, genetic transfer among bacteria. He meant to sort things out. There were three modes of such sideways inheritance, Zinder explained. The first was conjugation, as discovered by Tatum in collaboration with Zinder's mentor, Joshua Lederberg. The second mode was transformation, as discovered by Griffith and illuminated by Avery's team. The third mode was transduction, as discovered (though he didn't crow) by him and Lederberg. Conjugation was analogous to sex. But the other two were different, involving processes more like infection, as the Lederberg team had suggested passingly. These other two processes deserved their own descriptive category, their own metaphor. Zinder, picking up on the Lederberg's phrase, called them infective heredity.

52

Bacteria, bacteria. Bacterial sex. Bacterial transformation. Dead bacteria, live bacteria, virulent and mild. Bacterial DNA. A person might get the impression that this whole subject of horizontal gene transfer is merely a matter of bacteria, by bacteria, for bacteria.

But that impression would last only until a person came to the work of Tsutomu Watanabe, a bacteriologist at Keio University in Tokyo during the 1950s and 1960s. In 1963 Watanabe alerted his fellow scientists to an urgently human implication of the new bacterial discoveries: that resistance to multiple antibiotics among bacteria spreads horizontally. It can happen by conjugation. It can happen by transduction. It can happen in a sudden leap. Consequently, it has become a dire problem. And the problem is especially severe in hospitals, where such huge volumes and such variety of antibiotics are used, selecting for resistant bacterial strains that then infect people who are already ill. The World Health Organization now considers antibiotic resistance one of the biggest threats to global health in the twenty-first century. Watanabe saw it coming. He understood why such resistance would spread so fast and so widely. Adopting the terminology of Zinder and the Lederbergs, Watanabe called it "an example of 'infective heredity.'"

Antibiotic resistance was already a serious and growing problem, just after World War II, in Japan and around the world. People were dying again of diseases, such as bacillary dysentery, that the great revolution in antibiotics was supposed to have tamed. The first sulfa drugs had been introduced in the late 1930s, and immediately there were reports of bacterial strains with resistance. Penicillin was discovered in 1928 and developed for human use beginning in 1942; initially, it was a very potent weapon against *Staphylococcus* of various sorts; but by 1955, penicillin-resistant strains of staph were turning up, especially in hospitals, from Sydney to Seattle. Methicillin became available in 1959, valued highly as an answer to forms of staph—especially *Staphylococcus aureus*—that had acquired resistance to penicillin. But resistance to methicillin also appeared soon and spread fast, so that by 1972 methicillin-resistant *Staphylococcus aureus* (infamous now as MRSA) was a concern in England, the United States, Poland, Ethiopia, India, and Vietnam. And by the early twenty-first century, MRSA was killing more Americans per year than AIDS. Despite some success at reducing MRSA transmission in hospitals, by a more recent tally the numbers were still bad: more than 23,000 deaths annually in the United States and seven hundred thousand deaths globally from infection by unstoppable strains of bacteria.

What has driven this grim, costly trend is not just the use of antibiotics but also the reckless *overuse* of them for foolish or unnecessary purposes—doctors pandering to patients, for instance, by prescribing an antibiotic for those who want to believe it will cure a viral infection. (Antibiotics target bacteria exclusively and have zero effect on viruses. You might just as well try to hose the dirt off your driveway using a flashlight.) Another contributing factor is agricultural use: feeding low doses of antibiotics routinely to domestic livestock, because that somehow increases their rate of growth. In the United States during a recent year, more than 32 million pounds of antibiotics were sold for use in livestock, and most of that went for growth promotion and preventive dosing of food-animal populations, regardless of whether the individuals were sick. Globally, total consumption of antimicrobials (that is, drugs against dangerous microbial fungi as well as bacteria) by livestock was roughly 126 million pounds, with China using even more than the United States, and Brazil in third place. Most of that total goes into cattle, chickens, and pigs. A significant fraction of it involves drugs that are also important in human medicine.

So there's an extraordinary amount of evolutionary pressure, out there in the world, forcing bacteria to acquire resistance or die. But the most startling aspects of the trend have been how speedily resistance has spread and how many different kinds of bacteria have acquired multiple resistance—that is, resistance not just to one antibiotic but also to whole arsenals of different kinds. The danger of multiple resistance is that whatever antibiotic may be prescribed, whatever pharmaceutical may be thrown at it, the bacteria just keep eating a person's flesh or blood or guts, sometimes to the death. The death of the patient is no dead end to the strain of bacteria, of course, if it has managed to infect other victims in the meantime. And the appearance of resistance to each drug so quickly, in one strain of bacteria and another, as occurred in the 1940s and 1950s, was a phenomenon that couldn't be explained by the slow Darwinian process of mutation, natural selection, and ordinary inheritance, occurring independently in each case. Darwinian selection was certainly involved, but selection can act only on variation: genetic differences between one individual and another. What was the source of the variation? Mutation alone couldn't account for the appearance of so many new genes, so fast, in so many different organisms. It had to be something else—something that moved speedily and sideways, even between members of different bacterial species. Tsutomu Watanabe recognized the alternate explanation and, based on research by him and his Japanese colleagues, laid it out in English for the first time.

The Japanese work began after World War II in response to an increase in cases of dysentery, an intestinal affliction resulting in bloody diarrhea and other symptoms. Postwar deprivation, dislocation, and disruption of sanitary and health services probably exacerbated the problem, but its proximate cause was a bacterium, *Shigella dysenteriae*. The preferred treatment at first involved various forms of sulfa drugs, but *Shigella* strains soon showed resistance to those sulfas, so medical people turned to newer antibiotics, such as streptomycin and tetracycline. By 1953, strains of *Shigella* showed resistance also to both of those. Each bacterial strain, though, was resistant to only one drug. It could still be stopped by the others. Then in 1955 a Japanese woman returned from a stay in Hong Kong, sick with dysentery, and *Shigella* from her feces tested resistant to multiple antibiotics. From that point, resistance spread fast, shockingly fast, and during the late 1950s, Japan suffered a wave of dysentery outbreaks caused by *Shigella* superbugs resistant to four kinds of antibiotic: sulfas, streptomycin, tetracycline, and chloramphenicol. Could these strains have acquired such multiple resistance so quickly by incremental mutations alone—one misplaced A, C, G, or T at a time? The odds against that were so high you'd need a string of twenty-eight zeros to print them. But if not, what was happening?

The alarm bell rang louder when researchers discovered that this phenomenon wasn't confined to *Shigella*. Some cultures of *Escherichia coli*, taken from patients with resistant *Shigella*, showed resistance to the same drugs. *E. coli* had shared. A whole set of resistance genes had evidently moved sideways, in the depths of the patients' guts, from one kind of bacterium to another. Two teams of Japanese researchers then reproduced the phenomenon in their laboratories, showing similar transfer between bacterial strains cultured together in flasks or dishes, and concluded that the capacity for multiple resistance was passed by conjugation. Yes, a sizable packet of genes, not just one bit of DNA, was moving across. And the exchange wasn't limited to *Shigella* and *Escherichia*. Further research showed that the packet could cross boundaries between other species, even from genus to genus, among almost every group in the enteric bacteria, a large family of bugs that live within human bellies.

What exactly was this packet of genes that traveled so easily across boundaries? Watanabe and a colleague, Toshio Fukasawa, in earlier work, had offered a hypothesis: it was an *episome*, a sort of autonomous genetic element that floats free within a bacterial cell, unattached to the cell's single circular chromosome. An episome is sublimely selfish DNA. It carries extra information beyond what's absolutely necessary to assemble and operate the cell. It codes for traits that might be useful in emergencies. It can hold multiple genes, exist in many copies within a cell, replicate independently apart from the chromosome, and send a copy of itself into another cell during conjugation. It might be lost entirely from a strain of bacteria when its genes aren't needed, because of changing environmental conditions, and then, when conditions shift again, reacquired from other strains. Wow: wildly mobile DNA. Esther Lederberg's F factor was such an episome, though she didn't realize that at the time of discovery. The concept didn't exist until 1958. But now Watanabe declared to the scientific world in his 1963 paper what Fukasawa and he had already said in Japanese: multiple resistance, to streptomycin and those three other antibiotics, was coded on an episome. They gave the episome a name: resistance transfer factor. It became known as R factor, for short, in parallel to Esther Lederberg's F factor.

This R factor could be transferred by conjugation. It could be transferred (at least in lab experiments) by transduction. It explained how harmless bacteria such as ordinary *Escherichia coli* could convey genes for multiple antibiotic resistance, across species boundaries, into dangerous bacteria such as *Shigella dysenteriae*, in a blink. Its medical significance was "limited to Japan at present," Watanabe wrote, but R factor and episomes like it "could become a serious and world-wide problem in the future." That was prescient and understated.

53

ord of the Japanese discoveries spread, thanks to Watanabe's publications, though word didn't spread so quickly and broadly as bacterial resistance did. Unless you were a reader of *Bacteriological Reviews* or ate lunch with bacterial geneticists, you probably would have been unaware, in the early 1960s, of the way horizontal gene transfer was carrying this problem around the globe.

A young American named Stuart B. Levy, on leave from medical school, heard about it while on a research fellowship at the Pasteur Institute in Paris around that time. A Japanese researcher at the Pasteur gave an informal talk on multiple drug resistance, describing what his countrymen had learned, and Levy approached him afterward. Levy was fascinated by the work and especially interested in Tsutomu Watanabe. "Do you know him?" Levy asked. The Japanese colleague, a man named Takano, knew him very well. Watanabe was at Keio University in Minato Ward, part of Tokyo, and had hosted some of Takano's own work. "If you want, I'll write a letter for you," Takano said. That led to an invitation, and Levy finagled another getaway from med school so he could work for a few months in Watanabe's lab. It was a formative experience.

Stuart Levy MD is nowadays a professor at Tufts University School of Medicine and an internationally renowned authority on antibiotic use, overuse, and resistance. Old photos from fieldwork and conference events show him as a jaunty young man with a big, dark mustache and the good sense to smile and relax when the work was done. His twin brother, Jay, is also a medical researcher—one of the three scientists whose labs first isolated the causal virus of AIDS. While Jay stayed with viruses, Stuart focused on bacteria. He cofounded the Alliance for the Prudent Use of Antibiotics (APUA) in 1981 and still serves as its president. He's also a past president of the American Society for Microbiology, a huge and august organization with an international membership. He reminisced about Watanabe when I visited him in his office, on the eighth floor of a drab building just outside Boston's Chinatown. By this time, Dr. Levy was in his midseventies, clean shaven, with thinning hair and brown eyes that looked sad in their deep sockets above his mild, welcoming smile. He had seen a lot since Paris and Tokyo in the early sixties.

"We worked in the lab without air-conditioning," he said of his time with Watanabe. "It was very, very hot. Hot and humid." Levy's lab bench was on an upper level, with a sort of overlook from which, glancing down, he could see Professor Watanabe doing experiments in shirt sleeves, "because it was so hot." Periodically, someone would bring forth a hose and spray the professor with water to cool him off. He was a small man, an inch or two shorter than Levy, who spoke impeccable English and had a simple, straightforward manner toward students and postdocs. He would bicycle through the streets of Minato with his junior colleagues and sometimes took three or four of them out to a bar, Levy recalled, for an evening of karaoke. "We would be singing English songs, and he was reading, he was directing it. And it was . . ."—for a moment's pause, Levy left me to picture the professor, arms waving, as he joyously followed the bouncing ball and crooned something from Roy Orbison or the Honeycombs—". . . unbelievable moments." On a visit to Philadelphia for a scientific meeting, a few years later, Watanabe stayed with Levy's parents, nearby, in Wilmington, Delaware. "I was delighted that Watanabe would come," Levy said, "because I worshipped him, in a weird way." A lively mentor, a focused and dignified Japanese scientist. What became of him? I wondered.

"He passed away of stomach cancer," said Levy. "He probably was in his forties, early fifties."

Before that occurred, though, Levy himself had coauthored a paper or two with Watanabe on resistance factors. "One is in Japanese," Levy recalled. "Don't ask me what it says." So I didn't, and it has never been translated, but the title in English suggests they were investigating possible ways to fight resistant infections by preventing the bacteria from replicating their DNA. Levy had returned to the United States by then and resumed his medical studies, pointed toward a career that mixed research on bacterial resistance with some clinical practice and a sense of mission. His mission, pursued through publications, lectures, and APUA, was to protect the world against bacterial superbugs by devising defensive therapies and raising awareness about the scope and the consequences of needless antimicrobial excess.

In his research, Levy focused especially on resistance to tetracycline, and in the mid-1970s he led a pioneering study of how such resistance could be transferred from the gut bacteria of poultry to the gut bacteria of humans. That work, published in the *New England Journal of Medicine*, showed that intestinal bacteria of chickens, if the birds ate tetracycline-laced feed, acquired resistance to the antibiotic within a week. Less expected, more worrying, was that bacteria in the bowels of farm workers on the same site acquired the same resistance over a period of months. Soon after the early farm studies, Levy's lab also discovered just how those resistant bacteria evade tetracycline: by pumping it back out through the cell wall, with a sort of efflux mechanism. This mechanism is coded by a single gene on an episome (by that time, the word *episome* had been replaced by a synonym, *plasmid*) that sometimes travels laterally, carrying other genes for resistance to other antibiotics as well as the tetracycline-evader gene. A plasmid, as known ever since, is a short stretch of DNA, sometimes circular like a bracelet, that exists and replicates in a cell independently of the cell chromosome. That independence facilitates its lateral passage to other cells and helps explain how the efflux mechanism against tetracycline moved sideways so fast, from one bacterium to another, and from Leghorns into the bellies of people.

Levy's research continued, likewise his leadership as a voice of concern, and in 1992 he published a book titled *The Antibiotic Paradox*. The paradox is that these drugs, antibiotics, which made human lives so much better and longer during much of the twentieth century, have also been making our bacterial enemies so much more formidable. The book was updated in 2002, and in that edition he said:

"The discovery of transferable R factors, forty years ago, opened the eyes of microbiologists and medical scientists to a breadth of gene spread never before imagined. Transfer of resistance genes could occur among bacterial species more genetically and evolutionarily distant than a horse is from a cow."

The implications of those discoveries weren't fully realized at the time, Levy added, but they presaged the spread of antibiotic resistance, by horizontal gene transfer, around the planet.

54

All this talk about bacteria, which are invisible to the naked eye but nearly ubiquitous—on our skin, in our guts, throughout our environment, and spanning the history of life as we know it—made me want to see some. Gene sequences from bacteria and journal papers on bacterial dynamics are all very well, but I hankered for a glimpse of their corporeal presence. So I flew to London and connected to Porton Down, a high-security science compound located discreetly in the Wiltshire countryside, amid rolling fens and stubble fields, near the city of Salisbury in southwestern England. There, behind a tall chain-link fence, lies the National Collection of Type Cultures (NCTC), one of four major repositories of cell lines and microbes under custodianship of an agency called Public Health England. The other three, encompassing cell lines for medical research, and infectious viruses, and fungi, are tucked elsewhere around the country. NCTC, with its primary site at Porton Down, is devoted to bacteria.

I got through the security gatehouse because a genial press liaison, Isobel Atkin, had arranged for me to be welcomed. Inside the secure perimeter, I found a long, severe redbrick structure, looking like a Dickensian shoe factory, and a number of functional, metal-box modular buildings, some of them stacked two high for efficient use of space.

The Communications Office, including Atkin's desk and others in a single crowded room, was on the second floor of one stack, reachable by a metal stairway. At the far end of the row sat a larger metal box, formally known as Building 17. Less formally: the Warehouse, within which rests the Ultra-Low Temperature Storage Facility (ULTSF). That's where the bacteria sleep, and that's where I spent most of my day.

Building 17 is carefully compartmented, a repository of records as well as specimens, and as you move through it, you penetrate deeper into zones of secure containment, cryogenic preservation, and British medical history. Atkin and I were accompanied by Julie Russell, head of Culture Collections for Public Health England. Just inside, we passed among office carrels and then buzzed our way through a locked door into the storage area, the ULTSF, and its first big section, known as the Tank Room. This was a hangar-like space filled with dozens of freezer tanks—great steel barrels with tightly sealed lids, containing liquid nitrogen chilled to about minus 190 degrees Celsius. Within each tank rested about 25,000 ampoules, sealed glass tubes of frozen bacteria, neatly boxed and labeled and racked. These were production samples of various bacterial strains—including many strains of medical importance—grown and packaged at another facility, from historical strains kept here at Porton Down. They were available, at modest cost, for use by researchers around the world. (NCTC formerly gave away its bacterial samples upon request to reputable labs; nowadays the samples are sold for a reasonable recoup of expenses, but their genomic sequence data are given freely.) Producing such viable samples and making them available for all manner of scientific study is one of the chief purposes of the NCTC. Storage of the original strains, literally frozen in time—frozen in evolutionary stasis as they existed decades ago—is the other chief purpose. Those original strains have their own storage rooms, more like bank vaults than airplane hangars, deeper within the building.

Two staff people greeted us to give the tour. Steve Grigsby, supervisor of the ULTSF, with his chiseled features and vice-grip handshake, looked like an aging Daniel Craig in a black turtleneck. Jodie Roberts, senior cryostore technician, was a smart young woman with a multicolored ponytail who appreciated the subtleties of bacterial evolution as well as storage. Roughly a million ampoules of bacteria are stored in this room, Grigsby said. Roberts opened one tank, and we watched the frozen vapors rise. The tanks are on an alarm system, Grigsby explained; if one of them loses power and starts to warm up—to a dangerously balmy minus 153 degrees C. or above—an alert clangs and brings him running at whatever time of day or night. The room itself has a separate alarm for the safety of human personnel. If the ambient nitrogen level rises above a certain threshold, another alarm tells whoever is present that they are about to asphyxiate; windows open automatically, fans bring in outside air. He demonstrated for my benefit, with the windows and fans, a generous use of Public Health England electricity.

Beyond the Tank Room was another door, giving into a sanctum called Room 17/11. A special badge was needed here (Atkin's wouldn't open the lock), and Jodie Roberts buzzed us in. The temperature was just below 5 degrees C. (roughly 40 degrees Fahrenheit, like inside your refrigerator), which is much warmer than the tanks but cold enough for long-term preservation if the bacteria are lyophilized (freeze-dried) in their vacuum ampoules. The ceiling was low, and along the left wall stood eight storage cabinets holding many of the British collection's more notable samples. These included a sample of Oswald Avery's *Streptococcus pneumoniae* (known here by its accession number, NCTC13276), descended straight from the strain in which he discovered transformation, and another *Streptococcus* strain (NCTC8303) worked on by Fred Griffith. Freeze-dried securely, they were stored in the cabinets for future reference and research. There was also a specimen of *Haemophilus influenzae*, a bug that lives harmlessly in many people but sometimes causes nasty infections; this sample (NCTC4842) was isolated in 1935 by Alexander Fleming, the discoverer of penicillin, reportedly from his own nose. Eventually I got to see them all.

Roberts opened a cabinet, within which were nested a number of cartons. She lifted out one carton and raised its top, revealing dozens of delicate, balsa-wood boxes, each roughly the size to hold a single Montblanc pen. The preferred term for these little boxes is "coffins." It seemed nicely appropriate; but they were Dracula coffins, holding dangerous creatures quite ready to wake up. Tiny labels on the lids, hand lettered, told which coffin was which. Each coffin contained a glass outer tube protecting the delicate ampoule, which was a smaller tube, round at one end, the other end heat sealed to a needle point. Inside the ampoule, under vacuum conditions, was the bacterial sample. All you could see was a dab of yellow material, barely more than a smudge, at the ampoule's round end. The yellow stuff was mostly precipitate of a nutrient broth, a special mix used to protect the sample when it was freeze-dried. Within the smudge lurked dormant bacteria.

We test the integrity of the vacuum, Roberts explained, by sparking the ampoule. If it's good, it will fluoresce. Then she showed me, applying her electric spark-testing wand to one little tube. It did indeed fluoresce, with a gentle blue light, indicating a good vacuum. If an ampoule didn't fluoresce, that meant the vacuum was lost, she said, and the sample was spoiled. They would dispose of it.

I had asked in advance to see one particular sample, a historical treasure, and Roberts obliged. Wearing orange medical gloves, she gingerly lifted the coffin of a specimen known as NCTC1. Its number 1 testified that it was the very first sample acquired, back in 1920, when NCTC was founded. It was a strain of *Shigella flexneri*, a bacterium that causes *shigellosis*, a form of dysentery, similar to what Watanabe saw in Japan. Shigellosis manifests as bowel inflammation and diarrhea, and the infectious dose for a human is low—only a trace needed to take hold and bloom in someone's gut—meaning that it is highly transmissible through tainted water or food. The diarrhea can be bad, especially in the absence of decent medical care: shigellosis still kills more than a million people a year, most of them children in developing nations. In early 1915 it afflicted a British solider in France, and that's how the sample later known as NCTC1 came to be gathered.

The soldier was one Private Ernest Cable, twenty-eight years old, serving with a regiment out of East Surrey. No photo survives, and almost nothing is known about Private Cable or his next of kin; he had lodged with a family in England before joining the army, the family had a toddler, and to that toddler he wrote his will. Cable arrived sick at a military hospital in Wimereux, France, just down the coast from Calais, thirty or forty miles behind the trench lines of the Western Front. The hospital was in a converted hotel, formerly the Grand Hotel of Wimereux, and devoted entirely to infectious diseases, not battlefield injuries. Although his clinical records have disappeared, Cable probably suffered bloody diarrhea and gut cramps. The diagnosis was dysentery. On March 13, 1915, as the Allies pushed an offensive in Artois and Champagne, a hundred miles south, Cable died—but not before a stool sample was taken. The stool sample went to Lieutenant William Broughton-Alcock, a bacteriologist at the Wimereux hospital, who isolated a bacterium. That isolate was later placed to the species *Shigella flexneri* and identified as serotype 2A. Five years later, it went into the collection as NCTC1. What makes it interesting is not its numerical priority but what was found a century later about its biology and its genome.

Private Cable's *Shigella* was resistant to antibiotics that hadn't yet been invented. More precisely, to antibiotic substances not yet *discovered* by humans. More precisely still, Cable's strain was resistant in 1915 to penicillin and erythromycin, which went into use against human infections in 1942 and 1952, respectively. This seeming reversal of cause-and-effect order was deduced by a team based at the Wellcome Trust Sanger Institute, in Hinxton, a little village just south of Cambridge. Kate S. Baker was the first author on their paper, published in 2014.

Baker and her colleagues took an old ampoule of NCTC1 from Porton Down, broke it open, revived the bacterium, and cultured it, waking the old bug from its time capsule. They plated out samples of the growing bacteria onto separate dishes of nutrient agar, then challenged them with a whole list of modern antimicrobials. Baker's team found "intrinsic antimicrobial resistance" against those two drugs, penicillin and erythromycin. Intrinsic? It seems befuddling—like an intrinsic proclivity to wear bulletproof vests before guns were invented. How and why would anyone do that? In this case, such preparedness reflects the fact that resistance to antibiotics does exist in the wild because *antibiotics* exist in the wild. Some bacteria produce antibiotic substances as natural weapons, deterrents, for use in their competitive struggles against other bacteria. Resistance likewise arises naturally, slowly, as an evolved trait, for defense against such weapons.

Some evidence of this phenomenon came in 1969 from the Solomon Islands, a remote archipelago in the southwest Pacific. Modern medicine hadn't arrived, and the people of the Solomons were still "an antibiotic virgin population." Despite the absence of lab-produced drugs, a group of American researchers found bacteria, in a soil sample, that proved resistant to both tetracycline and streptomycin. How did antibiotic resistance arrive before manufactured antibiotics? Unknown, but again, the answer probably lay in the natural battles among bacteria.

In this Solomon Islands case, the same double resistance—against tetracycline and streptomycin—turned up also among human-dwelling bacteria. In a fecal sample from one native islander, a person "from the innermost bush country," the researchers found *E. coli* resistant to both the same drugs. Had it somehow been transferred sideways from soil bacteria to bacteria in the human gut? Although the American team couldn't answer that question, not in 1969, they did attribute the resistance in both bugs to R factors, transferable genes on little plasmids, of just the sort that Watanabe and Stuart Levy had deduced.

Another antibiotic, vancomycin, has an origin story that also traces to distant soils. Vancomycin was developed from a natural substance generated by bacteria found in dirt from Borneo. In 1952 a missionary sent this smidgen of dirt to one of his friends, an organic chemist in the United States. The friend worked at Eli Lilly, a pharmaceutical company then sorely interested in discovering antibiotics that would work against staphylococcus strains resistant to penicillin. The natural substance produced by the Borneo bug, known first as compound 05865, was purified and modified slightly, then named vancomycin because of what it could vanquish: the gonorrhea bug and other troublesome bacteria, including staph—even staph strains that resisted penicillin. But again defenses arose against the vanquisher, in the hospitals of Japan and the United States, if not naturally in the soils of Borneo. By the late 1980s, vancomycin resistance had shown up among bacteria of one genus, *Enterococcus*. The first identified gene for resistance in *Enterococcus* (eventually there were several such genes) was called *vanA*, and because enterococci contain plasmids and other systems for horizontal gene transfer, it was probably just a matter of time—and not much time—before *vanA* dodged sideways into other kinds of bacteria. Sure enough, it soon went across genus boundaries from *Enterococcus* into staph, including *Staphylococcus aureus*.

This was bad news for human medicine. Back in 1982, vancomycin had been one of the go-to weapons against methicillin-resistant *Staphylococcus aureus* (the dreaded MRSA), but by 1996, in Japan, there were staph infections with reduced susceptibility to the drug, and soon afterward vancomycin-resistant staph started showing up in the United States. A patient in Michigan, a poor soul with multiple health problems, including diabetes, kidney failure, and chronic infected foot ulcers, yielded the first American sample of vancomycin-resistant *Staphylococcus aureus* (a new demon to fear: VRSA), some of it isolated from the infected foot. That patient had been treated with vancomycin among other antibiotics for infections following an earlier toe amputation, so the staph may have acquired its resistance gene by horizontal transfer within the patient's own body. Oy vey. But the resistant staph might just as likely have been ambient in the hospital, having recently snatched up its new gene while abiding within somebody else's multiply infected body. Soon afterward, another VRSA strain was isolated from a woman in Pennsylvania, this time despite the fact that she herself *hadn't* recently been treated with vancomycin. The Pennsylvania patient, like the one in Michigan, was suffering an infected foot ulcer, and from that ulcer, again, came the resistant staph.

These two foot-eating strains of VRSA both carried the *vanA* gene, suggesting they had each gotten their resistance by transfer from *Enterococcus* bacteria, probably in separate events. The US Centers for Disease Control and Prevention promptly issued a warning that, because this gene was jumping around so deftly, from gut-dwelling bacteria into skin-rotting bacteria, "additional VRSA infections are likely to occur." No one's feet were safe, and especially not those of anyone foolish enough to walk barefoot in a hospital.

Private Ernest Cable's fatal dysentery, back in 1915, seems like a warning signal of all this travail, but only in retrospect. Its meaning couldn't be divined at the time because human-made antibiotics didn't yet exist—and therefore neither did the problems of their use, their overuse, and the resistance genes that spread so quickly from natural beginnings. Cable's case merely represents an obscure step along the way. Its particulars were unexceptional—a man's death in wartime, a bacterial sample archived routinely—and their implications weren't recognized until a century later, when Kate Baker and her colleagues regrew the bug, extracted its DNA, and sequenced its genome. They found genes that would have defended Cable's original bug from penicillin and erythromycin, if any such drugs had existed in that Wimereux hospital. Private Cable, even if given such treatment, probably would have died. But we'll never know.

He was buried in the Wimereux cemetery, and his *Shigella* strain, after lab culturing by Lieutenant Broughton-Alcock, plus a few other steps along the way, was buried in its own tiny coffin within a box, within a cabinet, within a very cold room, within Building 17/11, within the NCTC compound at Porton Down. The orange-gloved hands of Jodie Roberts now held that coffin for me to see. I bent close and read the label aloud: "*Shigella flexneri*, type 2A, strain Cable."

"We know that it's still viable," said Julie Russell, the head of collections, "because there was more than one." More than one such coffin, that is, each containing an ampoule of Cable's bug, taken from the same early batch, reawakened briefly and freeze-dried for better preservation, by the latest new method in 1951, then reburied in the cold. "And we've grown it from the other one," she said. That was for the Baker team's work, when they broke open one ampoule, cultured it, and looked at its deep identity. Now, besides this one, Russell added, there is one more ampoule of NCTC1, strain Cable, from the earliest batch, under storage here at NCTC. Two left, total.

Are they precious? I asked.

"That's right, yeah," she said casually, and what she meant was: they can be very damn useful.

55

During the late 1960s and early 1970s, scientists began to realize that the implications of horizontal gene transfer go far beyond the problem of bacterial resistance to antibiotics. Those implications include the whole matter of how evolution works—by Darwinian mechanisms, or otherwise?—and how it *has* worked for much of the past four billion years. A British bacteriologist named Ephraim S. Anderson hinted at this in 1968.

Anderson, born in 1911, came from an Estonian Jewish immigrant family in a working-class neighborhood of Newcastle upon Tyne, an unadvantaging start for an aspiring British scientist in the hard years between the world wars. He showed his brilliance in school, won a scholarship to study medicine, then struggled to find work, disfavored for at least some positions because he was a Jew. He joined the Royal Army Medical Corps and spent five years in Cairo, Egypt, tracing typhoid outbreaks among British troops. Returning to England, he took a research job at the Enteric Reference Laboratory, a national facility with a practical mission: identifying and characterizing strains of intestinal bacteria that threaten human health. Within a few years, Anderson became its director. By then, he was an expert on enteric bacteria such as the *Salmonella* group, which includes the typhoid bug, and during the 1960s he emerged as an influential voice on public health, warning early and loudly about the dangers of antibiotic resistance. He was known for his brusqueness, his feistiness, his "great gift" for rubbing people the wrong way, and for his strong opposition to the routine use of antibiotics for growth promotion in livestock. He was among the earliest bacteriologists in England who recognized what Watanabe and his Japanese colleagues had seen: that resistance genes could spread quickly, from strain to strain, from species to species, on plasmids. For that alone, he would be notable. But in one of his journal papers, Anderson went a step further, speculating that another important effect of such transfer factors "is their possible importance in bacterial evolution in general."

The fact that resistance genes could move sideways so easily was a clue that genes for other traits might be moving too. And so, Anderson wrote, "the temptation is very strong to suggest that the transfer factors may have influenced bacterial evolution." Maybe it wasn't all just a matter of mutation and natural selection after all. Maybe horizontal gene transfer also played a big role in the long history of microbial life.

This was a gentle statement of a revolutionary prospect. Was the history of evolution—at least among one group, the bacteria—really *that* different from Darwin's theory as we have come to embrace it?

Anderson's suggestion was echoed in 1970 by a pair of British researchers in a different area of bacteriology. Dorothy Jones and Peter Sneath were microbial systematists—namers and classifiers—at the University of Leicester. They worked in the same long tradition as Ferdinand Cohn, the great early classifier of bacteria, but with a concerted effort to make use of new data, modern methods, and fresh thinking. Their preferred method was known as numerical taxonomy, which at that time stood adamantly opposed by another school of classification, newer, known as cladistics. The numerical taxonomists classified creatures into species and higher categories by overall similarity, regardless of evolutionary history. The cladists argued that common ancestry—and therefore evolutionary history—is the only cogent basis for classifying. This was a bitter and arcane fight about which, trust me, you don't need the details. Suffice to say here that in 1970 Jones and Sneath coauthored an influential review paper titled "Genetic Transfer and Bacterial Taxonomy," the main purpose of which was to use horizontal gene transfer as a cudgel for bashing cladistics on its head.

Cladistics works poorly for classifying bacteria, they noted, because of the near-total absence of a fossil record. But even more damning for that evolution-based approach was the evidence emerging from studies in Japan and elsewhere of gene transfer between one bacterial species and another. Jones and Sneath proceeded, through the bulk of their long paper, to describe and document much of what was then known about horizontal gene transfer among bacteria. Then they speculated that one "foreign" gene, transferred into a bacterium and integrated there, might make that genome more able to accept other transfers. Barriers between species might begin to fall. "This in turn could favor extremely reticulate modes of evolution, with numerous partial fusions of phyletic lines." Reticulate modes? That meant weblike. Fusions? That meant genes jumping sideways, from one genome to another. Tree branches never fuse, never reticulate, so how do you draw this situation as a tree of relatedness?

"It may well be that gene exchange is so frequent," they wrote, "that the evolutionary pattern in bacteria is much more reticulate than is commonly believed." Weblike, not treelike. What they were saying, implicitly, was: Whew, all this gene transfer makes classifying bacteria tricky—and harder for us, but *impossible* for those poor obdurate cladists.

Jones and Sneath and their allies were destined to lose this battle. Cladistics would triumph and become the reigning approach to classification, at least among evolutionary biologists. But the paper by Jones and Sneath served other purposes. It broadened awareness of how horizontal gene transfer does complicate the enterprise of classifying organisms and portraying their evolutionary history. And it seems to have been one of the first scientific sources, if not *the* first, to float the concept of reticulate evolution—the idea that the limbs of the tree of life are intertangled.

56

A t this point, we come to the frangibility of another absolute. By which I mean, the ricketiness of another *apparent* absolute. The concept of "species" is commonly supposed to be secure. It isn't secure. It's especially insecure in the realm of bacteria and archaea, but it's even a bit blurry when scientists try to distinguish one species of plant or one species of animal from another. The boundaries blur. The edges are as porous as Goretex—or, in some cases, as cheese cloth. One reason for the blur, one symptom of the porosity, is horizontal gene transfer—genes moving sideways instead of just downward from parent to offspring. If genes cross the boundary between one species of bacteria and another, then in what sense is it really a boundary?

The conviction that bacterial species are fixed and discrete goes back, as you've read, to Ferdinand Cohn. Working in Breslau during the 1860s and 1870s, trying to put order into bacterial classification, contending against the notion of bacteria as shape-shifting creatures that change their forms according to conditions of environment, Cohn found ways to grow pure cultures of one bacterial strain or another on solid media. A pure culture had a continuity of identity and form that suggested giving it a species name—*Bacillus anthracis*, for instance—was a rational exercise. And the confidence of identity was very useful when you wanted to tell *Bacillus anthracis* (causing anthrax in humans) from *Bacillus subtilis* (causing rot in potatoes). Robert Koch helped this effort by developing the technique of streaking bacteria onto solidified gelatin surfaces, from which one tiny smidge of cells from a mixed sample could be teased out and grown again separately, yielding a pure culture. Koch's lab assistant Julius Petri helped further by inventing the Petri dish, in which a pure strain grown on gelatin or agar could be protected from airborne contamination with a glass lid. Ferdinand Cohn won the battle of bacterial classification, and his victory held good for about fifty years.

But this certitude of discrete identities began to weaken with Griffith's discovery in 1928 that one type of *Pneumococcus pneumoniae* could transform into another type. It weakened still further when Watanabe announced that *Shigella dysenteriae* could receive genes from *Escherichia coli*. Nowadays bacterial taxonomists recognize that *Shigella* genomes are very similar to *Escherichia* genomes—so similar that they should probably be lumped in a single genus. In fact, some strains of *Shigella* are more closely related to *E. coli* than they are to one another. And the crumbling certitude about how to sort bacteria into species, despite horizontal gene transfer, has progressed far beyond that little confusion.

Sorin Sonea, a Romanian-born microbiologist at the University of Montreal, took the conundrum of blurring bacterial boundaries to its logical extreme. First in a French edition, then in English, in 1983, he and his coauthor, Maurice Panisset, published a book titled *A New Bacteriology*, making the case that all bacteria on Earth constitute a single interconnected entity, a single species—no, wait, maybe even a single *individual* creature—through which genes from all the variously named "species" flow relatively freely, by horizontal gene transfer, for use where needed. This freedom of transfer, this universal interchangeability of parts, gives the bacterial entity "a huge available gene pool," Sonea and Panisset wrote, and thereby allows bacteria to adapt so well, and so quickly, to so many different environments and situations. The genome within an individual bacterium is typically small, much smaller than genomes of most eukaryotes, and contains relatively few genes—only the bare necessities for bacterial life and replication. There's little excess, redundancy, or emergency provisioning of intermittently useful genes for special circumstances. The advantage in such parsimony is that it allows bacteria to reproduce quickly. The disadvantage is a lack of versatility for special circumstances—but horizontal gene transfer, bringing in new genes from other strains or species as needed, compensates nicely for the lack, complementing the bare-bones endowment. As a result, bacteria get by with few genes, some of which (especially those on plasmids, unattached to the bacterial chromosome) are continually being lost or gained.

This is radically different, Sonea and Panisset claimed, from evolution as described by Darwin, who focused on animals and plants. Animal and plant species, as well as other eukaryotic species, arise mainly by genetic isolation. Bacteria are never so isolated. Instead of tortoises and mockingbirds marooned on islands, mutating and adapting, diverging slowly into distinct subspecies and eventually new species, finally reaching a point where they can't or won't mate with other populations—instead of that, you have relentless bacterial togetherness. You have genes oozing sideways all over the planet, from one bacterial strain to another, like pulsing juices inside a gigantic, invisible version of the Blob.

Okay, they didn't call it the Blob. They called it a "superorganism," which is almost as spooky. Also, to be clear: this superorganism concept of Sonea and Panisset was quite distinct from the superorganism concept that James Lovelock and Lynn Margulis framed, under the name Gaia, and applied to planet Earth itself. Earthly Gaia was a superorganism comprising all physical and living constituents of the planet, according to Lovelock and Margulis. Sonea and Panisset's superorganism was "just" the total global population of bacteria. The two ideas are related in spirit—they're grandiose and daring and woozy—but very different in their particulars and their purposes. Sonea and Panisset intended to depict the fluidity with which bacterial "species" exchange genes; but their superorganism did *not* encompass all other forms of life. It was not the Earth Mother. It was the world's biggest germ. What made it vivid, and more useful than the notion of Gaia, was that it starkly contrasted two things rather than uniting everything: the way bacterial genes move sideways versus the way tortoise genes and mockingbird genes generally don't.

After the death of Panisset, Sorin Sonea continued to argue their big idea, the bacterial superorganism, in English publications, with mixed response from the scientific community. Lynn Margulis liked it, not surprisingly. Ford Doolittle called it "bold if inchoate"—a fair judgment—and he recalled that Sonea's theory and others like it "were widely dismissed during the 1970s and 1980s—they were so hopelessly radical!" Doolittle enjoys offering a bit of radical provocation himself, so this was a nostalgic and friendly comment when he wrote it, in 2004, by which time horizontal gene transfer was a hot topic throughout molecular biology and had destroyed all the old notions of sorting bacteria into species with neat boundaries and placing them like fruit on a tree.

57

Other scientists besides Sonea and Panisset began noticing, back in the 1980s, that this odd phenomenon might have broad implications. Slowly at first, it became a favored research theme in more than a few labs. The phrase "horizontal gene transfer" had just been coined ("lateral gene transfer" became a variant, meaning the same) and "reticulate evolution" was also in the air. Journal papers and review articles appeared, still based mostly on tenuous data, raising questions in the same spirit as Ephraim Anderson had, about the significance of HGT and whether it demanded a new theory of evolution—a major supplement to Darwin's.

It did seem to be widespread and common among bacteria. Some researchers even saw, or thought they saw, evidence of it in other creatures. Eukaryotes. Animals, plants. A species of fish, which carried a bacterial symbiont, appeared to have passed one of its fish genes into the bacterial genome. How was that possible? Another bacterium had sent bits of its DNA into the nuclear genomes of infected plants. Bacterium to plant? A species of sea urchin seemed to have shared one of its genes with a very different species of sea urchin, from which its lineage had diverged sixty-five million years earlier. That was a stretch. Still another bacterium, the familiar *E. coli*, was found to be transferring DNA on plasmids into brewer's yeast, which is a fungus. Brewer's yeast is microbial, a relatively simple little creature, but nonetheless eukaryotic. This mixing of fungal host and bacterial genes happened via a smooching process that looked much like bacterial conjugation, the researchers reported, and "could be evolutionarily significant in promoting trans-kingdom genetic exchange." Trans-kingdom is a long way for a gene to go.

A 1982 essay in *Science* offered an overview, titled "Can Genes Jump Between Eukaryotic Species?," with the implicit answer: probably. Some of these cases of distant transfer later turned out to be illusory—disproved when better data became available—but the basic premise was correct, and the research agenda took hold. Genes *were* moving sideways, across boundaries, between very different kinds of creatures, to a degree previously unimagined. And the new recognition did raise a challenge to Darwin and Darwin's tree.

The notion that genes might be transferred sideways among complex eukaryotic organisms, as the *Science* essay noted, was a radical step beyond the established reality of horizontal gene transfer in bacteria. It was an "apparently fanciful and certainly unorthodox" idea, a perplexing anomaly that violated some axiomatic principles, and its investigation would need to progress through two stages. First, does this weird thing actually happen? Second, if it does, how common and how important is it?

New investigations, as time passed, showed that the weird thing does happen. For instance: there's a peculiar group of tiny animals known as rotifers, once studied only by invertebrate zoologists, including Leeuwenhoek, but now notable throughout molecular biology for their "massive" uploads of alien genes.

Rotifers are homely beyond imagining—so homely you almost (but not quite) have to love them. They live in water, mainly freshwater, and in moist environments such as soils and mosses. They live in rain gutters and sewage treatment tanks. Look at one through a microscope, and you'll see what resembles a maggot, with a lamprey mouth and a long, thin tail, but the tail isn't really a tail. Rotifer biologists call it a foot. Some kinds of rotifer can suck their foot back into their body when it's not in use. An extrudable and retractable foot. At the end of the foot is a toe, or maybe two toes, or four toes, depending on the species. Among rotifers that attach to surfaces or inch around, the toes have cement glands for getting a grip. Very handy; if you had only one foot, one shoe, you'd want a Vibram sole or cleats. But some of them float free as plankton. The lamprey mouth is circled by cilia, little hairs, that move quickly and set up a swirling flow to bring bits of food down the gullet. That circular swirling is what gives them their name, rotifer, derived from Latin syllables meaning "wheel bearer." They eat detritus, bacteria, algae, and other minuscule forms of digestible mulch. Some fish hobbyists put them in aquariums to help clean the glass. Captive rotifers will reproduce, if your tank is hospitable, and, as a side benefit, your tetras and swordtails can eat them. A bottle of pet rotifers, starter rotifers, can be had for $17 from a company in Nashville.

Invisible teeming maggots. A big rotifer might be a millimeter long, barely big enough to see, but small as they are, these are not single-celled creatures. They're multicellular animals.

The rotifers of one particular group are especially peculiar and interesting. They're called the bdelloids, a name I can type easily but not pronounce. Bdelloid rotifers tend to live in harsh, changeable environments that sometimes go dry. The bdelloids cope with such crises by hunkering into a dehydrated, dormant state, like instant coffee, in which they can survive for as long as nine years. When the water returns, they rehydrate and come alive. Another oddity of the bdelloids is that they reproduce without sex. Females give birth to females, no fertilization necessary. The fancy term for that is parthenogenesis. A male bdelloid has never been seen. Genetic evidence suggests that bdelloids have gone without sex for twenty-five million years—quite a period of celibacy by anyone's measure. Despite the absence of sexual recombination, which shuffles the genetic deck in a population and offers up new combinations of genes, bdelloids have managed to find newness somehow. They have diversified into more than 450 species.

That diversification without sex might be partly explained by the other bdelloid anomaly deduced recently from genetic evidence: their strong propensity for horizontal gene transfer. This was noted in 2008 by three Harvard researchers who sequenced sections of genome in one bdelloid species and found all sorts of craziness that shouldn't have been there. More specifically, they found at least twenty-two genes from non-bdelloid creatures, genes that must have arrived by horizontal transfer. Some of those were bacterial genes, some were fungal. One gene had come from a plant. At least a few of those genes were still functional, producing enzymes or other products useful to the animal. Later work on the same rotifer suggested that 8 percent of its genes had been acquired by horizontal transfer from bacteria or other dissimilar creatures. A team of researchers based mostly in England looked at four other species of bdelloids and also found "many hundreds" of foreign genes. Some of the imports had been ensconced in bdelloid genomes for a long time, since before the group diversified, while some were unique to each individual species, and therefore more recently acquired. This implied that horizontal gene transfer is an ancient phenomenon among bdelloid rotifers, and that it's still occurring. Genes going sideways among animals? That was definitely supposed to be impossible. It wasn't.

All those researchers from Harvard and elsewhere wanted to understand why. The best clues lay in aspects of bdelloid life history that I've already mentioned: their tolerance for desiccation, their reproduction without sex. Desiccation can be damaging to membranes and molecules, even when a creature survives the drought, and biologists suspect that such drying-and-rehydrating stresses cause bdelloid DNA to fracture and leave cell membranes leaky. Given that they're surrounded in their environments by living bacteria and fungi, plus naked DNA remnants from dead microbes, the porous membranes and fracturing could make it easy for alien DNA to enter even the nuclei of bdelloid cells and to get incorporated into bdelloid genomes as they repair themselves. Let me say that again: broken DNA, as a cell fixes it, using ambient materials, may include bits that weren't part of the original. If that mended DNA happens to be in cells of the germ line, the changes will be heritable. Baby rotifers will get them and, when the babies mature, pass the changes along to their own daughters. Thus a bacterial or fungal gene can become part of the genome of a lineage of animals.

Furthermore, the absence of sexual reproduction in bdelloids, the absence of recombination, might leave them especially needy of just such new genetic possibilities. Variation is the raw material of adaptation, as you know, and no lineage survives throughout time and vicissitudes without it. Mutation provides only tiny changes slowly, at the scale of one base in the DNA molecule replaced by another. Sexual recombination, by contrast, makes big rearrangements of what's already there. The tiny changes alone may not be enough. Omit sex, and you streamline reproduction but sacrifice adaptability. Parthenogenetic populations can thrive in the short term, but in the long term, they tend to go extinct. All this is relevant. Maybe bdelloid rotifers, reproducing asexually for millions of years—with no remixing of gene combinations, only mutations to supply incremental change—have gotten much of their freshening innovation from HGT.

If so, it's an aspect of evolution that was unimagined by Charles Darwin. And it goes far beyond the bdelloids.

58

It started showing up among insects. Again, this was supposed to be impossible. There were fervent doubters. Alien genes cannot move from one species to another, they insisted. The germ line of animals, meaning the eggs and the sperm and the reproductive cells that give rise to them, is held separate from such influences. It's sequestered behind what biologists call the Weismann barrier, named for August Weismann, the German biologist of the nineteenth century who defined the concept. (*Germline*, as now commonly composited, is a good word because it suggests the linearity of that lineage of cells.) Those cells and their DNA are isolated within the ovaries and testes, sequestered from genetic changes that may occur in the rest of the body. Bacteria cannot cross that barrier, Weismann's barrier—so said the skeptical view—to insert bits of their own DNA into animal genomes. Impossible. And again it turned out to be possible.

One of the big revelations came in 2007 from a team that included a young postdoc researcher named Julie Dunning Hotopp, then at the Institute for Genomic Research (TIGR), a private entity founded by the brilliant and audacious J. Craig Venter and located in Rockville, Maryland. Venter is the wildcat geneticist who competed against a huge and publicly funded international research initiative, the Human Genome Project, to assemble the first complete (or nearly so) sequence of the human genome. Dunning Hotopp joined TIGR after that ruckus, and her work is notable in its own right. She had come from Michigan State University with a newly minted PhD in microbiology but also an aptitude for what's called computational biology, meaning the analysis of huge amounts of biological data using computer and mathematical skills. It's essentially the same as what I've earlier mentioned as bioinformatics. Dunning Hotopp teamed with another postdoc, a fellow named Michael Clark at New York's University of Rochester (where she had gotten her own undergrad degree), and with their two mentors, for a study to see whether bacterial genes might be sneaking into the genomes of insects and other invertebrate animals, such as head lice, crustaceans, and nematode worms. The answer, for eight of the genomes they checked, was a strong yes.

These transferred genes came from bacteria in the genus *Wolbachia*, a group of aggressive intracellular parasites that infect at least 20 percent of all insect species on Earth. *Wolbachia* bacteria target the germline cells of the animals they enter, especially ovaries and testes, and, once established, a *Wolbachia* infection is passed from mother to offspring within her infected eggs. It does not usually pass within infected sperm. *Wolbachia* work around that constraint, the absence of sperm-to-offspring transmission, and proliferate themselves by manipulating the reproductive outcomes of their hosts. They do it in four different ways: killing male offspring before they can hatch, turning males into females, triggering parthenogenesis (virgin females delivering more females), and sabotaging the viability of uninfected eggs when fertilized with *Wolbachia*-infected sperm. The net result of their interference is to change the male–female ratio, shifting whole populations of insects quickly toward more *Wolbachia*-infected females producing more *Wolbachia*-infected offspring. These are evolutionary wins for *Wolbachia*. Given the broad range of insects infected (plus many other arthropods and nematode worms), and the sizes of those populations, *Wolbachia* is an extraordinarily successful group of parasites. "Arguably," according to one expert, "the spread of *Wolbachia* represents one of the great pandemics of life on this planet."

Being intracellular parasites means, of course, that *Wolbachia* bacteria take up residence not just inside the host but also inside *cells* of the host, closely adjacent in each cell to the nucleus with its DNA. Since they invade not just any cells but primarily the germline cells, that puts *Wolbachia* close to the very molecules of DNA that will be duplicated in the production of egg cells and passed to offspring. Such proximity seems to offer special opportunity for getting *Wolbachia* DNA spliced into the insect's DNA. Julie Dunning Hotopp and her colleagues discovered, by scrutinizing genome sequences from twenty-six different critters, that four insects and four nematode worms had taken aboard *Wolbachia* genes by horizontal transfer. The most dramatic case was one species of fruit fly, which had accepted almost the entire genome of *Wolbachia* (more than a million letters of code) into its own nuclear genome.

The fruit fly in question, *Drosophila ananassae*, is a favorite laboratory animal, and its genome had already been sequenced by other researchers. The sequence was publicly available. That published version omitted the *Wolbachia* genome, probably not because it hadn't appeared during sequencing but because the sequencing team assumed it reflected a bacterial contamination, an error, in their lab work. Researchers at the time were so reluctant to believe that bacterial genes *could* be transferred into animal genomes that, before publishing a new genome sequence, they routinely edited out the bacterial stretches. Dunning Hotopp and her co-workers took a different approach. Up in Rochester, Clark raised the flies in a laboratory and cured them of their *Wolbachia* infections, using antibiotics. Under microscopic inspection, their ovaries were clean; the only *Wolbachia* genes left behind, therefore, would be those actually embedded in the fly's own genome. Clark then mailed the fly DNA to Dunning Hotopp, in Rockville, where she handled most of the sequencing and the computational analysis.

"Anything having to do with the animal, he did," she told me, when I visited her lab. "Anything having to do with the computer, I did. And then some of the stuff in between, we both did." What they found amid the fly genome, to their own surprise, was almost the entire *Wolbachia* genome. Their methods were sound, and *Science* published their paper. It got attention in popular media, including the *New York Times* and the *Washington Post*, and was generally well received among the genome biologists.

Generally, but not universally. Their dramatic and well-supported findings were dismissed outright in certain quarters, Dunning Hotopp told me. She was in Baltimore now, not Rockville, when I called on her, at the Institute for Genome Sciences of the University of Maryland, and continuing her work on horizontal gene transfer in animals. Her career was thriving, she had received a prestigious grant of support from the

National Institutes of Health, and she was pushing the study of HGT in new directions. Her group had recently discovered evidence, for instance, of bacterial DNA transferred horizontally into the genomes of human tumors. What that dizzying revelation means is still unclear, but there's at least some chance that such insertions might play a role in causing cancer.

For the cancer-related work, she and her colleagues used bioinformatics to scan a vast number of human genome sequences, from several sources, looking for stretches that resembled bacterial—not human—DNA. One of their sources was a publicly available database called the Cancer Genome Atlas, containing genome sequences from the tumors of thousands of patients. The genomes of tumors are often different, in small but important ways, from the genomes of patients suffering the cancer, because tumor cells mutate as they replicate. Hotopp's team did find bacterial DNA lurking within some normal human genomes, an interesting result. More peculiar and disquieting, though, was that they found it 210 times more common in tumor cells than in healthy cells.

Human cells are continually exposed to bacteria—the ones that live routinely in our guts and on our skin; the ones that infect us sometimes. That intimate juxtaposition has consequences. One consequence, unsuspected before but suggested by this Hotopp study in 2013, is that bits of naked bacterial DNA, possibly from broken-open bacterial cells, may often get integrated into cells (not necessarily germline cells) of a person's body. Into cells of the stomach lining, for instance. Or blood cells. By "integrated," what I mean is, not just absorbed or injected into the human cell but patched into its DNA. The good news about any such horizontal transfer, bacterial DNA into nongermline human cells, is that the change isn't heritable. It won't be passed to future generations. The bad news is that it might trigger cancer.

How? By disrupting the cell genome in a way that allows runaway cell replication.

Hotopp and her colleagues looked especially at two kinds of human cancer, acute myeloid leukemia and stomach adenocarcinoma. In the leukemia cell genomes, they found stretches resembling the DNA of *Acinetobacter* bacteria, a group that includes infectious forms often picked up in hospitals. In the stomach tumor genomes, they found pieces suggesting *Pseudomonas*, the genus including *Pseudomonas aeruginosa*, a nasty bug that also inhabits hospitals and medical equipment, and is especially feared for its resistance to multiple antibiotics. Bacteria have been linked to human cancer in the past—for instance, *Helicobacter pylori*, an intestinal bug associated with gastric ulcers—and the simplest hypothesis was that by causing inflammation, the bacteria damage DNA and sometimes lead to cancerous mutations. The alternate hypothesis offered by Hotopp's team, supported by genome data and now crying for further investigation, is that horizontal transfer of bacterial DNA may discombobulate one human cell, in the stomach, in the blood, wherever, and turn it cancerous. Putting horizontal gene transfer on the list of suspected human carcinogens brings it out of the realm of microbial arcana.

Even before that provocative suggestion, while she was still looking at insects, Dunning Hotopp had faced adamant resistance among a few influential biologists, including some Nobel Prize winners, to her and her colleagues' discoveries of HGT in the animal kingdom. "No, that's all artifacts, there's no way that's true," was the tenor of these responses. An artifact, in scientific parlance, is an illusion produced by a methodological mistake. "I have biologists who come into my office," she said, "and it's just, like, 'No, it's got to be an artifact. You have to be able to explain it some other way.'" Animals don't experience horizontal gene transfer, period. Humans, certainly not.

"Do you ever say to them, 'Is that a faith-based statement?'" I asked. What I meant was: it seemed almost as though the Weismann barrier had become a theological dogma.

She mused about that for a moment and allowed that some scientists did appear to be more religious about science than about religion. A touch of faith-based genomics? "I think it is," she said.

59

And yet in the background of this situation, fresh in memory both to Dunning Hotopp and to those who doubted her findings, was an episode that helps illuminate why such critics were skeptical. It involved an embarrassing overreach of scientific claims—an overreach that had put large-scale HGT into human germlines, not just into tumors, and not just into fruit flies and other insects. This intersects with Dunning Hotopp's story for several reasons, one being that the correctors of the overreach included four of her former colleagues from Craig Venter's TIGR.

The race to sequence the human genome, a bitter contest between Venter's private team and the publicly funded effort, had ended in a negotiated, face-saving tie. The public effort had combined a huge group of government- and university-supported collaborators, known as the International Human Genome Sequencing Consortium. Lots of money had been spent, and no one wanted to admit that duplication of activities had caused waste. Public access to the data, versus private proprietorship, was also at issue. President Bill Clinton announced the brokered finish at a White House ceremony, a fancy press event, on June 26, 2000. Prime Minister Tony Blair was patched in by video, because British scientists and resources had played a sizable role, after which Venter and his counterpart from the public effort, Francis Collins, made polite remarks. What the two groups had to offer at that point, though no one stressed this qualification, was just a rough draft of the genome, consisting of two more-or-less overlapping versions.

Eight months later, on February 15, 2001, the Consortium published a provisional analysis of their human genome sequence in the journal *Nature*. (Venter and his group published their own analysis, almost simultaneously, in the journal *Science*. The full sequence itself was too long to print in any journal, since it ran to about 3.2 billion bases and would have filled many book-length volumes.) Listed first among more than two hundred coauthors on the Consortium's paper was Eric S. Lander, then of the Whitehead Institute for Biomedical Research, in Cambridge, Massachusetts. Lander's priority reflected the fact that the Whitehead's Center for Genome Research, led by him, had contributed more letters of code to the final assemblage than any other participating group. His authorial priority also entitled him to a good share of the discomfort when, soon afterward, one of the paper's major conclusions was shown to be credulous and overstated, if not outright wrong.

"Hundreds of human genes appear likely to have resulted from horizontal transfer from bacteria," the Consortium authors wrote. This had occurred not perhaps in the recent past, they added, but sometime during vertebrate evolution. Hundreds of human genes? More precisely, they put the number at 223. What Lander and his coauthors were saying was that these bacterial genes had come to our vertebrate ancestors not through parentage but on a shortcut, by infective heredity. What was the evidence? The 223 suspects matched bacterial genes very closely but weren't present in certain eukaryotic creatures outside the vertebrate lineage— not in a yeast, not in a worm, not in a fly, not in mustard weed. So the 223 genes hadn't come down to us vertically, throughout roughly a half billion years of evolution. They must have come across more recently by horizontal transfer. Yes?

No. Not necessarily, said a paper by Steven L. Salzberg and three coauthors that appeared soon after the Consortium's analysis. Salzberg and his coauthors all worked within Venter's TIGR, giving a certain nuance of rivalry to their response; but it stood on its own and was published by *Science*. Salzberg himself was TIGR's director of bioinformatics, so he knew a thing or two about crunching big biological data. (One of his coauthors, Jonathan Eisen, would later be a professor at the University of California, Davis, and write an influential blog titled "The Tree of Life.") The team of Consortium authors had made two simple mistakes, Salzberg's group argued. They had failed to look at enough other eukaryotic genomes outside the vertebrate lineage for the possible presence of the supposedly leaping genes, and they had failed to take seriously the chance that those ancient genes had merely been lost from the four genomes they did look at: the yeast, the worm, the fly, and the mustard weed. Salzberg and his colleagues examined additional data and found an interesting trend: the more nonvertebrate eukaryotic genomes they scrutinized, the fewer genes seemed uniquely shared by bacteria and humans. By the time they finished, the original 223 had been reduced to 41, with a steady downward trend suggesting that further genome sequences, if available, might drop the number to zero. HGT among humans began to look like an illusion.

Other scientists, even some deeply engaged with the subject of horizontal gene transfer, found the Salzberg critique persuasive. Ford Doolittle and two colleagues wrote a comment in *Science*, calling the original claim about 223 transferred genes "the most exciting news" from the Human Genome Project so far, but concluded that it was "probably over-enthusiastic." William F. Martin, a tall American biologist at Heinrich Heine University in Düsseldorf, known for his ferocious intelligence, his important ideas, and his bluntness, called the Consortium's claim "at least an overstatement, very probably a gross exaggeration and possibly altogether erroneous." The *New York Times* took note of this "fresh skirmish in the genome wars," under a headline about HGT into humans being "Hotly Debated by Rival Scientific Camps." To the *Times* reporter, Steven Salzberg said that he'd been surprised by the Consortium's report of 223 alien genes, but that as he read further, "I was immediately struck by the fact it was likely to be an error, because the method was simply wrong." Eric Lander, also reached by the *Times*, didn't admit being wrong but declined to insist he was right.

This is how science proceeds: by fits and starts, by claims and critiques, by new answers in the light of better data. The fuss over those 223 genes was no disaster for Lander and the Consortium, no scandal, and arguably not even a major embarrassment. It was a correction—a call for greater caution and broader thinking, the oxymoronic combination that makes for genuine scientific advance. It reminded everyone that the prospect of horizontal gene transfer from bacteria (or other microbes) into the human genome is a boggling thing, a trespass on our sense of identity, and an improbability against which there should ever be a high standard of proof. But the other point worth noting about the 223 episode is that it wasn't the end of that discussion. It was the beginning.

PART ✴ VI

Topiary

60

Taw wak

The town of Embarrass, Wisconsin, is halfway between Wausau and Green Bay. In 1907, on a plot of land thereabouts, a man named John Krubsack planted some box elder saplings in a careful arrangement and began growing them into the shape of a chair.

Krubsack was a banker who also farmed (or a farmer who also ran a bank) and built furniture from driftwood as a hobby. He had decided to cultivate his "living chair" on a whim that amounted to a self-dare. His son later remembered him telling a friend: "Dammit, one of these days I am going to grow a piece of furniture that will be better and stronger than any human hands can build." In 1908 he began bending, shaping, tying, and grafting the trunks and branches of the young box elders into the configuration he wanted. The grafts took. The stems grew together in a crisscrossing pattern. Krubsack pruned away whatever growth was extraneous to the blueprint in his mind. After four years, he removed all the rooted trunks except four—the four legs of his emerging chair—and, despite that truncation, the grafted-in sections continued to meld and grow. The legs and the crosspieces and the sweeping back and the arms thickened. The structure got stronger. In 1914 he cut it free of the ground. Presumably he sat in it and enjoyed a moment of satisfaction. Success. A year later, Krubsack's chair was on display in San Francisco at the Panama-Pacific

International Exposition—the 1915 world's fair. Robert Ripley, who wrote a syndicated newspaper column titled Believe It or Not!, eventually featured the horticultural chair. Someone offered Krubsack $5,000 for the thing, but he declined. It stayed in the family. In due time, it became the totem, in a Plexiglas case, of a furniture company back home in Embarrass, Wisconsin.

John Krubsack hadn't invented any novel or esoteric techniques in creating his chair, though the concept and the execution were clever. Grafting was routine in the horticultural realm, and it's still practiced today. A fruit tree is generally grown from rootstock of one type, onto which an upper section (or scion) of another type is grafted. The upper is fitted into the lower like a splicing of ropes. You make the cuts, insert one stem into another so that their cambium layers (containing the vascular plumbing) are in contact, wrap the area with tape, and wait. The rootstock might be selected for hardiness, to resist drought or disease, or maybe for dwarfism, so that the tree doesn't grow too tall. The scion is selected for the kind and quality of its fruit. Grapefruit may be grown on rootstock of orange; commercial pears are often grown on rootstock of quince. When the cambium layers make contact, and the vascular systems merge, the graft has been successfully achieved. Water and nutrients from the rootstock can now flow up into the branches. Carbohydrates produced by photosynthesis can now flow from the leaves to the rootstock. Two trees have become one.

A form of natural grafting occurs even in the wild—though rarely. It's called inosculation. From a Latin verb, meaning "to kiss." When the limbs or the trunks of two trees rub together, scraping away bark, creating two raw spots, cambium to cambium, sometimes those layers smooch and fuse. Dense growth, competition, and wind-driven rubbing can be the causes. It's unusual, but it happens. What doesn't happen, or only with very greatest rarity, is that two branches on the *same* tree inosculate. Branches on real trees diverge, reaching outward and away for light. Limbs diverge. I've said it already, and I'll repeat: not everything that rises must converge. Limbs on an oak don't. Branches on a cottonwood don't. Twigs on a sycamore don't.

That's the difference between actual trees and phylogenetic depictions. And so the tree of life concept became ever less satisfactory, ever more challenged, as new evidence of horizontal gene transfer continued to accumulate during the 1990s, because "tree" just didn't suggest the right shape. There's something spooky and unnatural about any tree whose limbs grow together rather than branching apart. Believe it or not.

Edward Hitchcock, a self-styled "Christian geologist," in the 1840s and 1850s offered pre-Darwinian, non-evolutionary trees of life. *The Edward and Orra White Hitchcock Papers, box 24, folder 21, Amherst College Archives & Special Co.*

Ernst Haeckel (seated), zoologist and artist, en route to the Canary Islands on a collecting trip, with an assistant, 1866. He drew robustly Darwinian trees. *Photos.com.*

Fred Griffith, a "shy and aloof" medical microbiologist, in the late 1920s discovered a mysterious transformative "pabulum." *Photograph by Alvin F. Coburn. Courtesy of the National Library of Medicine.*

Oswald Avery, "the Professor," along with his young colleagues Maclyn McCarty and Colin MacLeod, showed in 1944 that Fred Griffith's transformative "pabulum" was DNA. *Courtesy of the Tennessee State Library and Archives.*

At the National Collection of Type Cultures, in Porton Down, England, a microbiologist checks vacuum-sealed ampoules of freeze-dried bacteria. Within these glass tubes, a sample can remain viable for upward of fifty years. *Courtesy of NCTC, Public Health England.*

Joshua Lederberg found evidence of horizontal gene transfer among bacteria, through viral infection, and in 1952 gave it a name: "infective heredity." *Image S10965, University of Wisconsin-Madison Archives.*

Linus Pauling, Nobel Prize–winning chemist, with his coauthor Emile Zuckerkandl in 1965, tossed off a suggestion: that the tree of life could be discerned from information in long molecules. *Courtesy Ava Helen and Linus Pauling Papers, Oregon State University Libraries.*

Barbara McClintock, who studied the genetics of corn, won a Nobel Prize for her discovery of mobile genetic elements, genes that change positions on chromosomes— and, sometimes, leap between species. *Smithsonian Institution Archives. Image # SIA2008-5609.*

Carl Woese annotating an RNA "fingerprint" at the light board, a tedious but profoundly revealing chore. After long hours and many months, he sometimes said to himself: "Woese, you destroyed your brain again today." *Courtesy of Ken Luehrsen.*

Woese trusted his friend Charlie Vossbrinck, an entomologist, to take a series of portraits with an old-fashioned Linhof camera. One was memorable. *Courtesy of Charlie Vossbrinck.*

Woese, Ralph Wolfe, and Otto Kandler, after a landmark conference on the Archaea, celebrated atop a mountain in the Bavarian Alps. *Photo by G. Kandler.*

George Fox, a rangy young postdoctoral fellow in Woese's lab, coauthored the 1977 paper announcing a third kind of life. *Courtesy of Ken Luehrsen.*

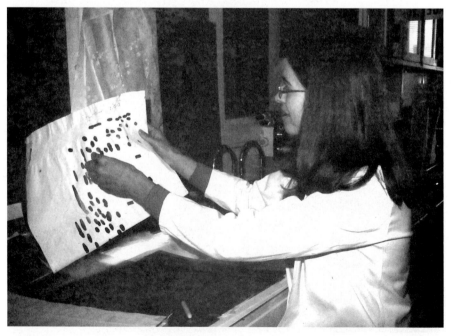

Linda Bonen, a technician in Woese's lab, who performed crucial tasks toward the RNA fingerprinting work, carried her skills to Ford Doolittle's lab in Halifax. *Courtesy of Linda Bonen.*

Ford Doolittle, who helped confirm the theory of endosymbiosis, later found that horizontal gene transfer vastly complicates Darwin's proposition that the history of life is shaped like a tree. *(Photographer/source unknown)*

Lynn Margulis paused at the Grand Canyon of the Yellowstone River, during a 2010 field trip, with Edward O. Wilson (in borrowed ranger hat), physicist Eric D. Schneider (wearing cap), and the author. *Courtesy of George Keremedjiev, Bozeman, MT.*

Thierry Heidmann, at the Gustave Roussy Institute, south of Paris, has led work revealing the vital role played by a viral gene captured within mammal genomes in enabling pregnancy. *Courtesy of Thierry Heidmann.*

Woese took great comfort and pleasure from a few close friendships. At the home of Harry Noller, a renowned expert on ribosomes and a onetime professional jazz musician, Woese sometimes relaxed by playing the piano. *Courtesy of Harry Noller.*

61

The earliest tree of life reflecting any such anomaly was probably Constantin Merezhkowsky's sketch of evolution via symbiogenesis, published as an illustration to one of his papers, in 1910. Yes, the crazy Russian pedophile again. This tree depicted the crossover of bacteria into the eukaryotic lineage, adding chloroplasts to complex cells, as a stream of dotted diagonal lines, from one limb to another. Lynn Margulis suggested the same thing, also with dotted lines, in the cartoony sketch that served as frontispiece to her 1970 book *Origin of Eukaryotic Cells*. Besides the chloroplast crossover, she showed two others, representing the bacterial origin of mitochondria (about which she was right) and of undulipodia (the little tails, about which she was probably wrong). Although these cases of endosymbiosis weren't horizontal gene transfers in the narrow sense—the sense in which HGT is now mainly understood—in a broader sense they were, carrying whole bacterial genomes into the eukaryotic lineage, where they assumed new functions and created new possibilities. But those fateful endosymbioses (whether you count two or three) represented anomalous events in the very distant past. They were rare conjunctions, not instances of an ongoing process; and whether for that reason or others, they didn't cause scientists to reconsider the whole tree metaphor.

Carl Woese, for instance, ignored the crossovers entirely in his "universal phylogenetic tree," offered with Kandler and Wheelis in 1990. Basing that tree on his favorite molecule, 16S rRNA, and on its eukaryotic equivalent, 18S, he felt no need to consider where and when other genes might have moved sideways. So the Woesean tree portrayed divergence without any hint of convergence. But then, during the later 1990s, ideas about horizontal gene transfer in evolutionary history changed drastically—and so did the illustrations by which those ideas were portrayed.

One factor driving such changes was the dramatic improvement in DNA sequencing methods and tools, which brought an explosion of new genome data. The dangerous, toxic, and laborious steps by which Woese and his team had deduced the sequences of a few handfuls of RNA fragments back in the 1970s, ingenious as they were, looked like a Stone Age campfire compared with the streamlined, automated operations of the mid-1990s. The Human Genome Project, begun in 1990 and conducted by that huge government-and-university consortium I've mentioned, helped catalyze the technological improvements with big money, medical incentives, and the endless fascination of us humans with ourselves. The race against Craig Venter's private group helped too, as competition put a premium on speedy and (for Venter's group) cost-effective production of genome data. This led to fancy new machines and clever shortcuts. (Carl Woese himself acquired one of the new machines—an ABI 370A, made by Applied Biosystems, state of the art in 1986—but Woese's lab people could never make it work.)

Applying those methods and tools to the sequencing of nonhuman genomes, both as practice exercises and for the sake of pure science, was a side benefit. And with every passing year, genome sequencing became not only faster and more accurate but also cheaper. Another constraint on the huge task of sequencing genomes, besides technical difficulty and cost, was computer capability. Speedy computers with lots of processing power were necessary for assembling a large genome and analyzing what was there. As computers became so much faster, and methods of applying them to genome assembly improved, that constraint fell away too.

Microbial genomes are much smaller than the human genome, and in those early days of automated sequencing, small genomes seemed less daunting but offered proof of principle. Venter's method, known as whole-genome shotgun sequencing, involved detecting the sequences of randomly grabbed fragments, enough when totaled to comprise the whole genome, and then putting those fragments together according to how their sequences overlapped. It was the jigsaw-puzzle approach: *Look, here's a piece of blue sky, and it seems to match* that *piece of blue sky, so let's see if they fit together. Yes!* This was faster than the Consortium's method, and so in 1995 Venter and his colleagues at TIGR, along with other partners from Johns Hopkins University and elsewhere, published the first complete genome of a free-living organism (that is, something larger and more complex than a virus). It was the bacterium *Haemophilus influenzae*, the same bug that I saw at Porton Down in a strain cultured from Alexander Fleming's nose. Venter and his group found that genome to be 1,830,137 letters long, and, to the best limits of their method, they identified every letter. Their report made the cover of *Science*.

Another big event came the following April, when a different team announced its success in reading the genome of brewer's yeast. Brewer's yeast may not sound thrilling to you or to me—it's not charismatic megafauna—but it is a eukaryote, and no complete eukaryotic genome had ever before been sequenced. The brewer's yeast genome was therefore closer to the human genome than any other whole-genome sequence yet produced. It was also somewhat larger than the average genome of bacteria. Those distinctions, plus the nervous atmosphere of the race toward the human genome, with Venter's gang in one lane and the Consortium in another, may account for why this third team, a far-flung international group, made its announcement by press release well before the formal report could appear in a scientific journal. *Hey look, we've done a eukaryote!—and we'll publish details pretty soon.* The pace of discoveries and the fervor of competition were picking up.

Just four months later, in August 1996, Venter and a large team of collaborators captured attention again, publishing the first complete genome of a member of the Archaea, the third of Carl Woese's three domains. This bug was *Methanococcus jannaschii*, a heat-loving and methane-producing microbe first isolated from a sediment sample taken at the bottom of the Pacific Ocean. That sample was scooped up by a robot submersible driven along the sea floor near a thermal vent, more than eight thousand feet deep near the Eastern Pacific Ridge. Like most other archaea known at that time, it was a strange little creature from an extreme environment. Woese himself appeared as an honored senior coauthor (second from last

in a list of forty, just before Venter) on the paper announcing the achievement, again in *Science*. He had persuaded Venter to undertake this piece
of work. Gary Olsen, one of his young and ingenious collaborators at
Urbana, was also among the coauthors, but otherwise the list comprises
mostly Venter people from TIGR. It must have been a bittersweet event
for Woese, when the first archaean genome came from Venter's institute,
not his little lab, with its nonfunctioning ABI 370A.

The *M. jannaschii* genome ran to 1,739,933 letters, including 1,738 sections that seemed to be genes. Of those genes, more than half were entirely
new to science, with no equivalents ever before seen in any other form
of life. That degree of uniqueness went far to confirm what Woese had
been saying since 1977 and what some hidebound scientists had resisted:
that these archaea were a separate form of life. The old two-kingdom
paradigm had now been "shattered," according to one eminent microbiologist, asked to comment by *Science* for a news story accompanying the
report. "It's time to rewrite the textbooks."

Ford Doolittle agreed. "This completes that basic set," he told *Science*,
meaning the triad of whole-genome sequences for a bacterium, a eukaryote, and an archaeon, "and so it will certainly have a major impact."

One impact it helped deliver, like the blade of an axe, was on the very
idea of the tree—particularly on the tree as Carl Woese had drawn it,
using ribosomal RNA as the definitive signal of life's ever-diverging history. Whole-genome sequencing of that first bacterium, then that first
archaeon, and then other organisms revealed more and more instances
of horizontal gene transfer, confusing the picture and inosculating limbs.
Within another two years, by 1998, more than a dozen microbial genomes
had been sequenced and another from a eukaryote, a nematode worm.

Scientists who inspected those genomes found puzzling mixtures of
bacterial genes and archaeal genes within single genomes, as though part
of a deck of tarot cards had been shuffled into a poker deck. Occasionally
a few bacterial or archaeal genes even turned up in eukaryotes. Another
reporter from *Science*, Elizabeth Pennisi, picked up the story and described the growing perplexity. By that point, even Woese was pondering
horizontal gene transfer, though in his view, it was a phenomenon limited
mostly to the earliest era of evolution, when cellular life was just taking
shape and there weren't yet any distinct lineages or species. A blurry time,
and HGT back then was part of the blur. Pennisi spoke with Woese for

her report, quoting him to the effect that "you can't make sense of these phylogenies because of all the swapping back and forth." He probably meant "all the swapping" during that earliest time, but the distinction didn't make it into Pennisi's article.

Another source, a molecular geneticist named Robert Feldman, who had just helped sequence another bacterium and found its phylogenetic relationships ambiguous, voiced his dawning distrust of Woese's faith in an rRNA tree. Feldman noted that "you get different phylogenetic placements based on what gene is used"—different genes yield different trees of relatedness—and he explained why: "Each gene has its own history." If each gene has its own history, Woese was wrong to draw such grand conclusions from a single molecule, however fundamental, and sketching the course of evolution as a single neat image was impossible. Pennisi saw that. Her piece ran in May 1998 under the headline "Genome Data Shake Tree of Life."

62

Ford Doolittle absorbed this new line of thought gradually. He was skeptical of the notion that horizontal gene transfer might have played—might still be playing—a huge, unsuspected role in the history of life. Yes, he could see, it was a mechanism for the spread of antibiotic resistance from one kind of bacteria to another. But beyond that? Did it explain why certain other genes, more basic to the functioning of single-called organisms, were turning up on what seemed the wrong limbs of the tree of life? Those anomalies multiplied as more individual genes were sequenced, and then whole genomes, but alternate explanations existed for why a bacterial gene, or what looked like one, might show up in an archaeon, and vice versa. The alternate explanations didn't require such extraordinary leaps. They weren't so dramatic and counterintuitive. Horizontal gene transfer still seemed deeply improbable, a rare sort of event, and Doolittle remembers calling it "the last resort of the impoverished imagination."

Unlike some scientists, Ford Doolittle abides in a zone of detached bemusement and pure curiosity that leaves him comfortable admitting when he was wrong. It's good scientific etiquette, arguably even the scientific ideal: you hypothesize, you test against data, you correct your views where necessary, regardless of ego; you hypothesize again. When you've goofed, and you need to backtrack, you admit it. Doolittle practiced that. He started revising his view of HGT under the influence of two colleagues and all the new data those colleagues helped him consider. One of the two was a postdoc in his own lab.

James R. Brown came east to Halifax after finishing a PhD at Simon Fraser University in British Columbia, where he worked on molecular evolution and the population genetics of sturgeon. Brown had always been a fish-loving kid, growing up in Ontario with aquariums full of cichlids and angelfish and an interest in marine biology. He snorkeled cold Ontario waters during summer vacations, watched Jacques Cousteau on TV, and read books about the sea. After his undergraduate degree in marine biology, he worked as a marine field technician and scuba diver in the Great Lakes and the Arctic for the Canadian government before going back to school. The group of sturgeon he studied for his PhD dissertation includes some fascinating fishes: long-lived animals with primitive traits and recognizable ancestors dating back more than two hundred million years. Brown focused his doctorate on mitochondrial DNA as a gauge of genetic diversity among their populations. In the process, he learned a bit about using molecular data to draw phylogenetic trees. Bringing those skills to Doolittle's lab, he joined his boss on a series of projects and publications, during the 1990s, that involved not sturgeon populations but the molecular phylogenetics of bacteria, archaea, and eukaryotes.

Brown's project involved the question of how to root the universal tree: Should it be midway between the bacterial limb and the archaea limb, with eukaryotes branching secondarily from the archaea? Or not? And could it be done on the basis of a single fundamental gene? They investigated one gene, for instance, that seemed to have gotten into certain bacteria by sideways transfer from archaea. Other genes, they knew, gave other answers to the rooting question, and they looked at several related genes for a possible clue. They recognized that horizontal gene transfer, with genes leaping sideways from limb to limb, might be what had made the rooting question (among others) so hard to answer. "Extensive gene transfer," Brown and Doolittle wrote, with two other colleagues, "may have played such an important role in early cellular evolution as to jeopardize the very concept of cellular lineages." Meaning: *Damn it, Maybe there is no single tree of life. Or if there's a picture of life's history, maybe it doesn't look like a tree.* Ford Doolittle was coming around.

Another influence on him was a colleague named Peter Gogarten, a German-born scientist who had trained in plant physiology, come to America in 1987, and shifted into molecular studies of early evolution. Gogarten, since his arrival in the States, had his own interesting experience with Carl Woese and the three domains. As a postdoc in California, he devised a method, in collaboration with his lab leader and others, of determining where the tree of life should be rooted. Their answer: at the base of a trunk that rose into two major limbs, one representing bacteria, the other leading to everything else. Either for that reason (two major limbs, not three) or others, Woese didn't care for Gogarten's paper and omitted it pointedly (despite its clear relevance) from the citations in his landmark 1990 paper with Kandler and Wheelis. Too bad for young Peter Gogarten, who was a new assistant professor at the University of Connecticut, needing some further publications and recognition if he were to get tenure.

Gogarten's relations with Ford Doolittle, as the two men converged on the subject of horizontal gene transfer, were more sunny. Their interaction began when Doolittle invited Gogarten to Halifax in 1994 for private discussions and to give a seminar. Gogarten talked about gene transfer from archaea to bacteria, among other things, and the idea that a "net of life" might better represent evolutionary history than a tree of life. Two years later, they both attended a big microbiology meeting held at the University of Warwick, in England. Doolittle's talk to that meeting stands less vividly in his own memory than the one by Gogarten, who spoke again about horizontal gene transfer: genes moving sideways across the great microbial divide between bacteria and archaea. There seemed to be so much of such transference, newly discovered, Gogarten said, repeating his Halifax point to a wider audience, that the phylogeny of species didn't look like a tree anymore. Not during the early phase of life on Earth, anyway. It looked more like a net. Evolution was "reticulate" as well as branching. Ford Doolittle was listening closely and inclined to agree.

Back in Halifax, Jim Brown coaxed Doolittle in the same direction, and the flow of new sequencing data tended further to tangle the tree. In 1997 Doolittle and Brown did a big tree-building effort together, looking at sixty-six different proteins that are essential to all forms of life, and at the different variants of those proteins as reflected in more than 1,200 different gene sequences, from a wide variety of bacteria, archaea, and eukaryotes. Most of these sequences were publicly available; Brown and Doolittle downloaded them from databases and subjected them to comparative analysis. They constructed an individual tree for each of the sixty-six proteins, showing how it had evolved into distinct variants within different lineages of creatures. Each protein had a name, of the sort that you or I could scarcely pronounce, let alone remember: tryptophanyl-tRNA synthetase, for instance. One version of that exists in humans, another in cows, another in the bacterium *Haemophilus influenzae*—each distinct but fundamentally the same protein. Why is the stuff so universal? Because it's a very basic tool, necessary to all forms of life for the role it plays in linking amino acids with code triplets during the translation process. The other sixty-five proteins chosen by Brown and Doolittle included some involved in DNA repair, some involved in respiration, some devoted to metabolism, structural proteins for ribosomes, and more. Brown and Doolittle compared the variants, constructing an independent tree of descent for each. They printed all sixty-six trees in their published paper, so that the paragraphs of text seemed to exist within a forest—or at least a very green suburb. This exercise yielded a telling point: the trees didn't match.

There was a lot of disagreement. Many of them were incongruent with one another—sprouting different branches in different places—and incongruent also with the supposedly canonical 16S rRNA tree of Carl Woese. The logical conclusion was that genes have their individual lineages of descent, not necessarily matching the lineage of the organism in which they are presently found. It was the same thing Robert Feldman would soon tell the reporter Elizabeth Pennisi: "Each gene has its own history." How was that possible? Horizontal gene transfer. While the creatures replicated themselves vertically—humans producing humans, yeast producing yeast, *Haemophilus influenzae* producing more *Haemophilus influenzae*—sometimes the genes went sideways. They had their own selfish interests and opportunities.

63

By 1998, Ford Doolittle had become one of the go-to experts for science journalists seeking comment on new developments in this field. He had a distinguished research record, he knew all the issues and most of the players, he would answer his phone, and he could turn a phrase. Reporters from the journal *Science* particularly favored him, including the one who covered Craig Venter's first whole-genome sequencing, of a bacterium, and the one who reported on the archaean sequencing a year later. Elizabeth Pennisi too, in her piece headlined "Genome Data Shake Tree of Life," quoted Doolittle several times. Then in 1999 he heard from the editors of *Science* with a slightly different request.

They were preparing a special issue on the subject of evolution. It would offer a package of articles, each one a wide-angle review of some realm of biology, seen from an evolutionary perspective, and the authors would include eminent figures such as Stephen Jay Gould and David Jablonski, as well as lesser-known scientists. Their reviews would cover a range of topics and scales, from nucleic acids structure to dinosaurs. The editors asked Doolittle if he would kindly recommend someone to write on evolution and microbiology. He replied: How about me?

They agreed, not knowing quite what would come. Doolittle himself thinks that the *Science* editors were caught by surprise, and that they said yes because saying no would have seemed rude. Grudgingly, they accepted what he calls "my self-promotion."

"I wanted to write about this topic," he told me years later—"this topic" meaning horizontal gene transfer and the tree of life. He felt impelled to produce a sort of manifesto. "And they weren't so keen on that. But they didn't really . . . I think they didn't see how they could back down." If he was authority enough to comment on articles and recommend authors, why not authority enough to write this review himself, and to choose those aspects of microbiology that were currently most interesting and important?

So the issue of *Science* for June 25, 1999, contained a review article by Doolittle entitled "Phylogenetic Classification and the Universal Tree." It became the most provocative paper he ever published. It put horizontal gene transfer in the center of a new discussion. And part of what made it arresting was not just Doolittle's words but also his drawings. He later told me that, while he was glad the editors of *Science* published his self-invited paper, he was pleasantly surprised that they accepted also those hand-sketched trees. "Normally, journals wouldn't do that. It must have been a flat day at the office, or something."

He began his text with a long view. "The impulse to classify organisms is ancient," he wrote, "as is the desire to have classification reflect the 'natural order.'" Histories of biology told that story, going back to Aristotle and forward to Linnaeus. But putting "natural order" in quotes was Doolittle's first hint of how ambiguous such an order could be. His purpose in this paper was to follow those ambiguities to their logical extreme.

He described the rise of evolutionary phylogenetics and reprinted the branching figure from Darwin's *Origin*, noting that this (and Darwin's passage about the simile of "a great tree") was what brought tree imagery into evolutionary thinking. He noted the big change in modern phylogenetics: the change from morphology to molecular evidence, yielding a whole new dimension of discoveries. He mentioned endosymbiosis theory and its two key tenets—about mitochondria and chloroplasts entering the eukaryotic lineage as captured bacteria—and that those tenets had been confirmed by molecular data. He highlighted the role of Woese, who seized on ribosomal RNA as the sole basis for a three-domain tree of life. Doolittle even drew a picture: a cartoon of the Woesean tree, with thick branches rising upward in three main clusters, labeled "Bacteria," "Eukarya," and "Archaea." Each branch was topped as an arrow, aimed vertically into the future. In addition to those vertical arrows, two diagonal ones jabbed sideways: from the early bacteria into the early eukaryotes, representing those two momentous instances of endosymbiosis, the origin of chloroplasts and the origin of mitochondria. He called it the "current consensus" model rather than simply "Woese's." Then Doolittle asked: How true is it?

His answer: probably not true enough. And the problem was horizontal gene transfer. Microbiologists had long been aware of HGT, going back through Joshua Lederberg to Oswald Avery and others; but for phylogeneticists, drawing their pictures and diagrams of vertical descent, it presented greater difficulty. Doolittle was largely addressing the latter group, scientists concerned with tracing phylogeny. If the new evidence proved correct, he wrote, and HGT was not a rarity but a rampant phenomenon, at least among bacteria, archaea, and the early eukaryotes, then Woese's tree—now the consensus model—was badly wrong and incomplete.

He mentioned some of that new evidence, citing Peter Gogarten, James Lake, Sorin Sonea, and others. For instance, two researchers looking at the "molecular archeology" of E. coli, the most-studied bug in biology, had just reported an unexpected finding: that its genome contains at least 755 genes acquired by horizontal transfer, accounting for 18 percent of its chromosomal DNA. Those transfers had occurred not during early evolution, furthermore, but more recently, giving E. coli adaptations it wouldn't otherwise possess. William F. Martin, the brilliant and blunt American based at Heinrich Heine University in Düsseldorf, had noted that there was "something quite ominous" about these E. coli results. If so many "relatively recent" transfers had gone into one bacterium, Martin wondered, then how many horizontal transfers had occurred among the entire bacterial domain throughout the depths of geological time? The rough answer was "countless." Martin made his comments in a paper titled "Mosaic Bacterial Chromosomes: A Challenge En Route to a Tree of Genomes," published not long before Doolittle's review. All this sideways transfer raised a challenge, Martin warned, to delineating a genomic tree of life. Doolittle, following similar logic and embracing the challenge, asked: What might the tree look like? Again he drew a sketch, and, to his surprise, the editors of *Science* printed it.

He called this one "a reticulated tree." It was a tangle of rising and crossing and diverging and converging limbs. It had its antecedent (as Doolittle acknowledged) in a somewhat similar figure offered by Martin in his "Mosaic" paper. Martin's tree resembled a sea fan, one of those delicate structures built by a colony of coral-like animals on the ocean floor. Its long limbs and branches rose wavily, diverging from a simple base. It was done in color, pastel shades, and some of the slim branches also converged: turquoise and lavender joining to form purple. Doolittle's sketch, by contrast, was in black and white and bulkier, as drawn freehand with a blunt pen; more robustly entangled from the bottom up, it resembled a mangrove thicket, if mangrove limbs could inosculate. It was so intricate and yet so fluid, so quirky, that it seemed almost funny. It looked like something John Krubsack might have grown, on a bet, from a cluster of box elder saplings in a field near Embarrass, Wisconsin.

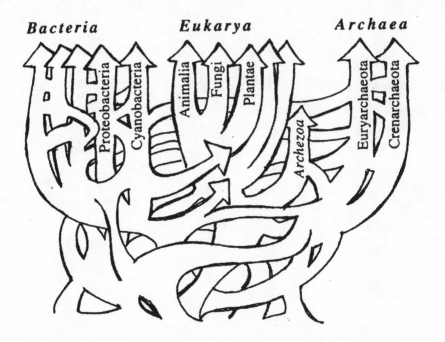

Doolittle's reticulated tree, as drawn by Doolittle, 1999.

And yet Doolittle's second figure was still much too simple: another cartoon. It rose by multiple trunks from multiple roots, then split into multiple limbs, but not so very many, and the roots weren't identified. It conveyed paradox, but it lacked detail. It was suggestive, not literal. It was eloquently weird. Maybe, Doolittle said in his text, as well as with this drawing, the history of life just can't be shown as an ordinary tree.

64

The 1999 paper was a landmark in several ways, one being its influence in getting horizontal gene transfer taken seriously. "That had a huge effect," William F. Martin said later. "It broke the dam." Suddenly HGT seemed a mainstream idea, an ongoing process of major importance at least in microbial evolution—something that had to be considered and discussed—rather than a hallucination, an artifact, or a quirk.

Another way of seeing Doolittle's 1999 paper, more personal, was that it marked the end of the friendship between him and Carl Woese. That friendship had already become strained several years earlier over a single word: *prokaryote*. Doolittle persisted in using it, to label the category in which Bacteria and Archaea could be grouped together, and Woese hated it, because his greatest discovery implied that those two never *should* be grouped together. So when Doolittle wrote about prokaryotes versus eukaryotes, in the old sense that Roger Stanier had used those terms, to Woese it seemed like an insult and a taunt. Then came this review paper of 1999, with Doolittle in his speculative tone posing an even more direct challenge to the central premises of Woese's career. Was it true that 16S rRNA (and its equivalent in eukaryotes, 18S) is a uniquely stable molecule, too fundamental in its role within each cell to be subject to horizontal transfer? And was it true, therefore, that those rRNA molecules constitute evidence for a uniquely definitive tree of life? Woese said: Yes. Doolittle said: Hmm, possibly not, and, in fact, maybe there *is* no definitive tree of life.

The historian Jan Sapp, so close to Woese in the later years, told me that Woese felt betrayed, harboring darkly imagined theories about why Doolittle had turned against him. Woese had known Doolittle when Ford was a young postdoc in Urbana, after all, and they had drunk beer together; Woese had sent Linda Bonen, with her crucial skills, to Doolittle in Halifax, making possible some of his best early work; the two men had shared many curiosities and ideas. This amicable history seems to have brought a bit of "Et tu, Brute?" into Woese's response. Science sometimes gets petty and emotive, as I've noted, and maybe especially so if you perceive yourself an isolated, neglected genius at a state university in downstate Illinois. Then again, if there was a "Woese's Army," in that phrase Lynn Margulis favored (and Woese detested), Doolittle now counted himself AWOL. But the reason for his departure from those ranks, notwithstanding Woese's dark theories about jealousy and betrayal, was simple: new data. New genomic evidence of horizontal gene transfer blurred the picture of Woese's three domains.

The rift between the two men was further exacerbated less than a year later, when Doolittle published a popularized version of his recent thinking in *Scientific American*. That magazine carried less authority than the journal *Science* and served a different purpose—not announcing discoveries but explaining them for a mostly lay readership. Doolittle's draft article was heavily edited and rewritten by *Sci Am* editors, so that it contained his ideas (and ideas from some of his colleagues, such as William F. Martin) but not his voice. And not his drawing hand: in place of Doolittle's pen-sketched cartoons of tangled limbs and roots, *Scientific American* offered smoothly professional (and more sterile) airbrushed figures. As for the title, that was iconoclastic even by Doolittle standards: "Uprooting the Tree of Life."

Horizontal gene transfer was "rampant," he told the readers, and it had affected the course of evolution "profoundly." Scientists already knew that bacterial genes sometimes move sideways, yes, carrying "the gift of antibiotic resistance" or other special adaptive traits. Bacterial geneticists such as Martin and his mentors were well familiar with that phenomenon. But most researchers concerned with evolutionary history and phylogeny, such as Woese and his followers, for whom HGT seemed rather startling, had assumed that a stable core of other genes—genes essential to cell survival, basic genes for metabolism and replication—remain firmly embedded within their original lineages and are inherited vertically, rarely if ever traded horizontally. "Apparently," Doolittle wrote, "we were mistaken."

"By swapping genes freely," he wrote, early cells "had shared various of their talents with their contemporaries." Eventually this ragout of changeable cells and interchangeable genes coalesced and differentiated into three major domains as we know them today. (He meant Bacteria and Eukarya and Woese's third, Archaea.) Even after that differentiation, horizontal transfer continued throughout another couple billion years of history and into the present, especially *within* each domain but even sometimes *between* one domain and another. "Some biologists find these notions confusing and discouraging," Doolittle granted. "It is as if we have failed at the task that Darwin set for us: delineating the unique structure of the tree of life. But in fact, our science is working just as it should." How so? Because the tree itself was always just an "attractive hypothesis," Darwin's hypothesis, for the shape of life's history. Scientists were now testing that hypothesis against fresh data, genomic data, and if need be, Doolittle concluded cheerily, they would reject it and find a new one.

The *Scientific American* article was published in February 2000. During the next few years, as evidence of horizontal gene transfer continued to pile up, Doolittle became ever more engrossed in trying to gauge its significance. He read the papers announcing new data, and the analyses, and talked with (or argued against) other scientists at meetings. He found common ground with three men in particular: Peter Gogarten, William F. Martin, and Jeffrey Lawrence, a genome biologist at the University of Pittsburgh. At the end of his 1999 paper, he had thanked Gogarten and Martin for "persuading me of the importance" of HGT. In 2002 he invited Gogarten and Lawrence up to Halifax for a weekend of brainstorming. While they were in town, Doolittle "locked" the three of them in his laboratory, where they sat at the big wooden table in his office and wrote a paper together.

All three believed that horizontal gene transfer was the new elephant in the room. They had agreed on that much during earlier conversations. If you weren't thinking about HGT, Jeffrey Lawrence told me, when I visited his lab in Pittsburgh, "you weren't thinking about what you were seeing." But no one had deeply considered its implications. "You have to move beyond the collecting phase," Lawrence said, meaning the phase of amassing data without analysis. "Or, 'Here's a case, here's a case, gosh, there's a lot of horizontal gene transfer.'" The big question, he said, was: What might it mean in the history of evolution?

I had asked Gogarten a few months earlier: How do three people write a paper in a weekend? Does one sit at the computer while the other two . . .

"No, we all sat at computers," he said. They had their laptops. "We discussed things. We had a kind of discussion outline, a draft. And then, yeah, we each went to our computers and wrote sections on it. Did some computations." Gogarten's computations led him to generate a few trees—trees of descent for individual genes, and one tree sketching the history of whole organisms. Lawrence nixed the tree of organisms. "We can't do that," he said, by Gogarten's recollection. "We can't write a paper on gene transfer and use a tree to illustrate that." Lawrence's point being that, with genes flitting sideways, there *was* no tree that could depict the history of whole organisms. It wouldn't just be wrong. It would be self-contradictory; it would be nonsense. So they included no tree in this manuscript—not Martin's colorful sea fan, not the *Scientific American* sort of thing, not even one of Doolittle's fluidly hand-drawn tangles.

They focused on "prokaryote" evolution, using the old word for bacteria and archaea, lumped together, that so aggravated Woese. Among such simpler microbial creatures, they wrote, horizontal gene transfer is far more important, in quantity and consequences, than imagined previously. Its impacts could be understood in four ways. First, new genes received by sideways transfer, from a different lineage or species, may allow a population of microbes (the recipient bug and its offspring) to colonize an entirely new ecological niche. Second, it may allow organisms to acquire a new sort of adaptation abruptly, without passing through the dangerous stage of being only half adapted to one situation or another. Third, this transformation happens fast compared with incremental mutation, which proceeds slowly. Fourth, HGT is a "font of innovation," bringing drastically new genetic possibilities, new supplies of variation, on which natural selection can act. All four kinds of impact are interrelated and represent overlapping perspectives on the same phenomenon.

Take those four together, and you have a strong case, they wrote, that horizontal gene transfer might be the "principal explanatory force" in prokaryote evolution. Darwin's natural selection is still there, but it's operating on a much different supply of variation from a much different source than imagined previously. What the three authors meant to show in this paper, they stated, was that recognizing the role of horizontal gene transfer led toward "a broad and radical revision" of the old paradigm. The old paradigm they meant was that microbes conform to Darwin's theory. But this should be a revision, they emphasized, not an outright rejection. They proposed "a synthesis" of old and new perspectives, acknowledging that gene transfer is horizontal as well as vertical, that life's history looks both "weblike and treelike," and that adaptations can evolve by "many modes," not all of which were discernible to Charles Darwin in 1859. Their paper appeared in December 2002.

In the meantime, that fourth figure, William F. Martin, had continued his own challenge to the conventional tree of life. Ford Doolittle sometimes speaks jokingly of "the four horseman" of horizontal gene transfer—the four scientists who most loudly declaimed its importance around the turn of the twenty-first century—and when he does that, he counts Bill Martin, along with Gogarten and Lawrence and himself. (Jeffrey Lawrence asks: Which horseman am I, Pestilence?) But for some reason, or no reason, Martin didn't participate in the "locked-in-a-lab" coauthorship weekend. Was he not invited, not interested, not available? Was there bad chemistry with one of the other horsemen? Martin, as I've mentioned, has a reputation for being smart, confident, forceful with his views, and sometimes startlingly brusque when disputing points of science in public. Then again, sheer geography might have been enough to keep him away from that Halifax confab. Martin lives and works in Düsseldorf, an ocean away. That's where I went to see him, curious for his thoughts, respectful of his work, but wondering whether I'd get a blast of his famous gruffness.

65

Bill Martin's lab is in the Department of Molecular Evolution at Heinrich Heine University, in the heart of Düsseldorf, not far from a great bend of the Rhine. I arrived early, was ushered through by his secretary, and sat waiting in his inner office while Martin finished participating in a PhD examination. That left me time to browse the books on his shelves, gaze at the arcane scribblings on a large standing flipchart, and notice the poems, cartoons, and other whimsical décor on the back of his door, including photographs of notable predecessors and colleagues, such as Constantin Merezhkowsky and Ford Doolittle. There were also framed photos of two smiling young daughters. Martin appeared at ten thirty, a large man, large enough to have played lineman during his time at Texas A&M University, though he did not. He greeted me with a robust Texas handshake, we sat at his table, and he talked for two hours, almost without need of prompting by questions from me.

The beginning of it all, for him, was endosymbiosis—hence the photo of Merezhkowsky. Martin was a young student in botany, hoping to run a nursery someday, and not yet aware that, as he told me, "if you want to have your own nursery, you don't study botany, you study business." He soon discovered, anyway, that he was more interested in research than in selling plants. "I just wanted to know about things." When he took microbiology, in 1978, the professor made two statements that gripped Martin's attention. "The first thing he said was, 'We used to isolate insulin from pig pancreas, and now we can take the gene, put it in *E. coli*, and make it in buckets.'" That was a glimpse of the prospects of genetic engineering. But the real fascination, for this student, lay in pure science. "The other thing he said is, 'And there are some people who believe that chloroplasts used to be free-living cyanobacteria.'" That was Martin's introduction to the theory of endosymbiosis—the idea that complex cells arose by capturing bacteria and converting them to internal organelles.

The theory was still in disrepute. Lynn Margulis had revived it, from the writings of Merezhkowsky and Ivan Wallin and others; Ford Doolittle and his colleagues had delivered some of the first molecular evidence to confirm it; further gene sequencing, of the slow and laborious sort, was adding support; but widespread acceptance hadn't yet come. As Martin reminded me, back in 1925 endosymbiosis had been called "too fantastic for present mention in polite biological society," and to many biologists, it *still* seemed too fantastic. Young PhDs of Martin's generation were warned not even to mention the theory when they auditioned for academic jobs. "It was just completely taboo," he told me. Until suddenly it wasn't—when the new molecular data assembled by Doolittle and others showed that those cell organelles, mitochondria and chloroplasts, carried genomes reflecting their bacterial origins. Within a few years, even polite biological society had accepted it. Around that time, in the middle 1980s, Bill Martin began his own work.

He had discontinued his studies at Texas A&M, worked as a carpenter, traveled through Europe, and become fluent in German, until his scientific interests resurfaced, and he enrolled at a university in Hanover. By 1988, he had finished a doctorate, in Cologne, on molecular genetics and plant evolution. During that period, Martin became familiar with bacterial genetics and bacterial geneticists, for whom HGT was common knowledge. His earliest research project, an investigation of chloroplast enzymes, led him straight into the endosymbiosis theory and the role of HGT within it. He already knew that the theory was probably correct, and he learned one other important thing: the genomes found in chloroplasts and mitochondria are tiny—far too tiny to code for all the enzymes and other proteins that those organelles need to function. They are only minuscule samples of their original bacterial genomes. The other genes, manufacturing those hundreds of other proteins, must be still present in the cell, somewhere, but not in the organelles. Where else could they be? In the one place where genes live protected: the cell nucleus. These missing organelle genes must have been transferred from the captured bacteria into the nuclei of their respective cells—maybe one gene at a time over the long eons of eukaryote evolution—becoming patched into the cell's own nuclear genome. "So gene transfer was a part of the picture from the get-go," Martin told me. Not just interspecies transfer, but transfer across the big boundary, between domains. "That is the world that I grew up in."

As his research progressed, he found more evidence that this sort of gene transfer—from the captured bacteria into the nuclei of complex cells—had been common and widespread throughout time and across a range of creatures. It occurred at different rates, with different results, in different lineages of animal, plant, and fungus. One plant that Martin looked at, a small flowering thing related to cabbage, has a nuclear genome of which 18 percent is bacterial. Another of his studies showed that yeast, a fungus, contains 850 genes from bacteria and archaea. The human nuclear genome includes more than 263,000 base pairs (letters of code) of what was originally bacterial DNA, transferred from our mitochondria. In these cases and many others, genetic material has escaped from the organelles, leaked into the cell nuclei, and gotten integrated into the chromosomes. This is a profoundly consequential process: the transit of DNA from organelles of bacterial origin into the chromosomes; alien genes becoming incorporated over millions of years into the deepest cellular identity of plants, fungi, and animals. And no one knows, not yet, just how it has occurred. Martin eventually coined a term for the phenomenon: endosymbiotic gene transfer.

It's horizontal in a slightly different sense: sideways passage of genes from one domain of life to another, yes, but within the confines of a single cell, and therefore even subtler than other forms of HGT. It began when our remote single-cell ancestors incorporated those fateful bacteria.

You could think of it this way: as domestication and transfer of duties. Wolves find their own food. Dogs, being domesticated, rely on humans to feed them. By mutual agreement, over the course of fifteen thousand years, wolf descendants have transferred their food-gathering functions (and responsibilities) to us. It began with bones and meat scraps at the edges of human campfires, probably. It progressed to all manner of extremes. In exchange for food and other emoluments, canines now offer love, bark at mailmen, herd sheep, point at pheasants, and chase Frisbees. Likewise mitochondria: they are domesticated bacteria in your cells. They have transferred many of their genes to your nuclear genome, and they rely on that genome to send back proteins enabling them to exist and do their work. Instead of chasing Frisbees, they manufacture ATP, that battery-like molecule I've mentioned before, the one that offers portable energy for fueling your metabolism.

Of course, all this discovery of endosymbiotic gene transfer, Martin explained, led to further problems for Carl Woese's tree of life. There were too many branches going every which way, including many that branched from one major limb and then inosculated with another. Further complicating matters was the fact, revealed as more genomes were sequenced and compared, that those bacteria captured early in eukaryotic cells—the ones that became mitochondria and chloroplasts—had themselves been recipients of horizontal gene transfer, from different kinds of bacteria, *before* their capture. This meant parts of genomes existed within other genomes before becoming parts of still other genomes, including yours. It was all a snarl. It was a mess. It was a plate of spaghetti. It was wonderful.

"It's not a tree, though," Martin said.

As we talked away the morning, he sometimes interrupted himself to jump up and grab a book, or to find the file of a paper on his computer and print me a copy, or he tilted back his head to ruminate and launch a new topic. Or he paused to refocus, asking "What were we talking about?" At one point, after a flurry of erudite speculation on the origin of photosynthesis, the varied methods of nitrogen fixation, and the adaptive value of sex, he said, "Sorry, I'm just rambling on."

"No, no. That's good," I said. But yes, I did want to get back to HGT and the tree. "How much time do we have?"

"We have all day. I planned you in all day."

"Bless you," I said, and then a half hour later, he suggested we break for lunch. Exhausted by the flow of ideas and information, I turned off my recorder.

Did I like Japanese food? he asked. Sushi? Yes indeed I did. Good, Martin said, because I'm supposed to be losing weight, but you're my excuse. We drove to his favorite Japanese restaurant, where he ordered us a huge and very excellent meal, ate voraciously, and allowed me to pay for nothing. Over the noodle soup and the combo plates, we talked more about the origin of eukaryotic cells, the origin of the cell nucleus, and the ultimate origin of life. I shared with him, somewhat shamefacedly, the fact that I had watched the Super Bowl broadcast from two to four that morning, streaming it live on my computer at the hotel, then went back to bed for a few hours' rest before meeting him. Patriots over Seattle on Malcolm Butler's last-second interception, worth every minute of lost sleep. I know, I watched it too, he said. All this time, I kept wondering: Where is the famously brusque, intolerant-of-fools, formidably combative Bill Martin?

After lunch and tea, we returned to his office for more talk. At the end of the afternoon, he offered to walk me back toward my lodgings, not far, so I wouldn't get lost. A little exercise was better than another taxi. It was a chilly winter day, and he wore a casual black jacket and a watch cap, striding the pavement like John Cleese disguised as a stevedore. At a major crossing, he pointed: straight down that road to your hotel. Himself, he would go back now to the lab for more work. We shook hands. Let's get together again and talk, Martin said. This was fun, he said.

66

Bill Martin did his part, back around the turn of the millennium, to integrate horizontal gene transfer into evolutionary thinking and to consider how it strained the idea of the tree. Although he didn't participate in the Halifax writing weekend with Doolittle and Lawrence and Gogarten, he published a series of his own papers, alone or with coauthors, in a similar vein. The first of them, "Is Something Wrong with the Tree of Life?," preceded Ford Doolittle's manifesto by three years. Another, as already mentioned, discussed the "mosaic" character of bacterial genomes. Given the evidence of relatively recent and abundant horizontal transfer in a familiar bug such as *E. coli*, with almost a fifth of its genome acquired from other bacteria, Martin found it "quite ominous" to contemplate how HGT must have played out across the depths of time. He meant "ominous" in a retrospective sense: a portent that the past, not the future, still held some wild surprises. This paper on mosaic genomes included Martin's distinctive tree illustration, the multicolored sea fan, showing HGT by way of merging lines and blended hues. "That's a nice tree. I mean, that's what it looks like," he told me with some small pride. "I drew that all by myself." Very different from Ford Doolittle's freehand style, but making the same three points: the shape of the tree is important, and counterintuitive, and here it is.

In a later coauthored paper titled "The Tree of One Percent," Martin described how endosymbiotic gene transfer, his favorite kind, continues to move genes of bacterial origin from mitochondria and chloroplasts into the nuclear genomes of complex creatures. He noted that only 1 percent of the genes in an average bacterial or archaeal genome—and maybe far less than 1 percent in the genome of a eukaryote—are so deeply and complexly essential to the organism that they *couldn't* be swapped by HGT. The 1 percent figure came from work by another group of scientists, who studied thirty-one select proteins of the roughly three thousand proteins coded in the genome of an average prokaryote, and offered those "universal" proteins as a basis for phylogenetic analysis. Martin, with a coauthor, turned that group's logic against them. Sure, you might well use your preferred, stable genes to define a single tree. (It's what Woese had done with 16S rRNA, though Martin didn't say so explicitly.) But if you did that, your tree of life would be really just "the tree of 1 percent" of those genomes—a small selection, unambiguous but not necessarily representative. And if your tree captured the story of only 1 percent of each genome, what was the point? You'd be better off depicting life's history with graphs and theories than drawing any such marginal tree.

Carl Woese, in Urbana, wasn't oblivious to all this. He saw the unfolding discoveries of horizontal gene transfer and grasped the challenge they raised against his Big Tree. While the four horsemen of HGT galloped across the field of battle, their banners flying, Woese took the phenomenon into account and made his own sense of it. He formulated a fallback position that allowed him to celebrate the importance of HGT, while reconciling—or anyway, seeming to reconcile—the new data on sideways genetics with his own older ideas and trademark discoveries. During the late 1990s and early 2000s, he published a series of conceptual articles on early evolution, the origins of cellular life-forms, and what he began calling "the universal phylogenetic tree." By that last phrase, he meant *his* tree of life, the one derived from comparing sequences of ribosomal RNA.

This group of papers, four of them, have been called his "millennial series," and as one expert commentator noted after Woese's death, they were peculiar in several ways. They didn't report any new research. But they weren't review articles in the usual sense. They contained almost no data from his own work or anyone else's. They were too serious and dogmatic to come across as essays or opinion pieces. The commentator, a brilliant Russian-American biologist named Eugene Koonin, who was quite

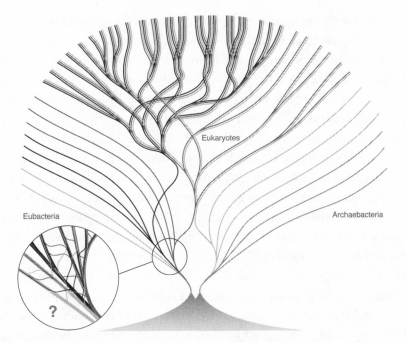

Eukaryotes

Eubacteria

Archaebacteria

?

Martin's reticulated tree, as drawn by Martin, 1999.

respectful toward Woese, found them hard to categorize except maybe as "treatises" or "tracts." Call them whatever, they were pronouncements on early evolution from a man with great personal confidence that he understood that vast, blurry subject better than anyone else. What makes the millennial series relevant here is how Woese tried to resolve the tension between rampant horizontal gene transfer and his beloved universal tree. He did it by pushing HGT into the very distant past.

Yes, it was a defining phenomenon back then, more than three billion years ago, Woese conceded, with genes or gene-like information leaping sideways between one sort of living entity and another. But that era of rampant HGT, in Woese's view, occurred at a time before the tree of life began to rise and branch. It happened before species existed, and therefore before limbs could diverge from a trunk, before branches could diverge from limbs. It happened prior to the origin of cellular creatures, as we understand cells today. Woese was alluding to another important idea, also unprovable but probably correct: that the origin of life, in some form, occurred a billion years or so before the origin of the first recognizable cell.

What did that early, inchoate period look like? Using a phrase coined by another scientist, he called it the RNA-world. In that world, RNA, not DNA, was the repository of complexity (nearly random complexity, at first); and RNA, not cellularity, was the basic structural unit of whatever passed as life. Certain chemical components, catalyzed somehow by energy and physical circumstances and chance, had assembled into "aggregates" of interactive molecules, with a capacity to replicate themselves. The crucial element of those aggregates was RNA, single stranded and not very stable. DNA, double stranded and stable, didn't yet exist. The aggregates differed from one another, there was a diversity of elaboration, and so as they self-replicated, drawing on environmental materials, they also began to compete. Some of them developed an inclination to translate their linear sequences of nucleotide bases (strings of A, C, G, and U) into another medium by linking amino acids together. Those linked amino acids constituted the earliest form of what a biochemist would now call peptides—short chains of amino acids—and then, as the chains grew longer and more intricate, proteins. These "aggregates" described by Woese in 2002 resemble the progenote concept he had offered back in 1987, with at least one signal distinction: in 1987 he hadn't yet become attuned to the importance of horizontal gene transfer, especially during the earlier stages of evolution.

Now he emphasized that phenomenon. As the molecular aggregates attained lives of their own, replicating, competing for survival, fragments of RNA from one aggregate would break off and attach to another aggregate. This was HGT in its earliest form. It happened often. The RNA-world was a great orgy of promiscuously shared materials. As time passed and possibilities burbled, some of the proteins proved useful in helping some of the aggregates replicate. At that point, RNA came to represent not just randomly assembled molecules but also potentially valuable information. But there were still no cells in this soup—nothing with a nice wall or a membrane bounding its internal substance, separating inside from outside, separating self from other—and certainly no division into groups that could be called species.

Complexity increased, under the impetus of competitive replication. Some of the aggregates brought proteins to their aid for slightly advanced structural purposes, and one possibility thus opened was self-containment. Packaging, with special advantages. Boundary protection. *Bingo*: the first cells.

These were primitive things compared with modern cells as we know them, probably leaky and clumsy and unstable; but cellular in some sense nonetheless. Each now enclosed in a wall or at least a membrane, they found ways to replicate themselves using the RNA (or maybe, by now, DNA) instructions they contained. The cell wall shielded them, to some degree, from the distractions and confusions of HGT—noisy information arriving from outside. Primitive though they were, the early cells replicated themselves more faithfully than could be done by a naked strand of RNA. They generated lineages. HGT became less rampant. Eventually the lineages yielded whole populations of cells that resembled one another. At this point, you might call those cells by a different name: organisms. The cells constituting a population weren't all identical, they encompassed some variation, but there was more similarity among them than between them and the cells of other lineages. *Bingo again*: species. A major boundary in the history of life had been crossed. Woese called that boundary the Darwinian Threshold.

"As a cell design becomes more complex and interconnected," he wrote, "a critical point is reached where a more integrated cellular organization emerges, and vertically generated novelty"—meaning parent-offspring inheritance, with modest variation, as distinct from horizontal gene transfer—"can and does assume greater importance." At that critical borderline, the Darwinian Threshold, evolution as Darwin understood it begins. This progression is all stated in present tense, you'll notice, as though Woese is sitting there, three billion years ago, and watching these things happen. In transcendent moments, he may almost have felt that he was.

None of it altered the shape of his preferred tree of life. In his 2000 paper on what he called the "universal" tree, and in his 2002 treatise on the evolution of cells, Woese included tree figures almost identical to what he had offered with Kandler and Wheelis in 1990. He ignored a decade's worth of discoveries in the realm of horizontal transfer. He omitted contortions of the sort shown in Ford Doolittle's drawings. His tree still had three major limbs. His branches diverged but did not inosculate. Woese seems to have felt that he had dismissed the challenge of HGT by pushing it far backward in time, down among the roots of the tree, before the rise and divergence of "modern" cellular life. Others disagreed.

67

half dozen years passed. Newer data, still more HGT. Eugene Koonin and two colleagues looked for horizontal transfers between major bacterial lineages and found plenty. The percentage of genes traceable to sideways transfer, they discovered, ranged from a small fraction in some bacteria up to a third of the entire genome in *Treponema pallidum*, the nasty spiral bacterium that causes syphilis in humans. As additional genomes were sequenced, Bill Martin and a pair of collaborators found even more HGT. Comparing a half million genes distributed among 181 different genomes of bacteria and archaea, Martin's team concluded that roughly *80 percent* of the genes in each genome had arrived by horizontal transfer at some point in evolutionary history. Peter Gogarten and a colleague detected HGT in plants, other scientists showed that it was important among fungi, and Julie Dunning Hotopp's work (which I mentioned earlier) revealed its widespread occurrence among insects. You could almost say that HGT finding became a scientific craze—except that *craze* doesn't capture the difficulty of such work, nor the depth of the implications. And few people outside the rarified realm of molecular evolutionary biology had heard about this phenomenon or the hubbub it was causing among experts.

Then, in January 2009, the British magazine *New Scientist* appeared with a tree on its cover and a big, incendiary headline: "Darwin Was Wrong." Beneath those words was a subhead in smaller type: "Cutting Down the Tree of Life."

The cover story, written by the magazine's features editor, Graham Lawton, began with Charles Darwin and his little sketch in 1837. A fruitful idea, depicting evolutionary relationships as a tree, Lawton noted, and by 1859, in *The Origin*, that spindly sketch had grown woody and tall, to a "great tree," Darwin's preferred simile. The article leapt from there to Ford Doolittle, citing his affirmation that the tree image was "absolutely central" to Darwin's thinking about evolution. It became the unifying principle for understanding life's history and, throughout the 150 years since first publication of *The Origin*, evolutionary biology had been largely devoted to learning details of the tree. "But today the project lies in tatters," Lawton wrote, "torn to pieces by an onslaught of negative evidence." He was talking about horizontal gene transfer.

Lawton's article was a fair introduction to the subject, despite the misleading provocations of the cover and the fact that those provocations were echoed inside by the very title of his piece: "Axing Darwin's Tree." Journalists don't often get the privilege of titling their own magazine stories (though a features editor might), so it's hard to know who was responsible. All that cutting down and axing—lumberjack imagery—and the claim of Darwin's wrongness were pointedly extreme. They may have helped to sell magazines and enticed readers to turn pages, but they caricatured the genuine challenge to Darwinian orthodoxy that the new discoveries raised. The article itself was more nuanced, but still drastic enough. Lawton quoted Doolittle's 1999 paper, the one that had splashed this debate onto the pages of *Science*, and also a more recent Doolittle comment, presumably gathered by email or phone: "The tree of life is not something that exists in nature, it's a way that humans classify nature."

Lawton cited Bill Martin's work—his critique of the tree of 1 percent—and mentioned endosymbiosis, but noted that, for a while, evolutionary biologists had taken HGT for a marginal phenomenon. No more. One researcher, Michael Rose at the University of California, Irvine, told Lawton flatly: "There's a promiscuous exchange of genetic information across diverse groups." Lawton mentioned endosymbiotic gene transfer from organelles as another version of that, making the point that HGT occurs in eukaryotes as well as in bacteria and archaea. He added a statement by

John Dupré, an English philosopher of biology: "If there is a tree of life, it's a small anomalous structure growing out of the web of life." Finally, the article came back to Ford Doolittle, who seemed uncomfortable with the vehemence of the attack he had helped launch. "We should relax a bit on this," Doolittle told Lawton. "We understand evolution pretty well—it's just that it is more complex than Darwin imagined. The tree isn't the only pattern."

Other scientists didn't want to relax. They wanted, Lawton wrote, mixing his metaphors, to see "the uprooting of the tree" as the start of something bigger and more radical. But wait, was the tree being uprooted or cut down? Never mind. "It's part of a revolutionary change in biology," said Dupré. Evolution as understood in the future would be much more about mergers and acquisitions and collaboration than about change within isolated lineages. It would be about not just the divergence of branches but also their inosculation. Another scientist, a one-time post-doc and coauthor of Doolittle's named Eric Bapteste, noted: "The tree of life was useful." In Darwin's hands, that image had helped people picture evolution, when evolution was such a startling new idea. "But now we know more about evolution," Bapteste said, and "it's time to move on."

The cover date of this *New Scientist* was January 24, 2009. The timing was significant: two weeks later came Charles Darwin's two hundredth birthday, on which biologists and historians around the world would be celebrating his life, his work, and his theory. That entire year, 2009, in fact, would be filled with Darwin events and retrospectives, from Cambridge to Mumbai to Albuquerque. The editors of *New Scientist* had planned their own observance, and this tree-cutting issue was it: an equivocal sort of birthday card for Mr. Darwin. Seizing the opportunity of the bicenten-nial, with its broad (if brief) public attention to all things Darwinian, they offered Graham Lawton's article as an alert that new discoveries—genes moving sideways, who knew?—continue to reshape our understanding of life's history. That was useful. But their cover headline was pure provo-cation: Darwin at two hundred, wrong! Promptly and stridently, some of the big names in evolutionary biology disagreed.

Daniel Dennett is a distinguished philosopher, author of *Darwin's Dangerous Idea*, among other writings. Jerry Coyne is an evolutionary bi-ologist at the University of Chicago, known for his work on the speciation process and his strong voice as a defender of evolutionary science against creationism. Richard Dawkins is author of *The Selfish Gene* and many other books, including *The God Delusion*, and probably the world's best known and most vociferously atheistic Darwinist. He's also an extremely bright man with a dangerous and confident wit. These three, along with one other biologist, Paul Myers, an influential blogger, coauthored an outraged letter to *New Scientist*, which the magazine published a month later. It began: "What on earth were you thinking . . . ?"

Their complaint was with the garish negativity of the tree of life cover, not the more nuanced explanations of the article. Announcing that "Darwin Was Wrong," Dennett and company argued, would give aid and comfort to the enemy, "handing the creationists a golden opportunity to mislead school boards, students, and the general public about the status of evolutionary biology."

On that point, they were probably right. Creationists did seize the opportunity, golden or otherwise. One instance turned up immediately at a creationist website called Apologetics Press, under the headline: "Startling Admission: 'Darwin Was Wrong.'" This piece, by one Eric Lyons, sketched the main points of Lawton's article, quoted Doolittle and others, and called the tree image just one more "iconic concept of evolution" that had "fallen on hard times." As decades passed and more data became available, Lyons claimed, yesterday's proofs of evolution were turning out to be mistaken. "One wonders what it will take," he wrote, "to convince evolutionists that it is not just Darwin's tree of life that needs to be cast aside, but the entire theory of evolution." Lyons's use of the *New Scientist* headline was exactly what Dennett and his letter coauthors had feared.

Amid the stink over the magazine's cover, and the essential or nonessential significance of the tree, a quieter comment from *New Scientist*—in the form of an unsigned editorial—went largely overlooked. This introductory note was meant to put Lawton's article in perspective and to negate, in advance, the sort of gleeful creationist misconstrual that came from Lyons. Biology, like physics, the editorial said, is a work in progress. Big upheavals in evolutionary thinking have occurred before. There was the original one, triggered by Darwin and Wallace in the nineteenth century. There was a second revolution in the 1930s and 1940s, when Mendelian genetics was integrated with Darwinian natural selection and mathematics. That came to be known as the Modern Synthesis of evolutionary theory, also known as neo-Darwinism. And now, in the era of molecular biology and genome sequencing, important biological topics were "turning out to be much more involved than we ever imagined." That was putting it mildly. "As we celebrate the 200th anniversary of Darwin's birth," said the editorial, "we await a third revolution that will see biology changed and strengthened."

Is that third revolution happening? If so, you could argue that it began with Carl Woese taking up a vague but astute suggestion by Francis Crick about doing "protein taxonomy" as a gauge of relatedness among different creatures. Woese's lonely work led to the now-crowded field of molecular phylogenetics and to the recognition that, against even so strong a signal as 16S rRNA, horizontal gene transfer has made the history of life unimaginably more complicated than Charles Darwin could have guessed. It also brought a dawning curiosity about what extraordinary things occurred during that distant time before what Woese called the Darwinian Threshold. It was all about evolution, and it went beyond Darwin's thinking without negating that thinking, just as Einstein and quantum mechanics went beyond Isaac Newton's. "None of this should give succour to creationists," the New Scientist editorial added. Right again, though, of course, it has.

68

By the time that ruckus broke in the pages of *New Scientist*, Ford Doolittle had begun what he half seriously calls "my retreat into philosophy." His talks at meetings and his publications, especially the 1999 paper in *Science*, had made him a leading spokesman for the view that horizontal gene transfer is vastly significant and that "the history of life cannot be properly represented as a tree." Tree or not tree, that was the question. It seemed very important to many biologists—not just Dawkins and Dennett and Coyne—because of their engagement in two parallel struggles: the struggle to understand life's history, yes, and also the struggle to defend the teaching of evolutionary theory against creationist rhetorical and political attacks. Now the tree image itself had become controversial, with "tree huggers" and "tree cutters" arguing pro and con.

"I'm not unhappy at having started that polarizing debate," Doolittle told me, "because I think it's been useful." The debate stimulated fresh thinking and work. It drove efforts to gather more data, and it generated new hypotheses, "some of which have turned out to be interesting and probably true, and some of which"—he chuckled—"have just turned out to be interesting. I guess I've always believed that if you're not wrong half the time," he said, "then you're not being brave enough."

But there came a point of diminishing returns. "I kind of got tired of opening every journal and saying, 'Oh, is this paper for me or against me?'" He realized, with increasing clarity, that this wasn't really a scientific issue. It was philosophical and representational and semantic. "Whether or not there is a tree of life," Doolittle told me, "depends a whole lot on what we mean by a tree of life."

It reminded him of a classic paradox that philosophers call the Ship of Theseus problem, recorded by Plutarch two millennia ago in his *Life of Theseus*. The legendary hero Theseus arrives back in Athens after an epic adventure in Crete, and the Athenians evermore preserve his ship as a historical relic. But it's a working relic. They don't mothball the thing, they don't put it on a pedestal; they use it for important ceremonial voyages. Some of the planks rot over time and must be replaced. Then more rotten planks, more replacement. At what point is the "Ship of Theseus" no longer the ship of Theseus? Hard to say. Well, the same problem exists, according to Doolittle, with the identity of an organism and its lineage after billions of years of horizontal gene transfer. If its genes are half bacterial and half archaeal, does that organism belong to the Bacteria or to the Archaea—or is the question impossible or meaningless to answer?

The best way to understand the tree, Doolittle decided, is that it represents a hypothesis about the history and relatedness of all life. He had said this back in 2000 in the *Scientific American* article, but passingly, without developing the thought. Now he did that. It isn't just an "attractive hypothesis" arisen from nowhere or by consensus; it's Charles Darwin's hypothesis, formulated in 1837, when Darwin drew that little sketch in his B notebook and scribbled above it: "I think." The hypothesis was meant to explain patterns Darwin had seen among living creatures and fossils, leading him to the conviction that evolution had occurred. If so, *how* had it occurred? From common ancestry, Darwin argued, beginning with a few original forms or maybe just one, passing slowly through changes, some forms diverging from others, differentiating into many species, those changes shaped primarily by the process he called natural selection. If all that were correct, what would life's history look like? Darwin hypothesized: it would look like a tree.

But the tree hypothesis works poorly for the history of bacteria and archaea, with all their sideways exchanges; and it works imperfectly for everything else. Darwin can't be blamed. He didn't have molecular phylogenetics to snarl his thinking. He didn't know about horizontal gene transfer. He did the best he could, which was exceedingly well, with the evidence he could see.

69

Axel Erlandson was a Swedish-born American farmer, brought to the United States as a child, raised in Minnesota, drawn to the Central Valley of California along with his family, in 1902, by the promise of irrigated land and extended sunshine. He married, had a daughter, and set to growing beans and other crops on his own place, near Hilmar, California, which is halfway between Modesto and Merced. One day he noticed a natural inosculation in his hedgerow: two branches from two different shrubs, kissing and fused. Inspired by this anomaly, as John Krubsack in Wisconsin had been inspired, Erlandson launched himself upon the same odd hobby: shaping trees into amusing, unnatural shapes. He learned by trial and error and achieved amazing results. His methods included pruning, grafting, bending, supporting his entangled saplings with external stakes and scaffolds until they were firmly grown into position, and, as he said teasingly when asked about his knack with bizarre trees, "I talk to them."

Erlandson favored limber species such as sycamore, willow, poplar, birch, and (like Krubsack) box elder. His creations became known by descriptive names. He grew a Needle and Thread Tree, featuring a crosshatch of diagonal limbs, some passing through holes in others. He grew a Cathedral Window, resembling a great composite of stained-glass panes, rising from ten small trees planted in a row. He grew a Ladder Tree and a Telephone Booth Tree and a Tepee Tree and a Double Spectacle Tree and an Archway Tree and a Revolving Door Tree and a Nine-Toed Giant. He grew six sycamores into a single Basket Tree. No one had ever seen such shapes—not among natural trees, wild or domestic, growing by their own arboreal rules. Erlandson, in his mind's eye, saw trees that couldn't exist, and he made them happen. The man was benignly demented.

Or maybe not. After twenty years of this, with World War II just over and tourism in California beginning its postwar boom, he got the idea to convert his arboreal sculptures into a roadside attraction. Maybe his hobby could yield a bit of cash.

He bought land on the old stage road between San Jose and Santa Cruz, a road that led from burgeoning suburbs toward the ocean, and he moved his strange nursery to the new property. He transplanted those novelty trees that were transplantable, cut the others and brought them like furniture, and began growing still others. In 1947 he opened for business. There wasn't much, but some. Visitors pulled over, stopped in, rang a bell, and Erlandson admitted them, for a small fee, to his collection—a steady trickle of customers, at least until a newer route, Highway 17, diverted most traffic from the old road. In a good year, he made more than $320. He called his place the Tree Circus.

The further history of Axel Erlandson is not what concerns us. I'm simply underlining a point about the mystifying unnaturalness of topiary. The other circus of peculiarly shaped trees, as you've read, grew from genomic data and phylogenetic analysis during the late 1990s and into the first two decades of this century. That topiary enterprise continues. Ford Doolittle has made his "retreat" into philosophy, to think about what it all means, but many of his former grad students and postdocs are now prominent researchers in this field, helping illuminate the deep history of life. Bill Martin and his colleagues continue doing important work, likewise Peter Gogarten and Jeffrey Lawrence and Eugene Koonin and many others, too many to count let alone list. Younger researchers, fresh voices—such as Thijs Ettema, a Dutchman in Sweden, studying novel lineages of archaea and the light they may cast on early evolution—have emerged to prominence in the past decade. The huge increase in sequenced genomes, available to all researchers through public databases, and of computing power and software tools to analyze those genomes, has pushed the waves of discovery and insight into high, breaking crests. No one can keep track of it thoroughly or reduce it to a linear narrative. But amid all the new data and phenomena and ideas, these scientists are addressing some very large questions.

Three in particular stand out. First question: Is it true that Charles Darwin was wrong? If so, about what? Has his theory of evolution been grievously challenged, or just amended? Second question: What are the origins of the eukaryotic cell? Endosymbiosis, and the ideas of Merezhkowsky and Margulis and its other proponents, only begin to answer this one. Mitochondria and chloroplasts as captured bacteria, okay. But what about the rest of the epochal transition, from prokaryote to eukaryote, that occurred about two billion years ago? How did the cell nucleus come to be? What was the identity of the host cell, the receptacle, within which all these fancy improvements occurred? What exactly were the materials, and what were the circumstances, that converged to become such a complex cellular entity, ancestral to all animals, all plants, all fungi, and other eukaryotic creatures? How did that complexity begin?

Third question, and the closest to home: What implications do these discoveries carry for the concept of human identity? What is a human individual? What are *you*? The reality here is more strange than you might think.

PART ❧ VII
E Pluribus Human

70

The realization that bacteria and other microbes inhabit the healthy human body goes back a long way, at least to the day when Antoni van Leeuwenhoek scraped some plaque off his teeth and looked at the stuff through one of his lenses. From there it was a great leap, about three hundred years, to the point where scientists could appreciate the extent of our colonization by other creatures and begin to census them. Modern microbiology arose in the meantime. But the shock that Leeuwenhoek experienced, finding alien creatures in his own mouth, would be matched later by the shock of finding alien genes in our own genomes, let alone whole menageries of intact microbes thriving amid the various compartments and surfaces of our bodies.

No doubt you've heard the term *microbiome*. It's a great scientific buzzword nowadays, in the news, in the magazines, in the grant proposals of many scientists. It's also an apt label for the phenomenon of tiny creatures as important constituents within big creatures, of which recent years have brought an exploding recognition. The word is old but Joshua Lederberg helped popularize it to mean "the ecological community of commensal, symbiotic, and pathogenic microorganisms that literally share our body space and have been all but ignored as determinants of health and disease." As that suggests, *microbiome* commonly serves to mean "human microbiome," the community of microbes resident within *us*. Every other sort of complex multicellular creature has its own version of such indwellers. The horse microbiome and the tiger microbiome each constitutes a different community living in its own little world. Ours is ours. These particular bugs have evolved as participants in peoplehood. And we have evolved with a deep dependence on them.

Of course, the human microbiome is not one invariant list of microbial species, common to all people at all times. It's a set of possibilities. It's kaleidoscopic. Just as "the human genome" is a single phrase suggesting a single entity but really encompassing much variation from one person to another, so is "the human microbiome." When I say "these particular bugs" are participants in peoplehood, I mean a whole roster of residents and combinations of residents within and upon the human body, differing somewhat from person to person and changing with circumstances and time.

That changeability of the microbiome is why it's relevant to a new understanding of human health. The composition of resident microbes, for each of us, is contingent—contingent on who we are, what we do, how we've been born and raised, where we go, what we eat. And its contingencies have impact.

The human microbiome has become a major focus of research, and of medical concern, for its involvement or suspected involvement in a broad array of unhealthy conditions: obesity, childhood diabetes, asthma, celiac disease, ulcerative colitis, certain kinds of cancer, Crohn's disease, and others. A healthy person, scientists have realized, carries a healthy and diverse microbiome; and if the microbiome is depleted or disturbed or left undeveloped, by one factor or another, there can be hurtful effects. Just as the presence of certain microbes in a human body can be problematic (we call that infectious disease), the absence of certain microbes can be problematic too. Take away *Bacteroides thetaiotaomicron* from his or her gut, and a person might have trouble digesting vegetables. An imbalance among microbes, too many of one kind, too few of another, can also cause trouble. Disrupt the bacterial community of a healthy person's colon by administering an antibiotic, and the small resident population of *Clostridium difficile*—surviving the drug, but now freed of competition, because so many other bacteria have been eliminated—may explode to a raging infection. It may weaken the bowel wall, cause fever and diarrhea, possibly even death.

A recent estimate suggests that each human body contains about thirty-seven trillion human cells. It also contains about a hundred trillion bacterial cells, for almost a three-to-one ratio of bacterial to human. (Another study proposes a lower ratio, roughly one-to-one, but that still means thirty-seven trillion bacterial cells in your body.) And this doesn't even count all the nonbacterial microbes—the virus particles, fungal cells, archaea, and other teeny passengers—that routinely reside in our guts, our mouths, our nostrils, our follicles, on our skin, and elsewhere around our bodies. These stowaways may represent more than ten thousand species (or "species," given the fuzziness of the category as applied to prokaryotes). About a tenth of that diversity, maybe a thousand species, are bacteria living just within the human gut. Because each of our trillions of microbiomic cells is generally much smaller than a human cell, the complete microbiome constitutes only about 1 percent to 3 percent of our body's mass. In a two-hundred-pound adult, that's roughly two to six pounds. In volume, maybe three to nine pints. Still, they're busy and they're consequential, those few pints of cells.

But that's not the half of what I'm getting at here. Other books, some of them technical, some of them popular, have recently described the nature and workings of the human microbiome. Among the best of the latter is Ed Yong's *I Contain Multitudes*, an encyclopedic and lively survey of microbial ecosystem dynamics in humans and other species. My purpose is different, and there's no need to cover much of the same ground. But since the microbiome has entered our cultural vernacular, it makes a good starting point toward something else, something even more fundamental and tricky: a new appreciation of the composite nature of human identity.

Leeuwenhoek reported his mouth-creatures epiphany in a letter to the Royal Society, dated September 17, 1683, along with further observations on similar samples from four other people. To his naked eye, this dental plaque appeared as "a little white matter, which is as thick as if 'twere batter." Magnified under his microscope, though, it was more: "I then most always saw, with great wonder, that in the said matter there were many very little living animalcules, very prettily a-moving." The biggest of them "had a very strong and swift motion, and shot through the water (or spittle) like a pike does through the water." That's fast: a pike is arrow-like compared with a bass or a carp. The smaller beasties "spun round like a top." It was a rich, busy ecosystem, especially as collected from one of Leeuwenhoek's subjects: an old man who had gone a lifetime without ever brushing his teeth. In the old man's savory offering of oral goop, he found "an unbelievably great company of living animalcules, a-swimming more nimbly than any I had ever seen up to this time." These bugs included largish ones that "bent their body into curves in going forwards." He was looking at microbes of various sorts, their identities now indeterminable from the sketchy evidence he left. But all the spinning and curving behavior suggests that spirochetes—corkscrew bacteria—may have been among them.

That possibility is supported, and set in broader context, by a report from a team of medical microbiologists more than three centuries later, in 1994. Their leader was Ulf B. Göbel of the University of Freiburg, in Germany. Searching for biodiversity in the mouth of a twenty-nine-year-old woman suffering severe periodontitis (gum disease, which was rotting her teeth away from the jawbone), these scientists found dozens of kinds of spirochetes that were assignable to a single genus. The genus was *Treponema*, notorious for *Treponema pallidum*, the little demon that causes syphilis. Other treponemes, it turns out, are less dramatic though still troublesome. They cause periodontitis of the sort seen in that woman. But what makes this case notable for us is not the diversity of bone-eating corkscrews that were rampant in one person's mouth. What makes it notable is that Göbel's group did its work using the methodology of Carl Woese.

More specifically, they identified those many different kinds of oral spirochete by sequencing and comparing samples of 16S rRNA. But they went beyond Woese, who had done his work by growing bacteria and other microbes in cultures and then extracting their ribosomal RNA. Sometimes the bugs can't be grown in captivity. They're too feral and sensitive for culturing in a lab. Under those circumstances, it takes a different method to detect new, unknown microbes. That method, as Göbel and his colleagues acknowledged, had been developed by Woese's great friend and foremost acolyte, Norman R. Pace.

71

Norman Pace grew up during the 1950s in a small Indiana town, a bright kid with an early taste for science, who blew out his left eardrum during a home-chemistry-lab experiment and, like Ford Doolittle, got serious and saw his career path in the aftermath of Sputnik. He attended a summer science camp during high school, then went off to college in the big city: Bloomington. In 1964 his prairie horizons widening, he moved to the University of Illinois as a grad student, attracted by the prospect of doing a PhD under Sol Spiegelman, an eminent figure by then, renowned for his work on nucleic acids. Carl Woese arrived in Urbana about the same time, thirty-six years old, recruited as an associate professor but largely unknown, and just beginning work toward his early book on the genetic code. Woese's lab was along the same corridor as Spiegelman's, on the third floor of Morrill Hall. Pace, again like Ford Doolittle but several years earlier, seems to have found the young Woese more companionable, and in some ways more inspiring, than the crepe-soled and spooky Dr. Spiegelman.

"We would talk and be friendly. He was just down the hall," Pace told me fifty years later, as we sat in his lab at the University of Colorado, in Boulder. "And he was doing some stuff I was pretty interested in." But maybe *doing* was the wrong word for such deep speculations, and

Pace corrected that to "—*thinking about* stuff that I was pretty interested in." Those interests went down through the mechanism of the genetic code, Francis Crick's favorite puzzle, to the evolutionary origins of the code, and beyond that to the earliest amorphous twitches of life itself. Woese was pondering big questions, pushing the boundaries of what seemed accessible to scientific investigation, not whimsically but with a fierce desire to know. "Smart, congenial guy on the corridor," Pace said, "thinking about origins of life, and those sorts of things that legitimate scientists didn't think about." He laughed. "We interacted a lot during the course of those years." Pace the grad student asked Woese the young professor to be on his dissertation committee. Woese agreed, and their friendship grew.

By now, it was the late 1960s. Pace was a fit and adventurous young man with a passion for caving, a fondness for motorcycles, and a lively wife, Bernadette Pace, who had her own doctorate in molecular biology. He favored the BMW R69S as a highway bike and rode one, with Berna-dette holding on behind him, from Germany to Turkey back in the day. Then she wanted one for herself. "That was before she took up trapeze," Pace told me. Bernadette became a high-flying trapeze performer of pro-fessional skill, good enough to perform with the Carson & Barnes Circus when they had an injury (which happens often in trapeze). Norm and Bernadette erected a high-flying setup for her in their yard, and some-times when the couple threw parties, and she performed for their friends, he would put on a top hat, tights, and a bow tie, and serve as ringmas-ter. None of that vitiated the seriousness of the science they did, with Bernadette his coauthor on more than a dozen arcane papers. Nor did it diminish him in the eyes of Woese, a conservative man though a radical thinker, who merely viewed Norman's more robust activities as "swash-buckling."

After getting his doctorate, then two years on a postdoc fellowship, then leaving Urbana for his first academic job, Pace stayed closely in touch with Woese. He established his own lab at the National Jewish Hospital and Research Center, in Denver, and continued research on what had been his dissertation topic, the replication of RNA. Ford Doolittle came out to Denver from Urbana and did a postdoc under his roof. Then so did Mitch Sogin, Woese's brilliant handyman for the early sequencing work, still interested in ribosomal RNA. Pace coauthored papers with both of them, and with Woese too, during the 1970s.

In the 1980s, his work focused more intensely on the application of an ingenious insight: that a variant of Woese's approach for identification of microbes and placing them on trees of relatedness could perhaps be used for detecting organisms never grown in a lab. Those unknown creatures, *whatever* they were, had 16S rRNA too, after all. If he could find it, extract it, sequence it, and sort one sample from another, he wouldn't need to grow the bugs themselves. He wouldn't need pure cultures. Or, better still, he could extract DNA instead of RNA from the samples, amplify that into workable quantities, and look at the genes coding for different versions of the 16S rRNA molecule. This might illuminate some wild, mysterious little creatures living in extreme but species-rich environments, as well as a lot of others, in places not so extreme but previously overlooked. If it worked, it would open a broad new vista on Earth's biological diversity, since most of the planet's life-forms are microbes, and most of those kinds of microbes had never been—possibly never could be—enticed to live and replicate in captivity. They were biological dark matter, vast and weighty, but unseen.

It did work. Pace and his young collaborators proved the principle with samples from three very different extreme environments. The first was a hydrothermal vent thousands of feet deep beneath the Pacific Ocean. The second was a leaching pond at a copper mine near Hurley, New Mexico. The third was a hot spring in Yellowstone National Park. From the hydrothermal vent, Pace's team got tissues of a giant tube worm, provided by colleagues at the Woods Hole Oceanographic Institution, who had presumably grabbed their worm samples using Alvin, the deep-ocean research sub. From the leaching pond at the mine, they scooped two pounds of toxic mud. Yellowstone was a different story— slightly more of a lark for the boys in Denver.

"I was sitting in my office, reading this book," Norm Pace told me, and he pulled a thick volume off his shelf: Thomas Brock's treatise on heat-loving microbes. Brock is the man, as I've mentioned, who discovered *Thermus aquaticus* in a Yellowstone hot spring, thereby opening the way to the polymerase chain reaction technique (using a *Thermus aquaticus* enzyme) for amplifying DNA and all the work in molecular biology that has followed from it. "There is an article in there about Octopus Spring, Yellowstone," Pace said, "with a whole bunch of these so-called pink filaments in it." The filaments had been "so-called" by Brock himself, who recognized the pink strands as bacterial agglomerations but never succeeded in culturing them in his lab. They were more untamable than *Thermus aquaticus*, possibly in part because they required very high temperatures. The upper heat limit for any form of life had been thought, by most microbiologists, to be about 73 degrees Celsius (163 Fahrenheit), until Brock went to Yellowstone in the mid-1960s. He found *Thermus aquaticus*, for instance, not at the very source of a hot spring but downstream along its outflow channel, at a relatively balmy 69 degrees C. But these pink filaments grew in the hottest part of Octopus Spring, at upward of 92 degrees C. "That's fucking hot," Pace reminded me. Water boils at 100. Reading the account in Brock's book, Pace had found himself wondering about the identity of those heat-loving microbes and how they made their living. This all occurred in 1981, but he remembered it clearly because it played a large part in setting the course of his career.

He had dashed from his office into the lab, where three of his grad students stood. "Hey, you guys, look at this," Pace recalls crowing. "This spring in Yellowstone, Octopus Hot Spring—supposedly kilogram quantities of high-temperature biomass!" Kilogram quantities, biomass: that's what "Eureka!" sounds like from the mouth of a microbiologist. He meant: *Whole shitloads of strange bugs that grow happily at near-impossible heat!*

Let's go get some, he said. He envisioned extracting ribosomal RNA from the pink filaments, sequencing it, and identifying entirely new creatures—creatures not just unknown but *unknowable* by the classic methods of microbiology. "My God!" he recalls thinking. "We can find out who is out there in the real world." Another quirk of microbiologyspeak: a bacterium or an archaeon is a "who."

The expedition took shape, and Pace invited Woese to join it. So the sage of Urbana rode with the Pace team to Yellowstone, camping amid the geysers, helping collect samples from Octopus Spring, a bemused participant in this "swashbuckling microbiology." What they found in the hot spring, thriving near the boiling point and characterized solely from their sequencing of ribosomal RNA, was a microbial community dominated by three forms. Two were bacterial and one was an archaeon. All three were new to science. Woese returned home to Urbana happy, but, he told Pace later, it took months for his back to recover from sleeping on the ground.

There is a grainy photograph, from almost precisely that time, snapped by someone in the lab and preserved in the memorial literature on Carl Woese. It shows a bearded Woese and a lean, young Norm Pace, both in T-shirts, Pace with longish hair and aviator glasses, left hand on his hip, the two toasting each other triumphantly with unidentified libations held aloft in laboratory flasks. The occasion is unspecified. Their return from Yellowstone with a harvest of strange microbial gunk? Maybe. The caption, as I've seen this photo reprinted, is simply "Friendship in the Kingdom." The kingdom referenced is really three kingdoms, or more accurately three domains: Bacteria, Eukarya, Archaea. The snapshot is a reminder that it was a lonely but thrilling time, and that Pace occupied a special position. None of Woese's long list of contacts—his departmental colleagues, his German admirers, his grad students, his postdocs, his lab assistants, his other coauthors and collaborators—ever did more than Norm Pace to confirm, and to defend for years after, the Woesean view of life's deepest history.

Pace was never a member of Carl Woese's lab. Woese had served on his PhD committee, but Spiegelman was the advisor of record. His relationship with Woese had begun casually, on the corridor, and grown deep over time because of shared interests and convictions. He had his own simple way of describing it, voiced during our conversation in Boulder: "I was his scientific son."

72

Microbes inhabiting people were never directly the focus of Norm Pace's work. He was more interested in the broader subject of microbial diversity on Earth. But the illumination of the human microbiome has followed in large part from what he and Carl Woese did. Woese pioneered the technique of using 16S rRNA to characterize and compare microbes. The microbes he worked on, such as Ralph Wolfe's methanogens, were isolated and cultured in the lab. They grew in captivity (not always readily, but Wolfe and Bill Balch solved that problem with the methanogens), and from those pure cultures Woese took his ribosomal RNA. Pace, on the other hand, with his own vision and skills, captured evidence of microbes that live only in the wild. He extended Woese's technique to raw samples from various environments, including extreme ones such as Octopus Spring. He used 16S rRNA and the genes that code for it, extracted wholesale from those samples, to detect and characterize new creatures that never had been—maybe never *could be*—grown in a lab.

Testimony to the importance of those two contributions lies in the literature of the microbiome, such as Ulf Göbel's paper on gum disease and many others. It lies also in the foundational history of what became a whole new branch of biology: *metagenomics*, the study of living creatures and communities that can be known only from their genetic material.

Research on the human microbiome has meanwhile led back to some new insights on horizontal gene transfer. That phenomenon, it turns out, occurs in our own bellies and noses and mouths. Bacteria are always evolving, and to evolve toward greater virulence, or antibiotic resistance, or transmissibility between human hosts, allows them to increase their Darwinian success. Fast evolution is generally better than slow evolution, and HGT represents the fastest way of adding major new possibilities to the heritable variation upon which evolution depends. So the acquisition of entirely new genes, by horizontal transfer from other kinds of bacteria with whom they happen to share habitat in a human colon or nostril, is an effective way for bacteria to improve their prospects of survival and abundance. This has been demonstrated in work led by Eric J. Alm at MIT.

Alm's group looked at 2,235 complete bacterial genomes, all available from an integrated, open database maintained by a federal institute. In other words, they downloaded this massive amount of sequence data gathered by other researchers the way you might download a season of *Stranger Things*. More than half of those 2,235 genomes came from "human-associated bacteria"—denizens of the human body, members of the human microbiome. Following Woese's method, Alm's team used the 16S rRNA gene as a standard of how closely or distantly those bacteria were related. They knew the geographical origin—at least by continent—of each of the bacteria represented. And they knew the ecological environment from which each came. With the human-associated bacteria, that meant which site within the human body. Stomach, mouth, vagina, armpit, skin? These three parameters, relatedness and geography and ecology, outlined the question they wanted to answer: Which factor is most significant in conducing horizontal gene transfer among bacteria?

They screened all the genomes for suspicious anomalies—cases where extremely similar versions of a given gene were shared by two very different kinds of bacteria. This was bioinformatics on a vast scale, but vast bioinformatics is now relatively quick and cheap. Their supposition, entirely reasonable, was that each close match of genes shared in two distant lineages signaled a relatively recent horizontal transfer event for that gene. Alm was hoping, in advance, that they might find a handful of such instances. "I was thinking five to ten," he told me by telephone from his lab. "If we could find five to ten genes, that would make it interesting." Instead they found 10,770. To find so much transfer, Alm said, was in itself "pretty shocking."

Of the three parameters, ecology was clearly important. They found twenty-five times more HGT among bacteria inhabiting the human body, for instance, than among bacteria inhabiting various other environments. But that wasn't all. Dicing their data at another angle, Alm's group saw that HGT between bacteria inhabiting the same *site* on a human body occurs more often than transfer between sites. To be explicit: bacteria that live in our guts tend to trade genes with other gut bugs. Bacteria of the gums, likewise. Of the vagina, likewise. Skin bacteria, likewise. These gene transfers mainly happen at short spatial distances, even when the phylogenetic distance (between two very different bacterial lineages) is great. The ecological circumstance of being resident in someone's vagina offers a proximity, a shared adaptedness to a very particular environment, and therefore an especially good opportunity, it seems, for two kinds of bacteria to share a gene by HGT.

So ecology matters more than phylogeny, in this transfer of genes within us. But there was still another message in the data. Alm's team also found that ecology matters more than geography. The microbes in their group of human-associated genomes came from people on several different continents. Not surprisingly, gene transfer between bacteria on the same continent occurred more often than gene transfer between continents. But the home-or-away difference at that scale wasn't as great as the home-or-away difference involving body sites. The shared ecology of the human gut, or the vagina, or the nasal passages, or the skin, was most conducive to horizontal transfer. The shared phylogeny of membership in the same bacterial lineage came second. The shared geography of the same continent was a weak third.

Alm's group wrote: "Taken together, these analyses indicate that recent HGT frequently crosses continents and the Tree of Life to connect the human microbiome globally in an ecologically structured network." In simpler language: genes go sideways, from limb to limb, even within our bodies.

73

Carl Woese followed the new discoveries by reading the journal reports, though he no longer stood at the forefront of lab work and discovery himself. He was now primarily a thinker.

In the early 2000s, as Ford Doolittle and other senior scientists published their influential alerts about horizontal gene transfer, as Eric Alm and other young researchers absorbed those alerts with high interest, Woese wrote his "millennial series" of papers on cell evolution and "the universal phylogenetic tree." He acknowledged in those papers that HGT must have been an extremely important process during the early stages of life on Earth, while making the rearguard argument, based on little offered data, that it had been far less significant since. This allowed him to continue insisting on the universality of his "universal" tree—the one drawn solely from the evidence of 16S rRNA, with three major limbs labeled Bacteria, Eukarya, Archaea. His pronouncement of the Archaea as the third great domain of life, revolutionary in 1977, had become an orthodoxy that he wanted to defend. This is what happens in science. Paradigms shift, but seldom twice in the life and mind of one scientist.

Woese wasn't alone (Norm Pace always stood with him, as did some others), but his adamant advocacy of a One True Tree left him increasingly in the minority, and he knew it. Many molecular biologists now seemed to feel, Woese admitted, that HGT over the ages had "erased the deep ancestral trace" of any uniquely valid, tree-shaped phylogeny. He disagreed. "These hasty conclusions are wrong," he wrote.

His own work life, meanwhile, changed in two ways. The first involved a new colleague. One day in September 2002 he reached out by email to a theoretical physicist in another corner of the University of Illinois campus. Nigel Goldenfeld was an Englishman, almost thirty years younger, who had arrived in Urbana as an assistant professor, risen to full professor, and spent his middle career studying the dynamics of complex interactive systems. That included topics such as crystal growth, the turbulent flow of fluids, structural transitions in materials, and how snowflakes take shape. The common element was patterns evolving over time. Goldenfeld had never met Woese but knew him by reputation. Later, he called that first ping "the most important email of my life."

Woese had contacted the Physics Department chairman and said: "I need a physicist to help. Who should I call?" The chairman had suggested Goldenfeld. In the email, Woese now explained that he wanted to discuss—with *someone*—the subject of complex dynamic systems. He felt that molecular biology had exhausted its vision, he wrote, and that it needed refocusing around drastic new insights. Those might come from recognizing that the cell itself is a complex dynamic system, and that cell evolution might have occurred through stages understandable only in that way. Properties sometimes emerge, when complex systems interact randomly, that go beyond what exists among unconnected elements. "My telephone is 3-9369," Woese wrote, "if you care to discuss the matter with me."

Goldenfeld replied fast. He was flattered, he was interested, but he didn't know much about biology, he confessed.

Woese emailed back: "You may not feel too much at home with biology as it now stands, but if I am any judge, the field is decidedly moving to meet you." What he had in mind, it turned out, was not just a conversation but a collaboration. Woese wanted a partner who understood complex interactive systems and could quantify their dynamics with brilliant math. Whether that partner knew a bacterium from an archaeon, or Darwin from Dawkins, mattered to him less.

"So began a scientific partnership and friendship that lasted more than a decade until his death," Goldenfeld wrote later. "During that time, we met nearly every day and talked on the phone or via email other-wise." Woese himself had trained as a physicist: math and physics as an undergraduate at Amherst, biophysics for his doctorate at Yale. Not only that, but also Woese prided himself on an academic lineage that went back through his PhD advisor, Ernest Pollard, to the exalted Cavendish Laboratory at Cambridge. Pollard's own PhD advisor at the Cavendish had been James Chadwick, who discovered the neutron; the lab leader at that time was Ernest Rutherford, who solved other mysteries of ra-dioactivity and the atom; before Rutherford's tenure, William Thomson (Lord Kelvin) had led the Cavendish, and before Thomson, James Clerk Maxwell. These are some of the giants of modern physics, and in addition to Chadwick, Rutherford, and Thomson, twenty-six other members of the Cavendish Laboratory (including two biologists named Watson and Crick) had won Nobel Prizes. The fact that Nigel Goldenfeld had done his own physics doctorate at the Cavendish may have added to his appeal as a youngish collaborator for the last phase of Woese's career. That career was now circling back toward biophysics and math, as Woese groped for deeper understanding of early evolution, the RNA-world that preceded his Darwinian Threshold, and the mystery of how so much complexity arose so quickly after the origin of life.

Woese and Goldenfeld wrote a number of papers during the decade of their collaboration, of which the most important and bold concerned the origin of the genetic code. This paper, coauthored with a young physicist named Kalin Vetsigian, appeared in 2006, bringing forward the subject with which Woese had grappled in his 1967 book. It offered the radical proposition that the universality of the code—all creatures using the same three-letter DNA combinations to call in the same amino acids—re-flects a dynamic evolutionary process during the early history of life, not a "frozen accident" traceable to happenstance in one small population of universal ancestors, as Francis Crick had suggested. That proposition was supported in the paper by a computer model and a tonic dose of math. The dynamic process involved "non-Darwinian" mechanisms that made innovation sharing between different lineages possible—and not *only* pos-sible but also so advantageous that a capacity for such sharing became mandatory. Among those "non-Darwinian" mechanisms, they argued, was horizontal gene transfer, at a level even "more rampant and perva-sive" than in later times. Horizontally transferred genes could be valuable to the recipient, by this view, even if the donor organism was still using a slightly different code. And the result of such rampant transferal, among other factors, was consensus on a single mode of coding and translation. Hence the universal code, by which a gene from one lineage of creatures can be optimally applied within another lineage. HGT is not just an ancient and widespread phenomenon, according to Vetsigian, Woese, and Goldenfeld; it was one of the preeminent factors in early evolution, shaping life as we know it and the informational system from which all life is built.

Another memorable Goldenfeld-Woese collaboration appeared in *Nature* a year later, under the title "Biology's Next Revolution." This one ran just a page. Its purpose, the two authors announced, was to explain why a radical transformation would soon sweep through the science, catalyzed by fresh thinking and "the coming avalanche of genomic data," and probably forcing biologists to revise some of their fundamental tenets, including the concepts of species, organism, and evolution itself. Their explanation started with horizontal gene transfer.

"Among microbes, HGT is pervasive and powerful," wrote Goldenfeld and Woese, acknowledging (as Woese earlier had been disinclined to acknowledge) that it wasn't just a thing of the dim past. "The available studies strongly indicate that microbes absorb and discard genes as needed, in response to their environment." Because of that genetic fluidity, the two men argued, the concept of "species" is useless among bacteria and archaea. With genes flowing sideways, information moving across boundaries, and energy flowing upward from cells through communities and environments, the concept of an "organism"—an isolated creature, a discrete individual—seemed less valid too.

And then there was "evolution" in its familiar Darwinian sense. That also seemed obsolete. Their newer idea—that evolutionary innovation might occur by means other than incremental mutation, and spread by means other than vertical inheritance—called the Darwinian model into question, they claimed. By this time, Woese had been reading some brilliantly unconventional scientific thinkers, such as the biologist Stuart Kauffman and the physicist Ilya Prigogine, associated with the swirl of ideas known as chaos and complexity theory, who proposed that certain "emergent properties," unpredictable and wondrously elaborate, could arise spontaneously within complex interactive systems. Kauffman particularly, in books such as *The Origins of Order* and *At Home in the Universe*, had suggested the possibility of "self-organization" emerging from biological systems. To some biologists, these ideas would seem dangerously metaphysical, steps toward rejecting Darwinian theory that, if misunderstood, as they surely would be, again might give aid and comfort to creationists. But to Woese, in his hunger for more, they appealed.

Somewhere amid the rich chaos of the RNA-world, Goldenfeld and Woese wrote, an "operating system" might have spontaneously taken form, by which the more promising innovations arising from random mistakes in RNA self-replication could be communicated and applied. They were alluding to the translation mechanism as seen eventually in cells, by which DNA information is turned into working proteins. At the core of that mechanism sits the ribosome, and at its core, Woese's beloved 16S rRNA molecule. That thought led to another: that early life evolved in what Goldenfeld and Woese called "a lamarckian way," meaning the inheritance of acquired characteristics, with vertical inheritance less important than horizontal gene transfer. "Thus, we regard as regrettable the conventional concatenation of Darwin's name with evolution, because other modalities must be considered." It was a fancy way of saying: *Let's pull Darwin down from his pillar. He may not have been wrong entirely, but his theory failed to cover the first two billion years.*

74

The other change in Woese's work life began earlier, back in 1997, when he arrived late to a meeting with the provost, several deans, some other faculty, and representatives from a charitable foundation that was considering a big grant to the University of Illinois. The grant, if it happened, would launch a center for comparative genome studies. The grant proposal came from a biologist studying mammal genomes and immunology, a young man who was friendly with Woese. The mammal fellow, Harris Lewin, had invented a fancy name for their enterprise: *phylogenomics*. Lewin would be lead researcher on this early effort. Woese would lend it star power, and the foundation's program officer badly wanted to hear from him before committing the money. But Woese had fallen from a ladder the day before, injuring his neck, and for that reason, among others (including his inherent crankiness), his colleagues didn't expect him to attend this meeting. Finally, he did show up, an "unearthly figure in a full neck brace," as Harris Lewin recalled later, and sat quietly at the end of the conference table. He looked uncomfortable. Probably he was in pain. The program officer asked him, "So, Professor Woese, what does phylogenomics mean to you?"

He took a deep breath, closed his eyes, and began to riff. Lewin remembered it as "an unrehearsed statement, a stream of scientific consciousness that blows the intellectual socks off everyone in the room. We all sit in stunned silence for a moment, knowing that Carl had just done something extraordinary." He had laid out a rationale for the whole, ambitious program: unraveling the true relationships among living organisms, using molecular evidence, and tracing those relationships backward in time to illuminate the true history of all life on the planet. Oh, only that? "I wish I had recorded it," Lewin said later.

Woese's performance especially surprised Lewin because he knew that Woese didn't even like the word *phylogenomics*. It seemed clever but undefined. But it was good enough, if one word be required, at least for the grant makers. It could pass as a label for Woese's own work, plus the work of Lewin and others among the interdisciplinary team they were assembling there in Urbana, and distinguish their enterprise from the efforts of others elsewhere. Woese hit his mark and made his speech, Lewin thought, out of loyalty to the university that had harbored him for a lifetime. "He did it for Illinois."

Whatever his motivation, whatever his unvoiced reservations, it worked. The foundation's check arrived within two months—seed money for a much bigger effort—and the project began. Ten years and $75 million later, the university opened its Institute for Genomic Biology, with Lewin as founding director and Woese as resident sage.

Harris Lewin considered himself an "improbable friend" to Carl Woese, because of wide differences in their scientific backgrounds, among other reasons. Lewin came from the field called animal sciences, meaning livestock and factors affecting their health, including genetics and immunology, and he studied topics such as bovine leukemia virus, a retrovirus that invades the genomes of cows. Woese didn't care much about animals, not scientifically, because the evolutionary questions that engaged him go back far earlier in time. Animals represent a small twig in the canopy of the tree of life, and he concerned himself with the big, deep limbs. Lewin had heard "some scary stuff" about Woese's unapproachability, but after a decade or so in Urbana, as his own work led toward comparative mammal genomics, he got curious enough to try to meet him anyway.

"He was perceived very much as a loner and a guy with a chip on his shoulder," Lewin told me during our conversation at the University of California, Davis, where he has lately finished a stint as vice chancellor for research. "I found a person that was quite the opposite of everything everybody said." He walked into Woese's lab, where yellowing sheets of paper showing the early trees, drawn with George Fox and others two decades earlier, still hung on the walls. Woese was seated, as usual, in his old swivel chair with his feet up on the lab bench. He stood, greeted Harris warmly, gave him a tour, and they talked for hours. Their friendship grew so trusting that, when the Institute for Genomic Biology (IGB) took shape, Woese lobbied university officials to appoint Lewin as director. And he lobbied Lewin, against a reluctance about giving up his own research, to accept. When the IGB opened its doors in 2007, in a new building just across from the university observatory, a sleek thing with big windows overlooking a stone plaza, they all moved in.

Woese left his old lab reluctantly. He had been up there on the third floor of Morrill Hall—site of his greatest discoveries, his hardest work, and many good times—for more than forty years. Lewin persuaded him to accept a nice office (he declined a grand one) in the IGB building, with a view of the plaza. That ended his "isolation" in Morrill, as Lewin saw it, and put him among "fresh young troops," Nigel Goldenfeld and others, "as well as his able and beloved assistant Debra Piper." Woese met Piper when he moved into the building, where she worked for the program that Goldenfeld called Biocomplexity, and she became a protector of Woese, as well as a helper and a friend. All seemed nicely hospitable to Woese, except that the office arrangement turned sour, and a small shadow came over his friendship with Harris Lewin, when Lewin commissioned for the plaza a certain three-piece sculpture meant to celebrate Woese's most famous achievement: the illumination of a third major domain of life.

"I wanted to do something to honor Carl's discoveries," Lewin wrote later, "with symbolic trees." A committee deliberated, and a statewide competition led to the choice of an "irreverent" artist from Chicago. The tree concept disappeared. The sculptor produced three abstract blobs, large things, molded from polyurethane but looking somewhat like mounds of cookie dough scraped off a spoon, all identical in their irregular shape but differing in scale and color. The biggest one was chartreuse. The middle-sized one was deep orange. The smallest was yellow. You can see them on your next trip to Urbana. Lewin accepted the blobs and renamed them, along with the plaza, Darwin's Playground. It struck Woese as an affront (to Lewin's enduring regret), and he wouldn't look at them. He started entering and leaving the institute by a side door, away from the plaza. He moved to an interior office with no windows. "Eventually, he forgave me," Lewin wrote, but it didn't come easily.

This marks the Late Woese period, when he was angry at Charles Darwin, disenchanted with molecular biology as practiced during the twentieth century, frustrated by resistance to what he considered his best ideas, annoyed by the very word *prokaryote*, embittered by what he considered his inadequate acclaim, disappointed in particular at not having won a Nobel Prize, and disinclined to appreciate whimsical modern art, outside his window, bearing Darwin's name. He had five years to live.

75

With the sequencing of the human genome, as a rough draft in 2000 and a better (but still provisional) version in 2003, by way of the competitive "collaboration" between Craig Venter's group and the International Consortium, scrutiny of the full sequence began to yield some surprises.

Those surprises related not just to the makeup of human DNA and how parts of it function, but also to the history of its assemblage and the sources from which some bits may have arrived. The first big revelation turned out to be a false alarm: the Consortium's announcement in 2001 that 223 human genes seemed "likely to have resulted from horizontal transfer from bacteria." The team had made some hasty assumptions, on insufficient data—their mistakes noted quickly by Steven Salzberg and other critics, as I described earlier—and that claim didn't stand. Further sequencing of genomes of nonhuman creatures put the human genome in better perspective. Meanwhile, another revelation from the sequenced human genome, also mentioned by the Consortium when it published its 2001 analysis, but given less public attention, was the vast amount of seemingly pointless repetition in the three-billion-letter sequence. There it sat, within the genome, like a mammoth landfill of what had been called "junk DNA."

So much redundant blather in the fundamental human blueprint seemed almost embarrassing. Most of it occurred in the form of relatively short bursts of code, each just a few hundred to a few thousand bases long, constituting units that reappeared thousands or hundreds of thousands of times. In total, those repetitive sequences accounted for half of the total genome. (Coding sequences that produced actual proteins comprised only 5 percent of the genome.) The repeats had been noticed years earlier, by other methods even before DNA sequencing became possible, and dismissed by some biologists with that "junk" label. Other scientists knew them, more descriptively, as "transposable elements." They were transposable in the sense that they seemed not just to have copied themselves many times, but also to have jumped around into different parts of the genome. These leaping, repetitive sequences weren't, in fact, useless; they were clues, and in some cases more. The Consortium scientists recognized them as potentially instructive—"an extraordinary trove of information" constituting "a rich paleontological record" about human evolution. What did that record reveal? Harder to say.

There's an ironic prehistory to the study of these transposable elements. They were first detected by the visionary plant geneticist Barbara McClintock back in the 1940s as she studied the genetics of maize (corn). At that time, McClintock worked at Cold Spring Harbor Laboratory, on Long Island, where she grew and tended her own maize, raising a few hundred plants on an acre or so of ground every summer. She looked for mutations, induced artificially by X-raying the kernels, and traced those mutation from chromosome to chromosome, from cornstalk to cornstalk, among genetic crosses she created by pollinating the plants by hand. Maize served well as a study organism for geneticists in the era before molecular biology, because many of its mutations show themselves clearly in the color of kernels within the variegated cobs. McClintock discovered that some of her induced changes took form as mobile entities, which could somehow bounce from one chromosome site to another in the course of plant development. She focused on two mutations in particular, observing the way they interacted to cause breakage in a chromosome. She called them "controlling elements" because they seemed to play a role in gene expression. What she had discovered, besides a gene-regulatory relationship, were the first transposable elements ever recognized. For that, almost forty years later, she received a Nobel Prize.

But the irony of her story isn't obscurity followed by acclaim. That's the mythic version, satisfying but inaccurate, and preferred by some tellers, who have made McClintock a great feminist hero. She was heroic, certainly, and she preferred the mythic version herself, though feminism was never her flag. The real irony is that McClintock always considered the controlling aspect of her elements far more significant than their jumping from place to place in the genome. By some testimony, she wasn't even much interested in transposition, at least later in her career. But the Nobel Committee was, and it gave her the prize "for her discovery of mobile genetic elements."

After McClintock's early work, as genetic research moved into the molecular dimension, other transposable elements turned up in the genomes of other creatures: bacteria, fruit flies, yeast, humans. They acquired a shortened label, transposons. And they acquired catchy names, some of them, just as genes are given names. There's a group of transposons known as *mariner*, which have sailed widely from place to place for millions of years and can be found in the genomes of fruit flies and many other animals, including humans. The original two *mariner* elements in humans arrived (from somewhere) during early primate evolution and copied themselves throughout our ancestors' genomes roughly fourteen thousand times. The most abundant human transposon is *Alu*. It's only about three hundred bases long, but that three-hundred-letter nonsense word recurs in the human genome more than a million times. Nature is wildly various, we know that, but nature under Darwinian selection is also thought to be sternly economical, and such redundancy has enticed some biologists to wonder what the devil has been going on. One of those wondering is Cedric Feschotte, a Frenchman from Toulouse.

Feschotte worked on transposable elements in insects for his doctorate at the University of Paris. He was hired as a postdoc at the University of Georgia, in Athens, to help study transposable elements again, this time in rice. As a side project, he worked on maize, crossing corn plants in a greenhouse, much as Barbara McClintock had done. Part of what makes maize still so apt for such research is that—unsuspected even by McClintock back in the 1940s and 1950s—transposons comprise 85 percent of its genome, and they jump around that genome frequently. From the Peach State, Feschotte went to the University of Texas at Arlington; and from cereals, he went to vertebrate animals. The constant again was these crazily mobile elements, bouncing and copying throughout genomes. By the time I caught up with him, he was a professor at the University of Utah School of Medicine, focused on transposons in humans and other vertebrates. On a shelf above his desk sat two variegated cobs of corn—mementoes of his sweaty days in Georgia, no doubt, and tokens of homage to Barbara McClintock.

Feschotte's first graduate student, in Texas, was a local fellow named John K. Pace II (no relation to Norman Pace), slightly older than his fellow students. Married, with kids, and ten years' experience as a computer programmer, John Pace just wanted to get a master's degree so he could teach biology at a community college somewhere. But then, under Feschotte's guidance, he made a major discovery. Putting his computer skills to work, scanning genomes in search of transposons, he found one in an East African primate known as a bush baby. The element ran to almost three thousand letters and repeated itself in the bush baby genome more than seven thousand times. That was notable enough—but what seemed more odd to Pace was finding virtually the same element in the genome of a very different animal: the little brown bat, native to North America. This time, nearly three thousand copies.

Pace and Feschotte, along with others from the lab, scanned genomes more widely, finding a close variant of the same transposon in a tenrec (a small mammal roughly resembling a miniature porcupine) from Madagascar. Recognizable parts of it occurred also in an opossum from South America, a frog from West Africa, and a lizard from the southeastern United States. Clearly, this thing—this fiercely assertive and agile stretch of DNA—had gotten around, both within and between creatures, within and between continents. But the fact of its complete absence from the genomes of many other vertebrate animals (including nineteen kinds of mammal) suggested strongly that it had gotten around horizontally, not by vertical descent throughout vertebrate ancestry. And once passed to a new genome, it replicated prolifically. The tenrec genome contained 13,963 complete copies. The bush baby held 7,145. Each version within the different animals was at least 96 percent identical to the other versions, giving full confidence that they shared a single source and a recent invasion history. Any set of transposons so invasive and strange deserved a vivid moniker, so Feschotte's team called it *Space Invaders*.

"Who coined the name?" I asked Feschotte.

"I did. You know, I do the marketing. They do the work." We both laughed at his professorial self-deprecation. "I coin the names of the elements," he said, still in joke mode about these crucial scientific chores: "I talk to the reporters and the writers."

Furthermore, he reminded me, these genomes of the tenrec and the opossum and the rest—they were all publicly available through an online database. In the current era, happily for science, people sequence whole genomes and share them. "This is the most democratic kind of research. Anybody could have found this at any time," Feschotte said. "You just need an internet connection, and you can do it." You can do it, that is, if you have that connection plus the biological knowledge and computational skills to ask the right questions in the right ways. But it's harder than Facebook. The paper on this work appeared in 2008, and John K. Pace II, the unassuming guy who just wanted a master's degree, left UT-Arlington with a PhD.

Among the big unknowns about all these transposons, intriguing to Feschotte, are (1) where they come from originally, (2) how they enter a new genome, and (3) why they copy themselves so profusely once they have gotten aboard. None of those three questions can be answered with certainty, but Feschotte has his preferred guesses. On the third, the matter of their busy self-copying, he favors the surplus DNA concept (non-gene DNA whose only "purpose" is to survive and proliferate) mentioned by Richard Dawkins in *The Selfish Gene*, his 1976 bestseller, and developed further by Ford Doolittle and a grad student in a 1980 paper. By this logic, transposons have acquired the capacity to self-copy because it improves their prospects of long-term survival. They replicate themselves more quickly than the host genome replicates, and they sometimes jump into other lineages, which enables them to evade extinction with the dying out of a single lineage. As a secondary effect, the redundant DNA that they add to a genome becomes available and might even prove useful, as it mutates, for cellular functions.

Gene regulation, for instance. That's how Barbara McClintock construed her transposons. Ultimately this may give the host organisms some survival advantage, and if the host lineage survives extinction, the transposon survives too. For the meantime, though, there was no proof of this. It was just McClintock's hypothesis.

The first unknown of Cedric Feschotte's three (ultimate source of any given transposon) remains mysterious, but on the second (mode of arrival), he has some ideas: parasites and infection. Viruses may occasionally carry such bits of selfish DNA from one species to another, just as viruses sometimes carry whole genes in the process of HGT. The parallel is close enough that Feschotte and his team began calling the process horizontal transposon transfer (HTT). That can be considered a subcategory of HGT. And they found evidence implicating one particular parasitic insect (*Rhodnius prolixus*) in cases of transposon transfer. It's a mean little critter, this insect, native to South America and Central America, that feeds on the blood of birds, reptiles, and mammals, including humans. It belongs to the group known as kissing bugs, because they tend to bite in the area near a victim's mouth.

Kissing bugs are despised in the American tropics not just for biting but also for their role in transmitting Chagas disease, a lingering and sometimes fatal affliction caused by a protozoan that replicates in a victim's blood and tissues. Charles Darwin encountered kissing bugs in Argentina during the *Beagle* voyage, when he made a horseback jaunt inland and slept in a bug-ridden village. He recorded in a notebook that it was "horribly disgusting, to feel numerous creatures nearly an inch long & black & soft crawling in all parts of your person—gorged with your blood." Typical of Darwin, when he was still young and robust, he shrugged that off with the aside "good to experience everything once." Whether the kissing bugs gave him Chagas disease is unknowable now (short of exhuming him from the floor of Westminster Abbey), but Chagas has been one hypothesis for the mysterious, chronic illness that punished Darwin throughout his middle years.

It turns out that this kissing bug, *Rhodnius prolixus*, carries more than the Chagas protozoan in its belly. Like the tenrec and the opossum and the frog of John Pace's study, it carries also a hefty dose of transposons in its genome. That genome was available—sequenced by others, presumably because of medical interest in Chagas disease—and Feschotte himself made the transposon discovery, he told me, while "messing around at home one night." By "messing around," he meant scanning a large number of published genomes, using sophisticated bioinformatics tools, to see where *Space Invaders* might turn up. He found it, unexpectedly, in the kissing bug. He already knew from the work with John Pace of its presence in several species of mammal, including the opossum, one of the bug's preferred South American hosts. That raised his suspicion, given the bug's blood-sucking habits, which seemed to offer opportunity for transfer of DNA as well as disease. It suggested that the bug might be an intermediary, a vector, for the transposon. Next morning, he alerted two of his postdocs and invited them to investigate. Further scanning the bug's genome, they found not just *Space Invaders*, represented in more than two hundred copies, but also three other transposons known previously in mammals. From the evidence of mutation rates, the transfers seemed to have happened within a time range of fifteen million to forty-six million years ago.

Let's pause briefly to appreciate just how odd this scenario is: selfish DNA passing from the genome of one species of mammal, through the belly of a blood-sucking insect, into another species of mammal, where it inserts itself into that genome. The transposed DNA becomes part of the second mammal's heritable legacy. And once the self-copying of the transposon begins, it adds masses of DNA to the genome. This could be bad or, much less probably, good. If bad, it disarranges the genome, destroys necessary gene functions, induces congenital diseases, and maybe even causes the mammal lineage to go extinct. Science will never see that transposon, because it has vanished with the unlucky lineage. But if the mammal is lucky, the new DNA brings no mortal harm, and some of it could even become useful. It adds possibility, it adds raw genetic material, it adds the chance of new genes taking shape from old transposon DNA. And new genes, as environments change, can mean the difference between survival and oblivion. If a new gene is distinctly valuable, it spreads through the population, it passes the test of time, and it enshrines itself in a lineage of opossums or monkeys or frogs or other creatures, such that Cedric Feschotte's group might find it there many millions of years later. Meanwhile, it could alter the course of evolution.

Including human evolution. Back in 2007, in a slightly different effort, John Pace and Cedric Feschotte assembled a list of transposons that have entered the primate lineage, most likely by horizontal transfer, over the past eighty million years. They found forty. Each has copied itself abundantly. Those copies now constitute about 98,000 distinct elements, 98,000 stretches of alien DNA, amounting to 1 percent of the human genome. They're still with us, changing slowly, and their effects too are largely unknown.

76

As his research career ended, Carl Woese assumed his new role as a much-honored but crotchety elder with strong opinions. He collected kudos, and he wrote. Having already received a MacArthur Fellowship, and an award from (in addition to his election to) the National Academy of Sciences, and the Leeuwenhoek Medal (microbiology's highest honor) from the Royal Netherlands Academy of Arts and Sciences, in 2000 he was announced as a winner of the National Medal of Science, bestowed by the president of the United States with advice from scientific counselors. Woese declined to attended the event in Washington because he didn't want to shake Bill Clinton's hand.

In 2003 came the Crafoord Prize, given by the Royal Swedish Academy of Sciences, as a complement to the Nobel Prizes, and presented by Sweden's king. Woese hated travel, but he did go to Stockholm for that event, inviting Harris Lewin and Gary Olsen (his faithful collaborator in Urbana) to accompany him, and he had no scruples about shaking the hand of King Carl XVI Gustaf. The Crafoord Prize is sometimes split between awardees (Edward O. Wilson and biologist Paul Ehrlich shared it in 1990), but Woese got it all, the $500,000 and the honor. "For Carl, winning the Crafoord Prize by himself was a tremendous vindication," Lewin wrote later, "and he joked that winning the Crafoord alone (especially without Craig Venter) was better than sharing a Nobel." Brave talk, but even Lewin didn't believe him. "In reality, I am certain that Carl coveted a Nobel Prize." He could console himself with the rationale that the Nobels recognize no biology category, but Barbara McClintock received one in the category Physiology or Medicine, for biological work, and so did Watson and Crick. Woese had been nominated for a Nobel, but maybe his discovery of the Archaea seemed a little too obscure, and maybe he just didn't live long enough.

One year after the Crafoord Prize, in 2004, he published another of his big, ambitious treatises. This appeared not in *Nature* or *Science* but in a narrower journal, *Microbiology and Molecular Biology Reviews*, the editors of which allowed him fifteen pages to vent. It was an appropriate outlet, not just a spacious one, because he wanted to tell the field of molecular biology just what he thought of it. He wanted to piss in the punch bowl.

He titled this essay "A New Biology for a New Century." His central point was that molecular biology had failed its early promise and declined to "an engineering discipline." By that, he meant it had come to concern itself with applications, such as genetic modification of organisms, for agriculture or environmental remediation, and the concerns of human health. Woese, for his part, wasn't nearly so interested in human health as in evolution. Back at the dawn of the molecular era—when Oswald Avery discovered transformation, when Watson and Crick solved the structure of DNA, when Crick suggested "protein taxonomy" as a way to discern the tree of life, when Zuckerkandl and Pauling proposed using molecules as an evolutionary clock—back in that glorious time, molecular biology had seemed a branch of science that might illuminate "the master plan of the living world." But then came a schism. Two biologies: molecular went one way, and evolutionary went another. Academic biology became divided, in universities all over America and much of the world. Two separate curricula, two separate buildings.

Worse, molecular biology took a "reductionist" perspective on what it saw as mechanistic problems, Woese argued, such as the workings of the gene and the cell. It lost sight of "the holistic problems" of evolution, life's ultimate origins, and the deepest mysteries of how life-forms became organized. It lost interest, or never had any, in the big story over four billion years. "How else could one rationalize the strange claim," Woese wrote, "by some of the world's leading molecular biologists (among others) that the human genome (a medically inspired problem) is the 'Holy Grail' of biology? What a stunning example of a biology that operates from an engineering perspective, a biology that has no genuine guiding vision!"

No one ever accused Woese of pulling his punches. And as he got older, ever more pugnacious, his disdain for Charles Darwin rose too, distinct from but alongside his disdain for molecular biology. The Darwin animus had kindled within him for a long time, other testimony suggests, but in impersonal and inconsistent form, an off-and-on resentment of the distant figure with the big name. According to his friend Nigel Golden-feld, Woese never read *On the Origin of Species* until around 2000, because he presumed it irrelevant to the evolutionary questions that interested him. What he knew of Darwin's theory, he knew (as most people, even biologists, know that theory) from secondary sources. Then he did read *The Origin* and a bit of Darwin's other work, and at first he reacted favorably. In 2005, responding to an interviewer's question about scientists who had inspired him, he mentioned Crick, Fred Sanger, and just a few others, including Darwin, "whose writings I encountered rather late in the game, but increasingly turn to as my foray into evolution deepens. How could he have been so right about so much? Astounding!" That interview appeared in *Current Biology*, a serious journal. He was on record.

But then something happened, something that radically changed his view of Darwin, or maybe just his bluntness in expressing it. He read Darwin's *Origin* more carefully—in fact, he pored through it, comparing the first edition with all five others variously revised by Darwin himself, according to Nigel Goldenfeld, who performed this exercise with Woese. He read the surviving correspondence between Darwin and Alfred Russel Wallace, the man with whom Darwin shared credit in 1858 for conceiving the idea of natural selection. Some of that correspondence bore on the question, persistently argued by several scholars, of whether Darwin seized an undue share of the credit. At their most extreme, those scholars charge that Darwin stole parts of his theory from Wallace and concealed the dastardly deed. It's a provocative accusation, with all the enticements of good slander, but its credibility evaporates (in my view) with a thorough reading of the Darwin-and-Wallace literature, which is abundant. Woese came to feel differently.

He also learned about precursors to Charles Darwin—such as Lamarck; Edward Blyth, an English zoologist working in India; and

Darwin's own grandfather, Erasmus Darwin—who had offered inchoate
notions of evolutionary change, or partial versions of what Darwin put
together so effectively, and to Woese, it all began to look like high intellectual larceny. His dark view was affirmed (if not triggered) by a book
titled *The Darwin Conspiracy*, a tendentious little volume accusing Darwin
of plagiarism and deceit. The author was a onetime BBC producer named
Roy Davies. Its subtitle: *Origins of a Scientific Crime.*

Discovering the Davies book, Woese took it as revelatory. He be-
friended Davies, ordered multiple copies, and gave them away. There are
kernels of fact and circumstance in the Darwin-stole-from-Wallace narrative sufficient to persuade the credulous and it makes a piquant tale.
The surprising thing in Woese's case is that such a deep thinker on other
matters would be relatively shallow on this. During the last phase of his
life, that piquancy seems to have helped soothe his own frustrations like
a menthol lozenge.

Mostly he expressed his grudge against Darwin in private to friends
and colleagues. On February 12, 2009, Darwin's two hundredth birthday, for instance, Woese noted the date by sending select friends a terse
message: "Let this be a day of rage." He began work on a book with Jan
Sapp, under the title *Beyond God and Darwin*, which Sapp conceived as a
sort of popularized version of his superb but dense tome *The New Foundations of Evolution*. This new book would offer a streamlined account of
the revolution in molecular phylogenetics, showing that the discoveries
of Woese and some of his colleagues transcended Darwinian theory, and
did it without giving any support to creationist ideology. That was the
point of Sapp's intended title. Such discoveries—of endosymbiosis, horizontal gene transfer, and the deeply contorted tree of life—go beyond the
God-versus-Darwin dichotomy, beyond creationism versus *The Origin*.
They go beyond Darwinian theory without undermining the reality of
evolution.

Sapp wrote an introduction and sent it to Woese for comment. To
this draft, Woese added his annotations in upper case, which gave the impression of a stern editor. "SHARPEN UP," Woese typed after one paragraph. "MORE PUNCHY." Most of the comments were small quibbles
and wording suggestions. But at the end of the draft, Woese wrote: "JAN,
YOU ACCORD DARWIN SO MUCH MORE SUBSTANCE THAN THE
BASTARD DESERVES."

Sapp dropped the project not long after this exchange, and *Beyond God and Darwin* never got written. Sapp had lost interest in the book effort, though not in the topic, he told me. He was also disheartened by Woese's increasingly needy ego. "Later in life, Carl thought he was bigger than life."

"And bigger than Darwin," I suggested.

We were discussing this over lunch in a noisy restaurant in Montreal. Sapp swallowed, let my comment pass, and said, "I didn't like that side of him." But their relationship survived to the end, despite Woese's disappointment about *Beyond God and Darwin*. Sapp didn't think Woese actually cared much about finishing the book, he told me later, but that nonetheless Woese might gladly have worked on it forever, because it gave them occasion for so many phone calls and emails, as well as discussion in person. "He was in it for the dialogue." Woese was a lonely man—though he had a wife and two children in a home just off campus—and he treasured his friends

This draft introduction with Woese's inserted comments resides among the Carl Woese Papers at the University of Illinois Archives. Another odd item preserved there (and called to my attention by the archivist John Franch, on the same day he showed me the X-ray films of Woese's earliest RNA fingerprints) is a cheap ring notebook, its yellow cover fading to cream, bought from a CVS pharmacy some time after 2006. It's unlabeled and untitled, but it contains a few pages of cursive scribble in Woese's hand. One page holds a single line, "Book 1: Growing up in Science," suggesting he may have intended an autobiography. The next page, headed "Preface," dates to the period of bicentennial celebrations of Charles Darwin's birth (just following the world financial crisis of 2008), when Woese hit his peak of disgruntlement.

Darwin was born, as I've mentioned, on February 12, 1809. "As I set pen to paper," Woese wrote, "the year is 2009. It's Darwin, Darwin everywhere and no one with a thought. Hopefully this year will be a nadir in biology—as it (hopefully) is in the world economy, and both biology and the economy will recover." The few paragraphs that follow are muddled and grumpy, inconclusive ruminations on evolution—"the heart center of biology"—and on the sad fact that society has confused evolutionary biology with Darwinism. The last page of the sequence is mostly crossed out with zigzag lines. One sentence is left to stand: "Science does not succeed by 'keeping secrets' and 'being diplomatic' (i.e., scheming); that's alchemy." The rest of the notebook is empty. The autobiography, like the book with Jan Sapp, never happened.

But there's more in this vein preserved in the memories of Woese's friends and colleagues. One of them, a chemist at the Institute for Genomic Biology, tells a story (though she wouldn't repeat it to me for quotation) reflecting Woese's late attitude toward evolution and his own place in its history. She went to see an important science bureaucrat in Washington—maybe a program officer at the National Science Foundation, someone like that—and noticed that high on his office wall hung portraits of two men: Charles Darwin and Carl Woese. Returning to Urbana, she mentioned that to Woese, presuming he might be flattered. Woese said: Why Darwin?

77

M olecular biologists now understand, Jan Sapp wrote in his draft introduction to the book that never happened, that "bacteria evolve by leaps and bounds through the inheritance of acquired genes." It sounded more Lamarckian than Darwinian, he well knew. HGT was what he had in mind.

And this wasn't true just of bacteria, evolving by leaps and bounds. Animals do it too, sometimes. And not just insects and bdelloid rotifers. Mammals do it too, sometimes. "The cells that make up our bodies have also not arisen gradually in the typical Darwinian manner of gene mutation and natural selection." Some of the changes occurred by quantum leaps. Our mitochondria came aboard suddenly, deep in the past of our eukaryote or pre-eukaryote lineage, as captured bacteria. Plants acquired their chloroplasts the same way. Our genomes are mosaics. We are all symbiotic complexes, even us humans.

"Consider too," Sapp wrote, "that a great percentage of our own DNA is of viral origin." The figure most commonly cited is 8 percent: roughly 8 percent of the human genome consists of the remnants of retroviruses that have invaded our lineage—invaded the DNA, not just the bodies, of our ancestors—and stayed. We are at least one-twelfth viral, at the deepest core of our identities. Consider that, Sapp urged.

So I did. And when I learned that few scientists if any have studied our viral genomic content more carefully than Thierry Heidmann, of the Gustave Roussy Institute on the south outskirts of Paris, it seemed worth a trip to see him.

78

T hierry Heidmann is another biologist who trained originally in physics and math. He grew up in Paris and got his education there at some of the best institutions in town: the École Normale Supérieure, followed by the University of Paris and then the Pasteur Institute. His father was an astrophysicist, and he considered astrophysics himself until he tipped toward neurobiology and the science of complex neuronal networks, roughly the same sort of complexity (with emergent properties) that intrigued Woese in his late collaboration with Nigel Goldenfeld. Heidmann, after finishing his doctorate in that area, hankered to do something more germane to human health, so he began to think about tumors and their relation to transposons and retroviruses.

A retrovirus is a virus that works backward relative to the usual DNA transcription process. Instead of DNA-makes-RNA-makes-protein, the normal path from genetic information to living application, a retrovirus uses its RNA genome to make DNA, the double-stranded molecule. That trick, plus a few others, allows it not only to invade a cell but also to enter the cell nucleus and patch a DNA version of itself into the cell's DNA, becoming a permanent part of the cell's genome. Whenever the cell or its descendants replicate, that alien stretch is copied too. If the retrovirus happens to infect germline cells—eggs or sperm or the cells in ovaries and testes that produce them—then the inserted viral sequence will be inherited, passed along as a permanent part of the genome. At this point, it's no longer alien to the organism. It's now *endogenous*, meaning native, inherent. Such viruses are known as endogenous retroviruses (ERVs) because they become endogenous to the lineage of creatures they have infected.

If a retrovirus inserts itself into the human genome, then that's a human endogenous retrovirus, or HERV. Those are the viruses, HERVs, that constitute 8 percent of the human genome. Trust me, this is leading to a point that will boggle you at the implications of Thierry Heidmann's work.

Some retroviruses cause cancer. Mouse leukemia virus, for instance. Heidmann studied that one. The most notorious of retroviruses, of course, is HIV-1, which causes AIDS. It would have been logical at the time, the late 1980s, when Heidmann established his own lab, for a scientist such as he with an interest in retroviruses to launch a research program on HIV-1. The money would have flowed, the significance was vast and urgent. Instead, he did something less obvious.

Tumor biology had its own urgency, and the connection of tumors with retroviruses continued to fascinate him. He knew of an old observation, made by more than one electron microscopist in the years before genome sequencing, that virus-like particles show up abundantly in placental tissue and in some tumors. Placental tissue? That seemed curious, a potentially fruitful digression, so Heidmann began looking for evidence of retroviruses in human placenta, freshly recovered from hospitals. He and his team found a new family of viruses, embedded in the placental DNA, that they named HERV-L. Further investigation showed that similar sequences, other ERV-L variants, exist in the mouse genome and the genomes of other mammals. "We made a sort of viral archeology," Heidmann said in a 2009 interview.

They found that the presence of ERV-L in animal genomes, including ours, dates back about a hundred million years, to before the big divergence of mammals. In primate genomes, this thing has copied itself, like a transposon, about two hundred times. Did it have a function? Did it serve as a gene? they wondered. Maybe and maybe not. It might just be selfish DNA, perpetuating itself by bouncing and replicating throughout various genomes. Heidmann's group didn't claim to know. But that was the beginning of Heidmann's decades-long exploration of the viral component of human genomes, human identity. What he found, in his long search, is that *some* HERVs, if not that one, have indeed acquired roles as human genes.

Thierry Heidmann is a generous man as well as a distinguished researcher. When he heard where I'd be staying in Paris, he said: Ah, very near my neighborhood. Forget the Metro, I'll pick you up. So I stood outside my little hotel in the 17th Arrondissement and, promptly eight in the morning, a small white Volkswagen pulled to the curb. Out stepped a man with a graying beard, robust eyebrows, and a blue blazer worn over his sweater, welcoming me to hop in. Instead of swinging north to the nearby Boulevard Périphérique, the great ring road, as I had expected, he drove southeast on a picturesque avenue toward the heart of the city. Traffic wasn't bad, and along the way, he pointed out some sights: there's the church de la Madeleine, this is the Place de la Concorde, the Louvre on our left as we cross the Seine, Notre-Dame, of course, just upriver, now here's Boulevard St.-Germain, and the Sorbonne, and over there the École Normale Supérieure, where I went to school, he said. This man, every day, makes the world's most elegant commute. We reached the Gustave Roussy Institute in less than an hour, talking the whole way about his background and his work, and then talked for another six hours in his lab office, interrupted barely by lunch.

The office, at the end of a lab corridor, is a small room with windows overlooking the treetops of Villejuif, a suburban village within which the Gustave Roussy Institute sits. The room is lined with shelves, the shelves filled with journal papers in piles, folders, and boxes, many of them printed out on colored paper—blue, green, orange, pink, as well as white—lending a nice sort of Mondrian décor to a very serious workspace. We sat at a table beside his MacBook Pro, on which he showed me slides of figures and graphs as he spoke, illustrating what he had been doing for twenty years. It began with the HERV-L work and proceeded into the story of another human endogenous retrovirus his group discovered not long afterward. What had changed between the two discoveries was that the human genome had been sequenced, and that sequence was publicly available. Heidmann's methods changed accordingly. His team screened the entire genome, looking for evidence of unknown ERVs. What they sought in particular were signs of one kind of gene: a familiar, recognizable gene that makes a viral envelope, which is a sort of sticky wrapping around the hard capsule of a viral particle. They saw twenty.

"There are all these envelope genes," Heidmann told me. "Among them, two are very important."

One of the two had already been discovered by other researchers and named *syncytin*. This gene expressed itself, turning out its protein, most notably in placental tissue. What was it doing there? Nobody knew, not at first, but its name derived from a capacity to cause cells (*cyt*) to fuse together (*syn*). This effect had been proven with laboratory cell cultures. Fusion of cells into aggregate cell masses with multiple nuclei instead of individual walls is a crucial step in building one layer of the human placenta. That layer, a sort of permeable protoplasmic cushion, is the part of the placenta that mediates between the maternal blood and the fetal blood. (Brace yourself for a fancy bit of lingo: it's called the *syncytiotrophoblast*. Okay, relax and forget.) So the hypothesis arose that *syncytin* might help assemble that layer. Heidmann's group found another envelope gene, left behind from an entirely different retrovirus, with a similar capacity to fuse human cells. They named it *syncytin-2*. (The first became *syncytin-1*.) Their lab testing revealed the same capacity for causing cell-to-cell fusion, adding support to the hypothesis that these genes help build placentas.

Soon after that, Heidmann's team recognized two other such genes in the common laboratory mouse. These were different enough from the human *syncytins* to get their own names: *syncytin-A* and *syncytin-B*. Further screening of genomes revealed close equivalents also in rats, gerbils, voles, and hamsters. So the two mouse genes were old, having entered the rodent lineage at least twenty million years ago, before those branches split.

"At this stage," Heidmann told me, "there was a question we asked. How is it possible in fact that a gene, which has been captured by chance, could be so important?"

Building placentas is obviously important, but the team needed more evidence linking *syncytins* to that function. They got it by experiments with genetically modified mice—"knockout" mice, in which the *syncytin-A* gene had been decommissioned by molecular manipulation. Breeding those mice, they watched the embryos all die in utero after no more than thirteen days of gestation. The usual gestation period in mice is nineteen to twenty-one days. These dead mice weren't coming close. Dissection revealed structural defects in the boundary between placenta and fetus, which constrained the fetal blood vessels, inhibited fetal growth, and killed the unborn mice. It was persuasive. But, of course, you can't do that experiment on humans.

Heidmann's curiosity spiraled outward. He and his team found a *syncytin* gene in the European rabbit. They found a carnivore *syncytin* in the genomes of dogs and cats. They found one in cows and sheep. They found one in ground squirrels. "We worked with many, many university and labs and zoos." They collaborated particularly with a zoo in France, "with a lot of animals, because we get placenta when possible." They even found one of these genes in marsupials.

Marsupials, they have a placenta?

"Very transient placenta. Because, in the marsupials there is opossum, or the kangaroo, wallaby, okay. So they have a very short-lived placenta." Some people assumed otherwise, he said—that a marsupial has no placenta "because the embryo is going to a pouch outside of the mother." So a marsupial female with her transient placenta and her gestation in an external pouch, yes, even she has a viral gene targeted to helping with the placenta. Heidmann's group named that gene *syncytin-Opo1*, for the marsupial in which they first found it: the gray short-tailed opossum.

All of these genes have four things in common. Each one derives from the envelope gene of a retrovirus that inserted itself in the mammal genome. Each one expresses itself as a protein suffusing the placenta. Each one causes cell-to-cell fusion (at least in lab cultures), suggesting that it can create that special fused-cell protoplasmic layer, which helps mediate between placenta and fetus, letting nutrients and gases seep in from the mother, letting wastes seep out. And each is an ancient gene, preserved for many millions of years in functional form (against the disarray of random mutations) by natural selection. Such preservation proves the genes have been useful. They aren't junk DNA. They are tools, not scrap. They have helped the fittest mammals survive.

Those four points represent the canonical standards, articulated by Heidmann's group, for what constitutes a *syncytin*. But equally notable, his team realized, is what these genes *do not* have in common: They all came from different sources. They represent independent captures, independent domestications, of viral genes from entirely different retroviruses. That independence, Heidmann suspects, accounts for the high diversity of types of placenta, which among mammal species is an extremely variable structure. He gave me a whole disquisition on placental structure and taxonomy, wondrous and arcane, which I'll spare you.

"They capture different syncytins," he said, "they" meaning the different mammal lineages throughout evolutionary history. "They capture different envelope genes, because they capture it from different viruses. And the difference in the capture is responsible for the variability of the structures. Okay." The captures had occurred, furthermore, at vastly different times. The primate *syncytin-2* dates back at least forty million years. The rodent version, as I mentioned, has sat in that lineage for twenty million years. The cow-and-sheep *syncytin* seems at least thirty million years old, and the one in marsupials may have entered that lineage more than eighty million years ago. What this reflected, Heidmann said, was a continuous bombardment of animals and their genomes by retroviruses. Most of those infections didn't result in the viruses inserting themselves into the genomes, but a tiny fraction did, and those few yielded the endogenous retroviruses. A still smaller fraction of those ERVs gave their envelope genes to be transmogrified into *syncytins*.

But wait. This whole pattern of different *syncytins* appearing in different mammal lineages at different times raises another question, a logical dilemma, which had struck me as I read through Heidmann's papers before ever boarding the plane for Paris. If some of these essential *syncytins* are twenty million years old, and some are thirty million years old, and some are forty million, how did the mammalian lineage arise at all—how did the first placenta evolve—before those gene captures occurred? They have been acquired intermittently and by chance, through the course of mammal evolution, but they have always been necessary. You can't *be* a placental mammal without a placenta. Which came first, chance or necessity?

"Yeah. Exactly," said Heidmann. "This is the paradox."

79

eidmann and his young colleagues answered the paradox with a hypothesis. Their hypothesis involves another nifty capacity of *syncytin* genes, probably also derived from the viral envelope genes from which they have been anciently modified. That capacity is immunosuppression.

Retroviral envelopes are complex and versatile structures, and the genes that code for them, equally so. Besides their ability to cause cell-to-cell fusion, they can also suppress the antiviral immune response of a host. This has obvious value to an invading virus. It has other value, less obvious, to a mammal for use in its placenta. Both the fetus and the placenta of an individual mammal carry a different genome from the mother's. Half their DNA comes from the father. If the mother's immune system were entirely on alert, her white blood cells might attack the fetus and reject it. Part of the role of the placenta, a uniquely adaptive organ among placental mammals, is to keep peace between the mother's immune system and the fetus by dampening that response. This allows for internal pregnancy and birth, an innovation back at the time when early mammals diverged from the reptile lineage, which evidently offered certain advantages over egg laying. Birds still exist, Heidmann reminded me, meaning that the advantage wasn't absolute. Birds don't have placentas.

They eject their embryos early within hard-shelled oval packets, sending them out self-contained, with an endowment of nutritious yolk and a mere promise to keep them warm. That is, they lay eggs. Crocodiles, likewise. But the placenta gave an edge to some vertebrate lineages in some situations. One lineage of mammals seized that advantage while another, the ancestors of what we now call monotremes (platypuses and spiny anteaters, egg-laying mammals) did not. What's so advantageous about pregnancy and live birth? Well, it allows a mother to walk around, for one thing, carrying the fetus to safety inside her body instead of squatting on it like a sitting duck.

What this all suggests, according to Heidmann's hypothesis, is that the earliest syncytial gene that was captured in preplacental mammals may have served such a purpose of immunosuppression toward the fetus, and then gradually acquired its additional role also as an intermediating layer in the evolving placenta. Later *syncytins* may then have been substituted into mammalian lineages as improvements upon the first one.

Heidmann and I went to lunch nearby, then came back for a final surge of talk. What does all this tell us, I asked him, about how evolution works? About the tree of life?

He sighed at the breadth and flat-footedness of my question. "That our genes are not only our genes," he said. Then he laughed, and I laughed too, but uneasily, because I was unsure I'd heard right. I asked him to repeat.

"*Our* genes are not only *our* genes," he said. "Our genes are also retroviral genes."

80

One step beyond Heidmann's sobering message—that a sizable fraction of our human genes have come from nonhuman, nonprimate sources—lies the brave new world of CRISPR. That crisp little acronym, as you probably know, refers to something very complex: a sensationally efficient system of genome editing, famed in newspapers, bruited in magazines, heralded by the journal *Science* in 2015 as Breakthrough of the Year, and widely expected to bring someone a Nobel Prize. CRISPR is more than a step beyond what Heidmann said. It's the latest big lurch toward genetically engineered futures. It's a technique that opens the prospect of cheaply, precisely altering genomes (including human genomes) in the laboratory and (eventually) in the clinic.

CRISPR stands for *clustered regularly interspaced short palindromic repeats*. A palindrome, of course, is a sequence of letters that spell the same words in either direction, front to back or vice versa. In the verbal realm, that's wordplay, yielding gambits of strained cleverness such as "Able was I ere I saw Elba," in the voice of Napoléon, and "A man, a plan, a canal: Panama," referring tenuously to Ferdinand de Lesseps. A verbal palindrome says something, but what it says needn't make much sense. For instance: "A Santa dog lived as a devil god at NASA." My own favorite is simpler: "As I pee, sir, I see Pisa."

Within a DNA genome, the available alphabet for palindromic patterns is only those four coding letters, A, C, G, and T. So a palindromic repeat of DNA coding would look more like GTTCCTAATGTA-ATGTA-ATCCTTG. Such DNA palindromes, it turns out, can be important functional markers. They were the first evidence that led scientists to discover the CRISPR mechanism where it exists in the natural world (more on that in a moment) and, within a few years, to develop it as an extraordinary new method of genetic engineering. The full system includes other molecular elements besides the palindromes—enzymes and forms of RNA—but CRISPR has become the informal label for the whole shebang.

CRISPR allows researchers to target any mutation, any single "wrong" letter, in the three-billion-letter human genome, and to send in biochemical tools that make a correction. It offers parents the hope that congenital defects, mutations that might kill or torment a child, can be not just detected by genetic screening but also reversed before a fetus begins to grow. Delete and replace a gene that would have brought muscular dystrophy? Wonderful. Erase a mutation that threatens cystic fibrosis? Heroic. By one count, there are more than ten thousand such heritable human disorders, each caused by a single bad gene, many or most of which could be fixable using CRISPR. Those fixes, furthermore, can be made not just to the somatic cells (the body cells) of the child but to the germline, the sacrosanct reproductive cells from which DNA passes onward to subsequent generations. How will that be done? By performing the edit very early during in vitro fertilization—one human egg in a dish, one human sperm, plus a dose of the CRISPR magic. Such germline engineering is especially powerful and controversial because it affects populations, not just individuals. A germline tweak may well become a permanent alteration within a lineage of creatures. It could change future lives, not just those in the present. It could change the evolutionary trajectory of a species—for instance, ours.

That hasn't happened yet. No test-tube baby, so far as we know, has been born with CRISPR modifications. Some prominent researchers in the field have called for caution, restraint, even a global moratorium on using CRISPR for human germline engineering. Others have noted that CRISPR, with its ultimate potential for inserting new bits of DNA as well as its near-term potential for fixing mutations, carries a threat of high-tech eugenics. Add a gene to your unborn child that promises higher intelligence, or athletic prowess, or first chair in cello at music camp? Ugh.

And then, what, we will live in a world like Lake Wobegon, where all the children are above average? This dreamy prospect is sometimes called "voluntary" genetic modification, as distinct from therapeutic modification, driven by urgent medical need. Like other sorts of "voluntary" parental nudging of offspring up the ladder of life, such as hiring a private tutor to coach your kid for college entrance exams, it might come to seem not just tantalizing and beneficent but necessary on competitive grounds. Still, what seems necessary to the wealthy and privileged is often unthinkable for everyone else. It would exacerbate gaps between the affluent and the struggling, between the optimized and the ordinary, between well-engineered children and those haphazardly conceived the old-fashioned way. And unlike SAT tutoring, or cosmetic nose surgery, or tae kwon do lessons from the age of five, it's a nudge whose effects, good or bad, would be passed to future generations.

At present, germline engineering of humans is still just a looming likelihood, not a runaway trend. But among the latest news, as I write, is a report in *Nature* from an international team of scientists, based in Oregon and elsewhere, who used CRISPR tools on single-cell human embryos to correct a mutation.

This particular mutation causes a heart disease called hypertrophic cardiomyopathy (HCM), which can show itself as sudden heart failure, sometimes in seemingly healthy young athletes. It kills some, and it shadows the lives of many. The work on HCM was a lab experiment, not an application of clinical medicine. It involved fifty-four human embryos, each treated with a CRISPR repair elixir, most of them successfully, some not, but none of them afterward implanted in a human womb or destined to grow into a CRISPR baby. Still, crossing that threshold is just a matter of time and, at the current pace of research, probably not much time. CRISPR is hot and democratic. It's inexpensive and relatively easy. In fact, one company is now selling do-it-yourself CRISPR kits (for bacterial, not human, gene engineering) online at less than $200. Around the globe, well-trained researchers and perhaps also some pedestrian scientists, more ambitious than judicious, now have the tools.

It's not my intention here to try to detail how CRISPR works or to explore the range of its ethical implications. You'll get plenty on that from other sources in coming years. The origins of CRISPR, not the repurposing of it by humans, are what link it to Carl Woese and the new tree of life. Those origins, seldom mentioned in press accounts of CRISPR's whizbang applications, are fascinating in their own right and very relevant to this book.

During the late 1980s and early 1990s, several teams of scientists discovered strange, repeating sequences in the genomes of certain microbes, including the familiar bacterium of the human gut, *E. coli*. The sequences were generally about thirty letters long, on each side of the midpoint, and situated like bookends around a short sequence of letters that didn't repeat. Think of it roughly like: "Able was I ere gesundheit ere I saw Elba." Remember that these words were all written in *As*, *Ts*, *Cs*, and *Gs*. No one knew what such odd stretches of DNA did—or whether they did anything at all. But in 2002 they were given their name and their acronym, CRISPR, and researchers continued guessing about their function. Three years later, that mystery was cracked by a Spanish scientist named Francisco Mojica.

Mojica grew up near the Mediterranean port of Santa Pola, knew the coastline, and did his PhD work on the genome of a microbe, one of Woese's cherished archaea, that dwelt in briny marshes of the Santa Pola area. The microbe was halophilic: salt loving. Examining its genome, he noticed an odd pattern: multiple copies of a nearly perfect palindromic repeat, with a spacer of other letters between the mirroring sides. Getting deeply curious, he spent much of the following decade looking for the same sort of pattern elsewhere. He found it—other versions, other palindromes—not just in archaea but in the published genomes of bacteria too. One case was that *E. coli* occurrence, reported back in 1987 by a Japanese team, who were mystified as to what it meant. By the year 2000, searching published genomes, Mojica had spotted CRISPR sequences in nineteen different microbes, a mixed list of bacteria and archaea. He suspected these similar sequences might have a common function. What particularly intrigued him were the stretches of code filling that space between the palindromic repeats—the gesundheit in my example, but slightly longer, a few dozen letters each—which became known as spacers. What was their meaning? What did they do? Why gesundheit in this sequence, maybe abracadabra in that one, and not damnyankees or rumplestiltskin? In 2003, during the heat of August, Mojica holed up in his air-conditioned office at the University of Alicante, just north of Santa Pola, and tried to understand it.

He typed out spacers from one CRISPR and another, using his word processing program, and entered them into a vast database of known genomes, looking for anything similar. He found some close matches: DNA stretches within certain viruses. Equally arresting or more so, he found other matches among bacterial plasmids, those infective little particles of horizontally transferable DNA. So it seemed that CRISPR might represent a record of past infections, during which bacteria and archaea captured fragments of foreign DNA and incorporated those fragments into their own genomes. But for what purpose?

Well, viral infection could kill the bacterium or the archaeon, and plasmid infection (horizontal transfer of DNA) could alter its genome, for better or worse. A microbe might acquire means to deter such intrusions. Maybe, Mojica speculated, these CRISPRs were some sort of immunity mechanism against reinfection by the bugs represented in the spacers. A memory of infection, as defense against future infection? There's a word for this in our own realm. We call it vaccination.

Mojica wrote his informed speculation into a paper, with three coauthors, and submitted it to *Nature*. Rejected. He tried several other leading journals in succession. Rejected. The editors didn't see anything that seemed very new or important. Months passed. Mojica worried that he would be scooped. Finally, he sent his draft to the *Journal of Molecular Evolution*, the same lively outlet in which Carl Woese and four colleagues had published their first hint about the existence of a separate form of life. Mojica's paper, suggesting that CRISPR might be involved in immune defense, appeared there in 2005.

Meanwhile, another clue was added in 2002, when a Dutch team reported finding an odd group of four genes that occurred adjacent to CRISPR sequences in various genomes. These CRISPR-associated genes (*cas* genes, for short) were conspicuously absent from microbial genomes that lacked CRISPR sequences. They seemed to have some functional relationship to CRISPRs—something beyond mere happenstance proximity. At first, neither the Dutch team nor anyone else knew just what that function might be. Then, rather soon, further insights on CRISPR and *cas* genes emerged from several sources, and it became clear that *cas* genes perform the function, guided by CRISPR spacers, of attacking and dismantling invasive DNA. The Mojica hypothesis was persuasively proven: CRISPR-*cas* among microbes, as it has naturally evolved, is a defense mechanism against infection and infective heredity. It's their version of an adaptive immune system. We have antibodies and white blood cells; they have CRISPR. It protects bacteria and archaea from killer viruses, and it serves as a barrier (sometimes useful, sometimes limiting) against horizontal gene transfer. It helps microbes maintain their health and their continuity of identity. HGT is still rampant among bacteria and archaea, but their CRISPR-*cas* genes protect them against at least some transfers.

This is the backstory of CRISPR. The more glorious and familiar part of the narrative began later, in 2012, when other scientists established how CRISPR sequences and *cas* genes could be repurposed for editing mammalian genomes—those of laboratory mice, endangered species, invasive pests, and us. That enterprise is what carries CRISPR forward into the present, the future, and the wondrous and fraught possibilities of human germline engineering, among other applications. When the Nobel Prize for CRISPR is announced, the names of the winners will probably not include Francisco Mojica, nor the Dutch team, nor any of their colleagues who have worked on CRISPR purely as an evolutionary phenomenon among bacteria and archaea. More likely you will hear the names of scientists who have made CRISPR a tool for human use: Jennifer Doudna, Emmanuelle Charpentier, Feng Zhang, or possibly others. People will be gladdened and concerned, variously, about the boggling new prospects that this particular Nobel Prize will celebrate. But it seems likely that Carl Woese, if he were alive, with his strong bias against "a biology that operates from an engineering perspective," wouldn't cheer.

In that cranky 2004 manifesto, "A New Biology for a New Century," Woese wrote:

> Modern society knows that it desperately needs to learn how to live in harmony with the biosphere. Today more than ever we are in need of a science of biology that helps us to do this, shows the way. An engineering biology might still show us how to get there; it just doesn't know where "there" is.

The proper purpose of biology is not to change the world, he added, but to understand it. Then again, this daring new century really wasn't *his* century, and he knew that.

81

Carl Woese's death, when it happened, would leave behind vivid memories in a wide range of people, some of whom published remembrances of him, some of whom kept their stories and views private. Part of my job has been to gather samplings of those memories and put pieces together. For four years I've felt a bit like the newsreel reporter in *Citizen Kane*, assigned by his producer to track down old friends and contacts of the protagonist, the newspaper magnate Charles Foster Kane, and try to solve the mystery of his character. What drove Kane to be such a ruthless success and a needy bastard? What was the meaning of his last spoken word, "Rosebud"? Was that word, that thing—that person, if Rosebud were a person—the key to his life and his character? Or was it just a false lead? Was *anything* the key to the man's life and his character, in the sense that others could find it and turn it and open him like a door? Orson Welles put the Rosebud device to good effect, and you probably remember his Kane (if you've seen the film, and if you haven't, you should) as a formidable, enigmatic figure. You probably don't remember the newsreel reporter, not by name anyway (trivia answer: Jerry Thompson), because he's just the fellow who travels from source to source and asks the questions. Thompson's back is always to the camera; he stands in the shadows or offscreen, and we never see his face. He's the proxy for the audience, to whom witnesses speak. That's how this job works.

Ralph Wolfe told me the story, and I've told you, about the insult Woese suffered at a big conference in Paris, amid important biologists, long before his own burst of fame. He presented his paper, no one commented, no one asked any questions, and they all walked off to lunch. "It was almost a mortal wound," Wolfe said. Woese resolved never again to let his work be ignored—and so he sanctioned a press release at the time of his Archaea discovery, instead of letting his journal paper speak for itself, and that approach backfired. He made the front page of the *Times* but got criticized by other scientists, the reality of his third domain doubted, on grounds that he had put scientific publication second to publicity.

Was that his Rosebud moment? I don't think so. Nor other frustrations, embarrassments, and perceived slights I've heard mentioned, some personal, some professional, including his lack of a Nobel Prize. I agree with Jerry Thompson's conclusion: no word, no wound, no grudge, and no childhood deprivation can explain a person's life. There are too many interactive pieces. Complexity theory offers better metaphors for human behavior than does this sort of mechanistic puzzle solving.

Woese left behind no personal journal or diary. His archives in Urbana are thick with scientific papers, drafts, and professional correspondence, but very thin on personal revelation. He published only one book, *The Genetic Code*, early in his career, and it was neither influential on its scientific merits nor illuminating of his character. In some of his review articles and contributed chapters, he included anecdotes and a little retrospection, as in the chapter he wrote for a big edited volume on the archaea; but those human details relate to laboratory moments.

For instance, when he recognized for the first time, from his X-ray films, that archaea represent a unique form of life: "I rushed to share my out-of-biology experience with George, a skeptical George Fox to be sure. George was always skeptical. That's what made him a good scientist." When the Nobel winner Salvador Luria called Ralph Wolfe, after publication of their first archaea paper, to tell Wolfe he should dissociate himself from Woese and his crackpot science: "How could this Luria fellow have the temerity to excoriate his friend and my colleague like that? What pedestal was he standing on?"

Woese wrote no autobiography. He kept his family life very private.

The closest he came to a self-portrait might be the formal, impersonal assessment of his own work and significance, written in third person, that he sent as a five-page email to Norm Pace in 1995, before his feelings toward Darwin went really sour, and probably on request at a time when Pace was nominating him for a Nobel Prize. It's Woese on Woese, this document, making a case for himself in scientific history, as you or I might make a case for ourselves in a job application. I've seen it but, at present, it's unavailable for quotation. The gist is that he saw himself as a peer of Leeuwenhoek and Darwin.

Larry Gold, now a distinguished molecular biologist and biotech entrepreneur, also based in Colorado, knew Woese from the early days in Schenectady, when they both worked for GE, and remained close to him through the years. Woese was a thirty-two-year-old biophysicist, hired at the General Electric Research Laboratory for purposes unclear both to himself and to his bosses; Gold was a nineteen-year-old Yale student with a summer job. Gold found himself set to a grim research project that involved dosing rats with a carcinogenic chemical, then trying to prevent the cancers from happening, and while he was lost without supervision amid these sick and dying rats, Woese came to his assistance. They worked on the poor rodents together, killing many. Did Woese get interested in the experiment?

"No, I don't think so," Gold told me fifty-four years later, as we sat on a bench in Urbana. "I don't think he cared about it at all. I think what he was interested in was spending time that made him happy." He had taken a shine to young Larry Gold. "He clearly got pleasure out of being around people who let him be exuberant." He loved to laugh. "He had an outrageous laugh. It was almost a cackle." They talked seriously too—or Woese talked, especially about the genetic code and the question of how it evolved, and Gold listened like a kid hearing a rabbi construe passages of Torah.

Some people took Woese to be dour, but Gold saw a different person: lonely, eager for good company, sometimes raucous, always interested in new ideas, and generous of spirit. Gold had a local girlfriend in Schenectady but nowhere private to go with her; Woese let them park for trysts in his driveway. He'd come out and sweetly say good night to Larry and his girlfriend in the car. "I just was lucky," Gold said, "and I stayed his friend forever."

Woese had a sort of bifurcated brain, Gold thought. On one side was his great depth of learning—acquired mostly by self-instruction, not formal training—and his relentless questioning. He was a biophysicist, Gold reminded me, not a biologist. "He didn't know any biology. He knew less biology by the time he died than *I* know," Gold said self-deprecatingly. "That's a terrible thing to say. But he didn't really think about biology. He was thinking about what happened three and a half billion years ago. That's not biology." It's more a gumbo of physics and molecular evolution and geology, Gold meant. The RNA-world as pondered by a consummate autodidact.

"The other part of him," Gold said, the other side of that bifurcation, "was he wanted to be around people that enjoyed living." For instance, Gold mentioned, two of Woese's oldest and best friends: Norm Pace, with his motorcycles and his trapeze-artist wife, and Harry Noller, with his jazz music.

So I went to see Harry Noller, now a professor emeritus at the University of California, Santa Cruz, and one of the world's leading experts on ribosomes. I drove the steep switchback road up the wooded bluff above Monterey Bay, to where UCSC campus buildings sit cantilevered over small canyons, shaded by redwoods and eucalyptus, and found Noller in his cheerful little office. He wore a black sweatshirt, jeans, and gym shoes, and with his oval face ringed by white hair and a beard, his quiet calm, he came across like a priest or an oracle. But he was genial and straightforward, not oracular. On one bookshelf sat two copies of *Ribosomes*, a definitive tome of which he was coeditor, and a near-empty bottle of Laphroaig.

He met Woese in the early 1970s, when Noller was a young assistant professor at UCSC, just beginning his own work on ribosomes. Noller's interest was structure and function, not deep phylogeny, but he used methods similar to Woese's, sequencing short stretches of ribosomal RNA, because he wanted to understand how those molecules contribute to function. Meanwhile he divided his time between the laboratory and playing jazz saxophone professionally, with various groups at a level sufficient to open for the Duke Ellington Orchestra. He shared a concert stage with vibraphonist Bobby Hutcherson; he sat in with the great trumpeter Chet Baker. And then, after a rehearsal or a gig, he would dash back to see what his latest electrophoresis run had revealed. Noller became troubled by disparities between his ribosomal RNA sequences and some published by a lab in Strasbourg. He called Woese, whom he had met when Woese made a short visit to UCSC, and Woese reassured him that his data were correct, the French versions wrong. Woese even sounded angry when he said: "These sequences are sacred scrolls. They should be entrusted only to those who appreciate what they mean."

Woese admired Noller's work, as it developed, and invited him to Urbana many times to give seminars or simply hang out and talk science. They collaborated on some papers. They drank a bit of scotch, sitting in the little study of Woese's house, and listened to jazz. Occasionally they even played some music together, Woese on the home piano.

"Was he a pretty good jazz pianist?" I asked.

"Um . . . he was not incredible," said Noller diplomatically. "He kind of noodled around at it, and he knew some tunes and . . ." There came a long, careful pause. "He was not an accomplished pianist, but he was fun to play with." Woese was a serious listener to jazz, if not a talented player. He loved Art Tatum. He loved Ella Fitzgerald and Gerry Mulligan. He had a black house cat that he named Miles. In one of his letters to Noller, he added a postscript, typing carelessly: "Milt Jackson makes Lionel Hampton look lilke [*sic*] Quasimoto's father." To Woese's credit, he recognized his limits—that he wasn't a musician in Noller's league—and for one visit, he surprised his friend by hiring a professional rhythm section, three guys, piano, bass, and drums, to play with Harry. They set up in the living room, and Woese invited some other friends over to share the fun. "I mean," Noller told me, "talk about hospitality."

Woese offered loyalty as well as generosity. During one of Noller's visiting seminars in Urbana, a professor from another department interrupted the question period to try to seize the room, which he claimed to have booked for that time slot—an intrusion that Woese considered "obstreperous, petty and insulting." Afterward, Woese wrote the man a curt note informing him that he owed Professor Noller an apology. Then he sent Noller a copy, after inking it gleefully with his favorite rubber stamp, walloped crosswise onto the formal and scolding letter: MAY A BAND OF NOMADIC BARBERS GANG-LATHER YOUR SISTER.

They talked often by phone, almost every day, even in the era of long-distance charges, before email, before fax. In the later years, Noller said, "I used to get calls from him in the middle of the night, after he'd had a few scotches, and he would go on and on." These were not discussions of ribosomal RNA but ramblings about evolution, about the universe, sometimes featuring Delphic pronouncements. "Time is the residual of being." "The ribosome teaches in silence." He liked the story of one lesser musician's reaction on listening to Art Tatum: "I hear it, but I don't believe it."

During one of his visits to Santa Cruz, for a small celebration with Noller's lab people, all the young students and postdocs, Woese happened across a plate of brownies in the kitchen. He ate four or five. One student noticed and, aware that these were *that* sort of brownies, alerted Dr. Noller with some alarm. Noller checked on Woese, Woese said don't worry about it, and for the rest of the party, Woese sat harmlessly in a corner, intermittently exploding with laughter while tears ran down his face. He had dabbled with stronger drugs in the 1960s, by some accounts, but this evening seems to have been more benignly joyous. In the morning, while Noller made coffee, Woese pinched his brow and delivered another of his pronouncements: "Last night I discovered humor."

"He was complex," Noller told me. "He was always a guru. He always saw himself as an unappreciated genius—or, *under*appreciated, unrecognized." But never too serious for too long. "After making some heavy pronouncement," Noller recalled, "he would then say something obscene." A dirty joke, or a crude malediction on people he resented, such as the great founders of molecular biology, or Charles Darwin.

At the end of his published remembrance of Woese, Noller wrote: "Carl was a profoundly creative and fiercely uncompromising scientist and thinker, who stood apart from the rest of his contemporaries." Harry Noller is a subtle man, candid while loyal, and it's easy to see that "stood apart" has two meanings. Woese was extraordinary and also, to many people, severe and remote.

That's why I found Charlie Vossbrinck's testimony so interesting. I heard about Vossbrinck by chance, not at the Carl Woese memorial symposium where Gold and Noller and Norm Pace and Nigel Goldenfeld and George Fox and others assembled, and not through the eulogistic memoirs that appeared in *Science, Nature*, and other publications. When the journal *RNA Biology* devoted an entire issue to Woese, Vossbrinck wasn't among the contributors. But he had met Woese, and they had gotten friendly back in the 1980s, when Vossbrinck was a PhD student in entomology, because the Entomology Department at the University of Illinois was just down the corridor from Woese's lab. The third floor of Morrill Hall, again bringing people together. I went to see Charlie Vossbrinck in New Haven, at the Connecticut Agricultural Experiment Station, where he sat in another third-floor office, this one tiny, his walls decorated with posters of insects and spiders, his desk cluttered with gypsy moth caterpillars in small plastic cups. The gypsy moth is an economically significant pest, sometimes defoliating whole forests of oak and other trees, and these larvae in cups were being reared for study.

Vossbrinck is a big, openhearted bear of a man with a Long Island accent, his hair thinning, his jowls covered with gray stubble. He was sixty-three, he confided, and he shouldn't have to be growing his own caterpillars, but his bosses didn't appreciate his work and refused to promote him. Never mind. He pushed aside the frustrations, if not the caterpillars, and reminisced fondly for an hour about his friend Carl.

They met because Vossbrinck heard about Woese's work in molecular phylogeny, the RNA catalogs, and Vossbrinck wondered how that technique might apply to insects or the parasites they carry. Also, he needed work, because he was an impecunious grad student, and his teaching assistantship had ended. "So I went over there and talked to Carl, and Carl has kind of a bawdy sense of humor. I don't know if you knew that." Vossbrinck had told a lewd joke, which I won't repeat, and Woese had laughed. "So we hit it off." Woese hired him on a project, and they became friends.

On a Friday afternoon, they would cross Goodwin Avenue to a place called Trino's for beer, or Timpone's, the Italian restaurant just beside it, Woese's favorite getaway. After the beer, half drunk, they might go to a movie: Cheech and Chong, or something offbeat in science fiction. Another of Woese's maxims, not quite so Delphic, was: "Beer opens the mind." They also talked seriously, about evolution, among other things, and Vossbrinck saw Woese's deep hunger to find "the big answer." He also saw Woese's ego. Vossbrinck himself became "kind of the moderator of his ego." Charlie wouldn't let Woese be too ponderous for too long.

"One time we were in his backyard—because he would invite students over for barbecues and stuff, and we'd get drunk, and he would stand up and start to pronounce something, you know," Vossbrinck told me. "And the first time I picked him up and I threw him in his bushes . . ."

"You threw Carl Woese in his bushes?"

"Yeah. And his wife and his kids would go, 'Wow, he's throwing Carl in the bushes.' And Carl is going, 'No, not in my bushes! In the neighbor's bushes!'"

Woese, full of beer and merriment, seems to have been more concerned about his hedge than his safety or dignity. His wife, Gay, evidently a woman of quiet resignation, seldom gets a speaking role or even a walk-on in the more public stories about Woese, but in this one, she was allowed to holler out. "Anyhow, I threw him in the bushes," said Vossbrinck. The friendship only prospered.

Occasionally they fermented their own champagne. They would buy cider from a local farm, Woese would get some champagne yeast from the Microbiology Department, and they'd let it bubble. "At the end of two weeks, it would be pee yellow, you know, and, a Friday afternoon, me and Carl would start drinking it." The more they drank, the better their homemade concoction seemed. Woese would say something to the effect that "those people, those connoisseurs of wine, they would turn their noses at this. But this is really good stuff." They coined a term for these happy states of inebriation: gooned up. "Let's get gooned up, Carl," Vossbrinck would say, and they did.

Woese was generous, lending his friend Charlie some cash when needed, and attentive to others. He had a Chinese student in his lab at that time, a humble young man who wore a Mao jacket. It was Decheng Yang, the same PhD candidate who wound up as first author of the 1985 paper on mitochondrial origins. Realizing that Yang had almost no money, Woese paid him extra to teach them tai chi. The odd thing about that arrangement was that Yang didn't know tai chi himself; he had to learn it in order to teach it. But there they were, on a courtyard outside Morrill Hall: a small, white-haired professor; a big, gentle bear; plus others from the Woese lab, and a Chinese student struggling to stay one lesson ahead of the class.

Vossbrinck saw the ambitious and competitive side too. Each year, with the announcement of newly elected members of the National Academy of Sciences, and the name Carl Woese not among them, Woese would say, "My friends are letting me down." One year he decided it was too late. If they ask now, Vossbrinck remembered him saying, "I'll turn them down." The following year, Vossbrinck heard he'd been elected. "You turned them down, right, Carl?" he teased.

Woese smiled sheepishly. "You know me pretty well, don't you." He was glad to be in. He needed to be in.

Yes, he felt underappreciated. He did such serious science, discovered the third form of life, became controversial, and throughout Vossbrinck's days in Urbana, Woese was "still sweating it out." He had no capacity for the gamesmanship, the career-building side of science—what Vossbrinck called "tap dancing." That was costly to Woese. Vossbrinck himself, among his caterpillars, knew something about such costs. And then the Darwin business. "Carl had this hatred for Darwin," Vossbrinck volunteered. It wasn't easy to explain—abstract, philosophical, part of his ego? "Every once in a while, he'd say, 'I'm more important than Darwin.'"

I've heard that sort of thing from others, I said.

"I would tell him I was going to throw him in the bushes."

At one point in those years, Vossbrinck acquired a fine, old-fashioned Linhof camera, 4 x 5–inch format, and asked Woese to sit for some portraits. They did a session around his house and yard. Vossbrinck showed me a selection on his computer. Woese in a flannel shirt, seated beside a hall table. The table lamp looks more comfortable than he does. Woese before a window, chiaroscuro, theatrically dreamy. Woese outside, in an aluminum lawn chair, squinting. Woese with head on fist, looking posed like a child. He didn't seem to have the knack for being photographed. But then came one in which Woese sat forward, wild haired, bearded, deep-set eyes glaring fiercely into the lens, with the famous bushes nicely blurred behind him. God, I thought, it's the best Woese image I've ever seen. So *that's* who he was. But who was *that*?

82

Among the biggest mysteries at issue when Woese died, and still at issue today, is the origins of the eukaryotic cell. That is to say, the deepest beginnings of us, among others. If there are three domains of life, as Woese proclaimed in 1977, and one of those domains is Eukarya, encompassing all animals, all plants, all fungi, and all microbial beings whose cells contain nuclei, then what is the foundational story of that lineage, leading eventually to humans and every other creature we can see? What made eukaryotes so different? What set them onto such a divergent course, away from the tiny size and relative simplicity of Bacteria and Archaea, toward bigness and complexity, redwoods and blue whales and white rhinos, not to mention humans and all our peculiar contributions to the planet, including major league baseball, iambic pentameter, and Gregorian chant? What were the pieces, and what were the processes, that came together to form the first eukaryotic cell?

Whatever happened that was so momentous, it probably happened between 1.6 billion and 2.1 billion years ago. The size of that window—a half billion years—reflects the current degree of scientific uncertainty. Several hypotheses exist, offered by bitterly divided camps. Fossil evidence in rocks, of early microbial forms, doesn't shed much light. Far better clues, precise and various, are mined from genomic sequence data.

Some of those clues still come from 16S rRNA. That's thanks to the insight of Carl Woese and four decades of work in his footsteps. But the meaning of those data is variously construed. All the experts agree nowadays that endosymbiosis played an essential role: somehow a bacterium got captured and domesticated inside another cell, a host, where it became a mitochondrion. Once present and abundant within early eukaryotic cells, mitochondria delivered vast quantities of energy, far beyond anything previously available, allowing increases in size and complexity among these new cells and the multicellular creatures that evolved from them. A salient feature of the increased complexity was containment—in particular, containment of genetic material. More specifically, that meant packaging most of each cell's DNA within an internal organelle: a nucleus, bounded by a membrane. So the mystery of eukaryotic origins encompasses three main questions. (1) What was the original host cell? (2) Did mitochondria acquisition trigger the most crucial changes—or, alternatively, did it result from them? (3) From what sources did the nucleus arise? A simpler way of asking all that: How did one thing get inside another thing to form a complex thingamajig, and what were the things?

New evidence regarding the first two of those three questions has lately arrived from an unexpected locale: the bottom of the Atlantic Ocean. It came up in marine sediments scooped from an area, almost eight thousand feet deep, between Greenland and Norway, near a field of hydrothermal vents known as Loki's Castle. Loki is a shape-shifting, devious god of Norse mythology; the Norwegian-led discovery team gave the vent field that name because the mineralized vent chimneys looked like a castle, and because the place was so hard to find. The marine sediments, shared with other scientists for analysis, contained DNA that revealed an entirely new lineage of archaea, a genome so different from anything known that it seemed to represent a distinct phylum. (Phyla are big divisions; all vertebrate animals, for instance, belong to a single phylum.) The biologist leading the genomic investigation, a young Dutchman named Thijs Ettema, based at a university in Sweden, named the new group Lokiarchaeota, after the deep castle and the devious god.

What made this find widely newsworthy, when Ettema's team published in 2015, was that the Lokiarchaeota genome seems such a near match to what must have been the host cell at the origin of our own lineage. One headline, in the *Washington Post*, said: "Newly Discovered

'Missing Link' Shows How Humans Could Evolve from Single-Celled Organisms." Were these archaea, pulled from the deep marine ooze, modern cousins of the creature that, two billion years ago, took a drastic divergence from its own lineage and became eukaryotic? Were they our closest microbial relatives? Maybe. That caught public attention.

But what made the Ettema work controversial among the mavens of early evolution were two other points. First, Ettema's group reported evidence that cells such as Lokiarchaeota had begun acquiring complexity *before* they acquired mitochondria. Important proteins, internal structure, the ability to bend around and gobble a bacterium, perhaps. If so, the Great Mitochondria Capture was an effect, not a cause, of the biggest transition in the history of life. Or, anyway, a later event within a cascade of changes. Certain people, such as Bill Martin, would strongly disagree.

Second, Ettema's team placed the origin of Eukarya *within* the Archaea, not beside it. If correct, that meant we were back to a two-limb tree of life, neither of which is the limb we have long cherished as our own. That meant we ourselves are descended from archaeans, a separate form of life, unimagined before 1977. (There are intricate complications to this scenario, involving horizontal gene transfer of bacterial genes into our archaean ancestors before our lineage even began, so that, yes, bacteria are blended into us too—but the essence is still: woops, *we* are *they*!) Certain people, such as Norm Pace, would strongly disagree. Carl Woese would disagree too, but he didn't live long enough to be aggravated by Ettema's 2015 paper in *Nature*.

On a June morning, in a conference room in Toronto, Thijs Ettema described this work to a roomful of rapt listeners, including Ford Doolittle and a few dozen other researchers, plus me. When I saw him later, Ford said, with his usual wry self-disregard: "I've drunk the Kool-Aid."

Still later, I sat down with Ettema. We talked about his newest work, at that point still unpublished, which pushes the same implications still further: mitochondria as secondary to the big transition, and human ancestry rooted within the Archaea, on a two-limb tree of life. He was quite aware of the opposing views and of how fervently they would be argued. He said: "I'm really sort of preparing for some wind."

83

In late spring of 2012, Woese's health began to fail. He was eighty-three, still going to his office at the Institute for Genomic Biology every day, still with the big questions, uninterested in retirement. On a morning in May, he spoke by phone with his friend Harris Lewin, by then at UC-Davis, congratulating Lewin on his election to the National Academy of Sciences. He warned Lewin to beware of the "shitkickers" in Section 61, the Academy's section to which Lewin had been elected. Section 61 is Animal, Nutritional, and Applied Microbial Sciences, appropriate to Lewin's background in agricultural biology. Hard to say whether Woese's "shitkickers" was a genial allusion to scientists wading in cow manure or reflected his grittier feelings toward applied biology generally. Lewin mentioned the call in his memoir of Woese, then wrote: "Sadly, over the next few months, it seemed that Carl's physical vigor was waning and his mental state was slipping." Lewin stayed in touch, as well as he could at a distance.

In early summer, Woese's health worsened, involving some sort of intestinal blockage, while he vacationed with his family on Martha's Vineyard. The family took him to Massachusetts General Hospital in Boston, where imaging diagnostics revealed the problem: pancreatic cancer. It was bad, a tumor wrapped into one of the arteries like a strangler fig.

He underwent emergency surgery to relieve the blockage, but surgery couldn't dice out the tumor, not one so closely entangled with arterial walls. What happened next, his final six months, is best seen through Debbie Piper, the administrative assistant who became his factotum and friend during his years at the Institute for Genomic Biology. She helped Woese die as he preferred to, which was neither amid "heroic measures" nor in Massachusetts.

Debbie Piper's bond to Woese had grown like an accidental flower in a vacant lot. He didn't hire her at the IGB or bring her with him from the old lab. She transferred, from a different job, when the institute opened in 2007, and got assigned to the Biocomplexity program, under Nigel Gold-enfeld. "I was just sitting at my desk one day, and here comes this little white-haired guy, carrying a bag of books," Piper told me. "And I said, 'Do you need some help?' And he was like, 'Sure.'" They clicked, and became close. "I think he trusted me because I didn't want anything from him."

Piper met me at a coffee shop in Urbana, just down the block from Timpone's restaurant. She was fiftyish, with graying hair scooped down around her face, and a soft voice, and a firm manner. "He told me once, he said, 'You know what, Debbie, we're like family.'" But almost better than family, he added, because "we don't have all that mess."

She spoke about how Woese had reacted to his own renown, back in the late 1970s, after he had his fifteen minutes of Warholian fame, in the *New York Times* and elsewhere, and then suffered what he took as rejection and backlash by the community of his colleagues. "I wasn't there at that time, but we talked about it a lot." Thirty years on, Woese's aggrievement remained. "He had made this amazing discovery," as she took it from him, "and either people didn't think it was significant, or they didn't think it was real."

But it was the work, not himself, in Debbie Piper's impression, for which he still craved renown. He detested (or professed to detest) what he called "the cult of personality" in science. Darwin had been canon-ized by it. Woese didn't want that—at least, so he persuaded Piper, who knew him well. Others, who had known him longer and as a scientific colleague, disagree; but she, supportive and forgiving, was there at the end. "He didn't want his work to be about him," she told me. "He thought the work should stand by itself." During the later years, the years of her protectorate, other academics visiting campus would want to meet him, the famous Carl Woese, regardless of their own research fields, their comprehension or incomprehension of his work. Sometimes he told her, "I am not a dancing bear." He didn't want to meet people, to bask in adoration. He didn't want to travel. "He just wanted to be left alone to think." Piper's own role at the IGB evolved, with the full blessing of Goldenfeld, toward being Woese's personal assistant. Whatever he wanted came first. It was odd, she said, because at the beginning, she had no idea who he was, what he did, the extent of his reputation.

"He was just this guy with the white hair," I suggested.

"Just this lone wolf professor with white hair, yeah," she agreed, "in the office across from me." As they got acquainted, she had learned he was funny. He was kind. He was private. He was the most intelligent person she ever met, though on quotidian matters, he could be clueless. His mind roamed, who knew where. He was one of the best friends Debbie Piper ever had, and she still missed him terribly. Unpretentious? I wondered. Yes, unpretentious. Then she qualified that.

"He was not pretentious at all. He was really pretty humble. But then he would say, 'I have a lot to be humble about.'" She laughed.

A few minutes later, we turned to the hard part. Woese got his cancer diagnosis at the very beginning of July 2012. He had the emergency surgery on July 3, she recalled. "He called me and asked me if I would come out there." On July 4 she flew into Boston.

84

mong the essential points of the upheaval that Carl Woese helped initiate, and of this book in which I try to sketch that upheaval, are three counterintuitive insights, three challenges to categorical thinking about aspects of life on Earth. The categoricals are these: species, individual, tree.

Species: it's a collective entity but a discrete one, like a club with a fixed membership list. The lines between this species and that one don't blur.

Individual: an organism is also discrete, with a unitary identity. There's a brown dog named Rufus, there's an elephant with extraordinary tusks, there's a human known as Charles Robert Darwin.

Tree: inheritance flows always vertically from ancestor to descendant, always branching and diverging, never converging. So the history of life is shaped like a tree.

Now we know that each of those three categoricals is wrong.

Biologists have argued for a long time, long before molecular phylogenetics began complicating matters, about how to define *species*. The concept dates back at least to Linnaeus and, in looser forms, to Aristotle. Him again! But never mind the deep philosophical and etymological history. As used by Linnaeus in his classification system, during the eighteenth century, a species was an entity (an aggregation of creatures, but still an entity) that had constancy and essence. Darwin in the nineteenth century, with help from Wallace and others, dismissed that sort of idealism, persuading people that species change, species originate and depart, species consist of individuals that vary from one another, sharing a certain degree of similarity but no ineradicable common essence. In the twentieth century, a clarified definition of *species* was offered by Ernst Mayr, whom I mentioned earlier as one of the founding neo-Darwinists. Mayr's view carried weight because, besides being an eminent evolutionary theorist, he authored books on the history of biology, often writing himself into the saga in respectful third-person singular. Mayr's famous 1942 definition was: "Species are groups of actually or potentially interbreeding natural populations, which are reproductively isolated from other such groups." You know enough by now to see two problems with that definition.

First problem: it's inapplicable to bacteria and archaea, which don't "interbreed" in anything like the way implied by Mayr. Second problem: How can "reproductively isolated" be an absolute standard if genes are continually transferred horizontally (by viral infection and other mechanisms) and if, furthermore, members of one species sometimes breed with members of another, producing new lineages of hybrid offspring? (Such hybridizing happens often among plants and sometimes among animals.) Answer: reproductive isolation is a useful and intuitive standard, yes, but not an absolute one.

Take *Homo sapiens*, the species most dear to our hearts. In the era of DNA sequencing, scientists have recognized that the human genome contains evidence of hybridizing events. *Homo neanderthalensis*, Neanderthal man (and woman), was discovered in 1856, named in 1864, and for many decades considered a discrete species, closely related to us within the hominid family, but distinct. Some experts now consider the Neanderthals to have been a subspecies of *Homo sapiens*, more properly called *Homo sapiens neanderthalensis*, but others argue that *Homo neanderthalensis* is still the right label, representing the group as a full species. In any case, our lineage diverged from their lineage sometime between about three hundred thousand and six hundred thousand years ago, maybe more, when pioneers left Africa and colonized Eurasia. From those pioneers would descend several non-African species, including *Homo neanderthalensis*. Our own lineage, known as "modern humans," sent another wave of dispersers out of Africa and colonized Europe again, but much later, around fifty thousand years ago. Then, for one reason or another, the Neanderthals disappeared.

Paleoanthropologists have long speculated that either our ancestors killed off the Neanderthals, by direct aggression, or forced them to go extinct, by competition, or else absorbed them to some degree, by interbreeding. But there was no conclusive proof. Nowadays, since the recovery and sequencing of Neanderthal DNA by a team including the Swedish biologist Svante Pääbo, analyses indicate that hybrid matings did occur between Neanderthals and modern humans. The human genome, especially as found among non-African peoples descended from those hybrid matings, now contains about 1 percent to 3 percent Neanderthal DNA.

And it's not just Neanderthals in our genome. The human lineage diverged from the chimpanzee lineage seven million, or ten million, or maybe thirteen million years ago—nobody knows the timing with any precision. But recent genomic analysis suggests that, sometime well after the big split, hominid ancestors and chimp ancestors came back together for hybrid matings, and that those hybrid matings have left genuine chimpanzee genes (not just close human equivalents) in parts of our genome. The imprecise dating of the full divergence, in fact, may owe to the ratcheting stages by which it happened. As a consequence, some parts of our genome even today look more chimp than human. This knowledge tends to blur our prideful and categorical confidence that *Homo sapiens* is a discrete entity, produced through gradual evolutionary processes but now standing alone in space and time. We're not so discrete, not so alone. Svante Pääbo calls our genome a mosaic. He wasn't the first to use that metaphor in the realm of genomics, as I've noted, but when applied to us, it carries a peculiarly strong challenge to selfhood.

Of course, the presence of chimpanzee genes, or Neanderthal genes, isn't the half of it. There's also that viral DNA—including *syncytin-2*, a gene co-opted from a retrovirus, repurposed to enable human pregnancy. The fact that endogenous retroviruses constitute 8 percent of the human genome certainly complicates our sense of *Homo sapiens* as a species of primate.

It complicates our sense of human individuality even more. So does the recognition that each of us contains, as a necessity for health and digestion and other aspects of our physiology, some hundred trillion bacterial cells, representing thousands of different bacterial "species." And so does the realization that within every one of our human cells reside captured bacteria, long since transmogrified into mitochondria, without which we couldn't exist.

Biologists and philosophers of science have struggled for a long time, and continue struggling, to define and clarify the concept of an "individual" in biological terms. Some have argued that it's crucial to have such a definition, because the logic of evolution by natural selection—Darwin's core principle—depends on the differential survival and reproduction of . . . individuals. If so, what *is* an individual? Is a single bacterium an individual? Carl Woese and Nigel Goldenfeld teased at that question in their 2007 paper "Biology's Next Revolution." Sorin Sonea, the Romanian who argued that all earthly bacteria constitute a single "superorganism," a single interconnected genetic entity, would say no, bacteria considered one by one are not individuals. Is a worker ant, incapable of reproducing itself, living its life to maximize the reproductive output of the queen ant, an individual? Or is the ant colony itself an individual? Is it another "superorganism"?

What about a Portuguese man-o'-war, that peculiar relative of jellyfish, floating the ocean surface like a swim bladder with stinging tentacles? An individual? It seems so, but biologists who study these things tell us that, no, a Portuguese man-o'-war is not. It too, like an ant hill or a termite community, is a colony of individual creatures (in this case, small multicellular forms known as zooids), aggregated for a common purpose and variously performing specialized functions. Likewise that very strange thing known as a cellular slime mold, which during one phase of its existence looks and behaves like a garden slug, but at another phase reveals itself to be a fine-tuned team of individual amoebae. When food is scarce, the amoebae aggregate into the slug, unified in their effort to crawl toward better habitat, raise a stalk atop which sits a fruitlike body, and, when that opens, disperse spores. If the spores land in a place where food particles (bacteria) are available, they awaken as new amoebae.

Likewise again with aspen trees in a grove. They may look like individuals, but, in fact, aspens grow as clonal eruptions from underground rootstock, all interconnected, all sharing the same genome, sometimes including hundreds of trees across a wide area. The grove is the individual.

By one accounting, the largest organism on Earth may be a single aspen clone composed of thousands of trees spread across more than a hundred acres in Utah's Fishlake National Forest. It weighs about thirteen million pounds, this aspen individual, and is roughly eighty thousand years old.

These cases and others illustrate what the philosophers of science confront in their erudite papers: The meaning of "individual" is hard to define, except on a case-by-case basis, and not so easy even then. Coral might be ambiguous. Lichens might be ambiguous. Everyone agrees that puppies are individuals, owls are individuals, humans are individuals, until you consider the disquieting molecular facts. We are mosaics, as Pääbo noted, as Bill Martin said, not individuals.

And then there's that third challenged categorical; the tree of life. You've read here the reasons why it doesn't, in fact, look like an oak. Why it doesn't look like a Lombardy poplar. Even aspens in a grove make an unsuitable metaphor because, though interconnected underground, they don't reconnect above. Their roots form a network, but their limbs and branches only diverge, growing away from one another, seeking open space in which their leaves may harvest light. They don't converge, they don't inosculate, not in the wild—not like John Krubsack's grafted box elders or Axel Erlandson's grafted sycamores. The tree of life is not a true categorical because the history of life just doesn't resemble a tree.

Carl Woese knew that, though it wasn't among his highest priorities to say so. He interested himself in big limbs, not small branches. And of big limbs, in his view, considering the past four billion years, there were three: Bacteria, Eukarya, Archaea. Those three diverged from the last universal common ancestor of all life as we know it—life on Earth, life using one common genetic code, life that began with the RNA-world and then yielded cells and passed through the Darwinian Threshold and became very complex. One of the ironies of Carl Woese's career, it seems to me, is that although profoundly interested in complexity and how it arose—infatuated with complexity theory and emergent properties late in his life, when he collaborated with Nigel Goldenfeld—he was also deeply entranced by simplicity. Three living domains, from which all else results: that's simplicity. His holy trinity. It's almost religious.

Woese was a theist, not an atheist like many other scientists. "He said he believed in a deity," Debbie Piper told me. She is an atheist herself, with no reason to romanticize his beliefs, like the pious liars who fabricated a deathbed reversion to Anglicanism for Charles Darwin. Woese made no deathbed lurch toward organized religion. He was steady, though vague. There was a deity. Sometimes in an email, he would say to Piper: "May the God you don't believe in bless you." She laughed, telling me that, as though it had been a sweet joke between them.

On July 4, 2012, arriving in Boston, she found him in bad shape, not just medically but also mentally. Stuck in Massachusetts General Hospital, surrounded by family, he was very unhappy with his treatment. He wanted out. "They were giving him Haldol, and it was just making him crazy," Piper told me. Haldol, trade name for a drug called haloperidol, is an antipsychotic medication used to treat schizophrenia, delirium, psychosis, and other forms of agitation. Had Woese become psychotic? No. Was he agitated? Yes indeed. He had torn the IV tube out of his arm. His wife and two children were concerned but cautious, disinclined to challenge medical authority. Piper felt no such constraint. "Why are you giving him Haldol?" she asked the doctor who seemed to be in charge.

"Well, because he was upset."

"He's upset because you're giving him Haldol. He wants his mind to be clear."

Woese hated the fogginess, she told me. He wanted to be able to think. There was nothing—no medical fate, no waves of pain—worse to him than deprivation of his ability to think. That was his life. So the doctor took him off Haldol, and then Woese refused even Tylenol, despite the major surgery he had just undergone. They had cut open his belly, rearranged his gut, a day or two earlier. But he wanted clarity more than he wanted comfort.

"When we got that all straightened out," Piper told me, "then he asked me if I would help him die with dignity." She paused. "Which I did."

The doctor recommended he stay in Mass General for three weeks. Woese didn't want that. He declined chemotherapy. Piper helped get him aboard a medical charter flight, costing $16,000, and back to Urbana. "They rolled him into the house," she said, and he was home.

Home but not done, not quite. People wanted to see him and hear from him. Piper helped him dissuade most such well-wishers, campus dignitaries, and others with no vital claim on the last of his energies and his privacy. Although his relationships with his wife and his son and his daughter had been attenuated, he wanted them close to him now. Few others. Piper brought food, which sometimes he could eat. Larry Gold visited. Nigel Goldenfeld made himself helpful. In August Woese consented to endure a series of video interviews for the historical record. Jan Sapp came to town for that purpose, Norman Pace also, and Woese did his best to respond to gingerly questioning, mainly by Sapp and Goldenfeld, eliciting reflections on his work, his discoveries, the science of his time.

These recordings would be archived at the IGB. (The IGB itself would later be renamed, becoming the Carl R. Woese Institute for Genomic Biology.) Pale and manifestly uncomfortable, seated before bookshelves and an ivy plant, he spoke to the camera for more than six hours spread across two days, laboring to remember facts and names, to express ideas, frustrated when he was unable, often saying "Cut" or "Hold" when he couldn't summon a focused response. The camera operator didn't cut. Woese seemed unaware of that, or unconcerned, and, in a moment, he would start again. There was so much that still needed saying. Now it was too late. He took long pauses. He blinked back his own mortality. At one point he said, "My memory serves badly, badly, badly." Meanwhile the camera captured it all. At the time of his memorial service, months afterward, someone raised the idea of playing some of this video to bring his voice and image into the event.

Debbie Piper, when we spoke, recalled her reaction to that thought: "Oh, please don't. Because he just looks and sounds like a sick old man."

But inside the sick old man was a multiplicity of other realities. Some had arisen straight and some had arrived sideways.

Acknowledgments

This project began with my reading of the work of Ford Doolittle, particularly his 1999 paper in *Science*, which I discovered belatedly in 2013. Doolittle's writings led me in several directions—most importantly, to the work of Carl Woese, who had died on December 30, 2012. From those leads, the broader subject of molecular phylogenetics and the radical rethinking of the idea of the tree of life opened out to me like a vast limestone cavern, filled with astonishing Neolithic rock art and lit suddenly by flashlight. My first active step was to make contact with Doolittle, and, from the beginning, he has been extraordinarily helpful and generous to this project, without ever trying to influence unduly its shape or direction. He sat for days of interviews on several different occasions, in Halifax and elsewhere, and he read the entire book in draft, offering corrections toward greater accuracy, again without trying to lobby my subjective judgments or conclusions. Thanks, Ford.

The historian Jan Sapp has also helped me in several ways: through his published work, most notably his superb book *The New Foundations of Evolution*; by sitting for interviews at great length; and by sharing with me not only his memories of Carl Woese and of Lynn Margulis (both of whom he knew well) but also some private email correspondence. Although we write for very different audiences, and in very different ways,

Sapp was always generously supportive and candid when I called on him for insight or clarification.

Two scientists, both outside the field of molecular phylogenetics but keenly familiar with the pageant of biology, and both personal friends of mine, read the entire book in draft and offered advice: Mike Gilpin (my faithful consulting biologist since *The Song of the Dodo*) and Dave Sands.

Another group of people did me double favors: sitting for extended interviews, or answering email and phone questioning over the years, then afterward reading short portions of the draft book for accuracy, giving me notes and crucial corrections: Linda Bonen, Jim Brown, Julie Dunning Hotopp, Thijs Ettema, Cedric Feschotte, George Fox, Larry Gold, Peter Gogarten, Nigel Goldenfeld, Mike Gray, Jonathan Gressel, Thierry Heidmann, Jim Lake, Jeffrey Lawrence, Stuart Levy, Harris Lewin, Ken Luehrsen, Bill Martin, Harry Noller, Norman Pace, Debbie Piper, Julie Russell, Dorion Sagan, Mitch Sogin, Jake Turnbull, Charlie Vossbrinck, Blake Wiedenheft, and Ralph Wolfe. George Fox also shared with me the sequence of working drafts of the 1980 "Big Tree" paper he coauthored with Woese and others.

Among the many scientists who indulged my intrusive curiosity, I want to thank four in particular, because these men gave so generously of their time, their thinking, and their patience, and yet purely for reasons of structure and focus, their work is mentioned little or not at all in this book: John McCutcheon, Gary Olsen, Jonathan Eisen, and Eugene Koonin. I spent ten days in Chile with John McCutcheon and his colleagues, for instance, shadowing fieldwork (that is, chasing cicadas with butterfly nets) led by his postdoc Piotr Łukasik toward a study of bacterial endosymbiont genomes within these particular cicadas. That work was part of the overall focus of McCutcheon's lab: nested genomes and gene transfer within endosymbionts of certain insects, and what such nesting and genome reduction may suggest about endosymbiosis generally—perhaps even about the endosymbiotic origin of mitochondria. Although McCutcheon's work is deeply fascinating and important, I found that its relationship to my subject was just too complicated for me to ask readers to follow me where I had tried to follow him. Too bad: Chile is picturesque. And McCutcheon's company and conversation were wonderful, as were the Chilean steaks and beers.

Likewise, I spent a week in Davis, California, auditing an introductory biology class taught by Jonathan Eisen—in a large lecture hall, to hundreds of students—under the title "Biodiversity and the Tree of Life." After the classes each day, Eisen and I talked about phylogenetics and evolution and baseball and books, and then, on my final day, he took me birding at his favorite protected wetlands nearby. Look, he said at one moment, there's a white-faced ibis! So I had reassurance that this biologist, whose lab website bears the slogan "All microbes, all the time," cares about macrofauna too. Gary Olsen, among the closest working partners of Carl Woese during their shared years in Urbana, walked me patiently through ideas and memories that similarly never made it into the book. Eugene Koonin's wide-ranging thoughts on microbial genomes and evolution intrigued me so much that, after a first interview at his office in Bethesda, I said I'd like to read his book *The Logic of Chance* and return for a second session. I did, months later, and though I don't portray either visit in this book, those conversations with Koonin were among the great ancillary privileges of doing the whole project.

The list of others who helped my research, welcomed my visits to their labs and their offices, and responded hospitably to my pestiferous questioning is much longer and best organized geographically. I will repeat a few names to put them in this context. In the United States and Canada: Eric Alm, John Archibald, Jillian Banfield, Linda Bonen, Austin Booth, Seth Bordenstein, Jim Brown, Tyler Brunet, Ford Doolittle, Laura Eme, Mark Ereshefsky, Cedric Feschotte, Greg Fournier, George Fox, Bob Gallo, Peter Gogarten, Larry Gold, Nigel Goldenfeld, Mike Gray, Jacob P. Johnson, Patrick Keeling, Jim Lake, Jeffrey Lawrence, Harris Lewin, Stuart Levy, Linda Magrum, Joanne Manaster, Carlos Mariscal, Harry Noller, Maureen O'Malley, Norman Pace, Debbie Piper, David Relman, Andrew Roger, Mitch Sogin, Ray Timpone, Charlie Vossbrinck, Blake Wiedenheft, and Ralph Wolfe. In England: Tom Cavalier-Smith, Matthew Cobb, Martin Embley, James McInerney, plus many people at the National Institute for Biological Standards and Controls, and the National Collection of Type Cultures, including Isobel Atkin, Miles Carroll, Ana Deheer-Graham, Steve Grigsby, Ayuen Lual, Hannah McGregor, Jodie Roberts, Jane Shallcross, and, again, Julie Russell and Jake Turnbull. In Germany: Christa Schleper and, of course, Bill Martin. In France: Thierry Heidmann. In Israel: Jonathan Gressel. In Sweden: Thijs Ettema. In Chile: Piotr Lukasik and Claudio Veloso as well as John McCutcheon.

In Champaign-Urbana, at the University of Illinois, I was welcomed and helped by Christopher Prom and his colleagues, especially John Franch, at the university archives. At the Carl R. Woese Institute for Genomic Biology, Director Gene Robinson and his assistant Kim Johnson arranged access for me to events, contacts, and materials. And through such an event at the institute—a memorial symposium—I met Donna Daniels, Carl Woese's younger sister. Mrs. Daniels later answered a list of my questions by email and graciously shared memories of her beloved brother and their family history. Gabriella Woese and Robert Woese, Carl Woese's widow and his son, kindly allowed me to quote from his unpublished writings.

Colleagues are important for writers too. I'll mention just four whose works, knowledge, and friendship specifically helped me in this effort: Carl Zimmer, Ed Yong, Dorion Sagan, and Barry Lopez.

Bob Bender, at Simon & Schuster, gave the book a wonderfully astute and valuable edit—the old-fashioned kind, guiding the author to make his meaning more clear, his pacing more steady, and his relationship to the reader more friendly. To Bob and his colleagues, from Jonathan Karp to Johanna Li, I'm very grateful for vital and genial collaboration. Philip Bashe did a keen copyedit.

Amanda Urban, my agent at ICM, played an enormous role by way of advocacy and counsel, again, in helping me choose the right project and find the right place.

Emily Krieger has again served as my chief defender against error of my own making, as she fact-checked this book with tireless rigor. Gloria Thiede has again—thirty years now—transcribed long recordings of arcane, mumbled conversation into pages a writer can dice and use, and, at risk of strabismus, she has typed the bibliography. Together these two women have saved me from working even slower and appearing more foolish than I do.

At home in Montana, my wife, Betsy, remains my first advisor, my most trusted sounding board, my exemplar of strength and love. She's also matriarch of our adopted brood of other mammals. Harry and Nick and Stella, old canine souls, saw the beginning of this project but not its end. Steve and Manny, youngsters, now chew the shoes. Oscar the cat abides.

Notes

This book contains a Montana blizzard of facts, and though all of them are keyed to sources in my private annotated draft, I won't burden readers, or the printer, with full citations of every one here. Anyone fervently compelled to know the provenance of a particular assertion is welcome to contact me through my website, www.davidquammen.com. The notes following here pertain only to the most crucial sourcing: for quotations from published works or archival sources. Spoken quotes, from my interviews, are set in their contexts within the text. Complete citations are supplied in the bibliography. In cases where a source is quoted several times within a single paragraph of the text, I've noted that source only once, confident that the especially curious reader will be able to find the full passages easily.

THREE SURPRISES: An Introduction

xii *"a separate form of life"*: New York Times, November 3, 1977.

PART I: Darwin's Little Sketch

5   *"have arisen from one living filament"*: quoted in Browne (1995), 84.
5   *"Why is life short"*: Barrett (1987), 171–76.
6   *"organized beings represent a tree"*: ibid., 176.

7 *The tree is "irregularly branched"*: ibid., 176–77.

11 *"proceeds" from lifeless things*: Archibald (2014), 2, guided me to this passage from Aristotle.

11 Ladder of Ascent and Descent of the Intellect: Pietsch (2012), 4–6.

11 *"Scale of Natural Beings"*: Archibald (2014), fig. 1.4.

12 *"very wee animals"*: Lane (2015), 4.

13 *"a figure like a genealogical tree"*: Stevens (1983), 206.

13 *"according to the order that Nature appears"*: ibid., 203.

13 *That is, a "natural order"*: ibid., 205.

13 *"It appears, and one can hardly doubt it"*: ibid., 206.

13 *classified animals as "bloodless" and "blooded"*: Mayr (1982), 152.

16 *"Stamen number is a striking character"*: Stevens (1983), 205.

16 *"This figure, which I call a* botanical tree": ibid., 206.

17 *an American who prided himself a "Christian geologist"*: Lawrence (1972), 21, 23.

17 *"of insects, of worms, and microscopic animals"*: Packard (1901), 37.

18 *rather quickly "forgotten and unknown"*: ibid., 56–57.

19 *He argued that "subtle fluids"*: Mayr (1982), 354.

19 *"the true order of gradation" . . . a "counterpart" arrangement*: Pietsch (2012), 36–37.

21 *called his illustration a "Paleontological Chart"*: ibid., 81.

22 *"nothing was before me but a life"*: Hitchcock (1863), 282.

22 *"I gave myself to this labor"*: ibid., 284.

23 *"a hypochrondriac of the first rank"*: Lawrence (1972), 21.

23 *"dismissed" from the Conway pastorate*: ibid., 24.

24 *"exclude a Deity from its creation"*: ibid., 25.

24 *"We know nothing of Mr. Lyell's religious creed"*: ibid., 29.

25 *"a higher organization" had been inserted*: Archibald (2009), 573.

25 *"is perfectly explained by the changing condition"*: ibid., 575.

27 *Darwin read it in early autumn 1838—"for amusement"*: Darwin (1958), 120.

27 *"the warring of the species as inference from Malthus"*: Barrett (1987), 375.

27 *"One may say there is a force"*: ibid., 375–76.

27 *"the grand crush of population"*: ibid., 399.

28 *repeatedly to what he now called "my theory"*: ibid., 397–99, 409.

31 *It presented the theory as "one long argument"*: Darwin (1859), 459.

31 *"Natural selection . . . leads to divergence of character"*: ibid., 128.

31 *". . . have sometimes been represented by a great tree"*: ibid., 129.

32 *"The green and budding twigs may represent"*: ibid., 129–30.

34 *"It is well to take heed to the opinions"*: Archibald (2009), 575–76.

PART II: A Separate Form of Life

38 *"It has not escaped our notice"*: Watson and Crick (1953), 737.

40 *"his method of working was to talk loudly"*: Ridley (2006), 86, quoting David Blow.

40 *His talk "commanded the meeting"*: Judson (1979), 333.

40 *"probably his most remarkable paper"*: Ridley (2006), 104.

40   *"Biologists should realize that before long"*: Crick (1958), 142.

41   *"vast amounts of evolutionary information"*: ibid.

42   they called it *"chemical paleogenetics"*: Zuckerkandl and Pauling (1965a), 97.

43   *"Why don't you work on hemoglobin?"*: Morgan (1998), 161–62.

44   *"most influential of Pauling's later career"*: ibid., 172.

44   what you have is *"a molecular evolutionary clock"*: Zuckerkandl and Pauling (1965a), 148.

45   *"one of the simplest and most powerful concepts"*: Morgan (1998), 155.

45   Crick himself later judged it *"a very important idea"*: ibid., 155–56.

45   *"branching of molecular phylogenetic trees"*: Zuckerkandl and Pauling (1965a), 101.

46   and not just what he called the *"cryptographic aspect"*: Woese (1965a), 1546.

47   *"I differed from the whole lot of them"*: Woese (2007), 2.

48   *"A universal tree would therefore hold the secret"*: Sapp (2009), 156.

49   *"A slight diversion in my research program"*: Woese (2007), 2.

50   *"Dear Francis,"* he wrote, *"I'm about to make"*: Woese to Crick, June 24, 1969. Woese Archives, University of Illinois, Champaign-Urbana.

50   *"unravel the course of events"* leading to the origin: ibid.

51   *"backward in time by a billion years or so"*: ibid.

51   *"There is a possibility, though not a certainty"*: ibid.

51   allow him to deduce the *"ancient ancestor sequences"*: ibid.

51   *"The obvious choice of molecules here"*: ibid.

52   under their previous name, microsomal particles: Crick (1958), 147.

54   *"I feel . . . that the RNA components of the machine"*: Woese to Crick, June 24, 1969.

54   *"What I propose to do is not elegant science"*: ibid.

55   *"Here is where I'd be particularly grateful"*: ibid.

56   distinguishing variant forms of a molecule by *"fingerprinting"*: Morgan (1998), 161, n. 34.

57   *"My work had sort of come to a climax"*: Browntree (2014), 132.

57   *"A knighthood makes you different, doesn't it"*: "Frederick Sanger: Sequencing Insulin," Wikipedia, https://en.wikipedia.org/wiki/Frederick_Sanger#Sequencing_insulin.

59   *"It was routine work, boring, but demanding"*: Woese (2007), 1.

63   *"There were days . . . when I would walk home"*: ibid.

66   *"other professors just liked to hear"* . . . *"he felt it took him away from his real love"*: Luehrsen (2014), 217.

66   *"he plopped me down in his office"*: ibid., 218.

68   *"What a mess that often was!"*: ibid.

68   Woese *"just chuckled and said not to worry"*: ibid.

73   *"disentangling almost everything that was correct"*: Bulloch (1938), 192.

73   *"entirely modern in its character and expression"*: ibid.

74   *"Chaos"* was the name of the group: Breed (1928), 143.

76   *"the abiding intellectual scandal of bacteriology"*: Stanier and van Niel (1962), 17.

76   *"probably represents the greatest single evolutionary discontinuity"*: Stanier et al. (1963), 85.

77   *"Any good biologist finds it intellectually distressing"*: Stanier and van Niel (1962), 17.

77   *"elaborate taxonomic proposal"* they had published: ibid.

77   *"Many, many years ago I often went around"* . . . *"During those periods"*: Sapp (2005), 295.

79   dismissed that as *"a fourteen-syllable monstrosity"*: Woese (2007), 3.

79 *"screamed out" their membership in the prokaryotes*: ibid.

79 *a "signature" sequence in all prokaryotes*: ibid., 6.

79 *"What was going on?"*: ibid.

80 *"Then it dawned on me"*: ibid., 7.

80 *his "out-of-biology" experience*: ibid., 4.

83 *"burst into my room in the adjoining lab"*: Sapp (2009), 166.

83 *"proclaiming that we had found a new form of life"*: George Fox, "Remembering Carl," "Carl R. Woese Guest Book" (of posthumous remembrances), Carl R. Woese Institute for Genomic Biology online, last modified January 13, 2013, www.igb.illinois. edu/woese-guest-book.

83 *"George was always skeptical," Woese himself wrote*: Woese (2007), 4.

83 *they seemed to "jump off the page"*: George Fox to Jan Sapp, January 24, 2005, quoted in Sapp (2009), 167.

87 *a wonderfully named substance called "mine-slime"*: Wanger et al. (2008), 325.

89 *"Carl's voice was full of disbelief"*: Wolfe (1991). 13.

91 *"We went into fast-forward mode"*: Woese (2007), 4.

91 *"a rare opportunity to put the theory of evolution"*: ibid.

92 *"Testing these two main evolutionary predications"*: ibid.

93 *"should in principle be definable"*: Zuckerkandl and Pauling (1965a), 101.

93 *didn't look much like "typical" bacteria*: Balch et al. (1977), 305.

93 *"the most ancient phylogenetic event"*: ibid.

94 *"These organisms . . . appear to be only distantly related"*: Fox et al. (1977), 4537.

94 *"There exists a third kingdom"*: Woese and Fox (1977a), 5089.

95 *"These organisms love an atmosphere of hydrogen"*: *Washington Post*, November 3, 1977.

97 *a paper on what Woese called a "ratchet" mechanism*: Woese (1970).

100 *"Scientists studying the evolution of primitive organisms"*: *New York Times*, November 3, 1977, 1.

102 *"Ralph, you must dissociate yourself from this nonsense"*: Wolfe (2006), 3.

102 *"I wanted to crawl under something and hide"*: ibid.

103 *"Ralph marched him into my office"*: Woese (2007), 5.

104 *"In my whole career I had never paid attention to lipids"*: ibid., 6.

105 *"if unusual cell walls meant anything"*: ibid., 5.

107 *"A Third Reich?" he snapped*: Sapp (2009), 210.

109 *"We are about to embark on a scientific meeting"*: Woese (1982), in Kandler, ed. (1982), 2.

109 *"Generations of failure had discouraged the microbiologist"*: ibid.

109 *"hopefully divert biology to some extent"*: ibid.

110 *"Woese and especially Wolfe were not in top physical shape"*: Wolfe (2006), 7.

PART III: Mergers and Acquisitions

113 *rejected by "fifteen or so" other journals*: Margulis (1998), 29.

114 *"On the Origin of Mitosing Cells"*: Sagan (1967).

115 *"probably represents the single greatest evolutionary discontinuity"*: ibid., quoting Stanier et al. (1963), 85.

115   *"This paper presents a theory"*: ibid., 226.

117   *a "bad student," by her own account*: the quoted words come from Lake (2011), an obituary, but Margulis herself gives a similar and fuller account in Margulis (1998), 15–16.

118   *"I was a scientific ignoramus," she would recall*: Margulis (1998), 16.

118   *"a fine teacher—the best of my whole career"*: Eric Goldscheider, *"Evolution Revolution,"* *On Wisconsin* 110, no. 3 (Fall 2009): 46, https://onwisconsin.uwalumni.com/features/evolution-revolution/6.

119   *"endosymbiosis must again be considered seriously"*: Ris and Plaut (1962), 390.

120   *"I was interested in evolution"*: quoted in Keller (1986), 47.

120   *"At Berkeley," she recalled, "there was absolutely no relationship"*: Margulis (1998), 26–27.

121   *called Merezhkowsky's proposal "an entertaining fantasy"*: Wilson (1925), 738–39.

121   *she would call Sagan "unbelievably self-centered"*: Goldscheider, "Evolution Revolution," 46.

121   *"a torture chamber shared with children"*: Poundstone (1999), 47.

122   *to her it was "a move of convenience"*: ibid., 70.

122   *"Enforced home leave permitted uninterrupted thought"*: Margulis (1998), 29.

122   *"sprouted, expanded, and eventually was pruned"*: ibid., 29–30.

122   *"I typed late into many nights"*: ibid., 30.

125   *"Merezhkowsky's career was unsettled"*: Sapp et al. (2002), 416.

126   *"the origin of organisms by the combination"*: Merezhkowsky (1920), quoted in ibid., 425.

127   *"would be somewhat reminiscent of a symbiosis"*: quoted in ibid., 419.

127   *in what he called "a completely spontaneous way"*: quoted in ibid.

127   *"organs" of each cell, which had "gradually differentiated"*: Martin (1999) translation of Merezhkowsky (1905), 288.

127   *they are "foreign bodies, foreign organisms"*: ibid., 289.

128   *"Let us imagine a palm tree"*: ibid., 292.

128   *those docile "green slaves," the chloroplasts*: ibid.

128   *"I have no doubt that it would immediately lie down"*: ibid., 292–93.

129   *featuring a "seven-dimension oscillating universe"*: Sapp et al. (2002), 432.

129   *"The Plant as a Symbiotic Complex"*: cited in Khakhina (1992), 48.

130   *It warned: "Do not enter my room"*: Sapp et al. (2002), 435.

132   *just a bit of money given "from time to time"*: Wallin (1927), ix–x.

132   *an intimate and "absolute" symbiosis*: Wallin (1923b), 68, 71.

132   *"the fundamental principle controlling the origin of species"*: Wallin (1927), 146–47.

132   *a third force, an "unknown principle"*: ibid., 147.

133   *"Dr. Wallin's writings stirred up much interest"*: Eliot (1971), 138.

133   *Ivan E. Wallin "asks us to believe"*: Gatenby (1928), 165.

133   *called them "certainly defunct"*: Lange (1966), quoted in Margulis (1970), 45.

135   *may have come straight from the spirochete*: Margulis (1981), 16.

136   *"Every major concept in this book"*: ibid., 67.

142   *"advice, encouragement, and much unpublished data"*: Bonen and Doolittle (1975), 2314.

145   *That is, he said: "AAA, UUG, AAG"*: the full sequence appears in Carbon et al. (1978), 155, fig. 2.

153   *"Has the Endosymbiont Hypothesis Been Proven?"*: Gray and Doolittle (1982).

153 *In 1985 his lab published a paper titled "Mitochondrial Origins"*: Yang et al. (1985).

156 *calling it a "false-flag operation"*: "College and University Professors Question the 9/11 Commission Report," http://patriotsquestion911.com/professors.html.

157 *"We don't ask anyone to accept Williamson's ideas"*: "Butterfly Paper Bust-up," *Nature* online, last modified December 24, 2009, www.nature.com/news/2009/091224/full/news.2009.1162.html.

157 *called her "science's unruly Earth Mother"*: Mann (1991), headline.

157 *"I quit my job as a wife twice"*: quoted in Martin Weil, "Lynn Margulis, Leading Evolutionary Biologist, Dies at 73," *Washington Post*, November 26, 2011.

158 *"Rather," they wrote, "the important transmitted variation"*: Margulis and Sagan (2002), 12.

158 *they become in effect "plant-animal hybrids"*: ibid., 13.

158 *"The evolutionary biologists believe the evolutionary pattern"*: Dick Teresi (2011), "Discover Interview: Lynn Margulis Says She's Not Controversial, She's Right," *Discover*, April 2011.

160 *"There's a role in science for iconoclasts"*: quoted in Mann (1991), 4.

161 *"I greatly admire Lynn Margulis's sheer courage"*: John Brockman, *Third Culture: Beyond the Scientific Revolution* (New York: Touchstone, 1996), 129.

161 *"If I hear her say it again, I'm going to sue her"*: he told Sapp, and Sapp told me: interview, July 6, 2015.

161 *what he preferred to call the three great "domains" of life*: Woese et al. (1990).

161 *"If you wish merely a complimentary letter"*: Carl Woese to Dean Nicholas at Chicago, January 14, 1991, Woese Archives, University of Illinois, Chapaign-Urbana.

PART IV: Big Tree

166 *"one of the most magnificent works which I have ever seen"*: Darwin to Haeckel, March 3, 1864, in *The Correspondence of Charles Darwin*, vol. 12, 61.

167 *"proper evaluation of nature required"*: Richards (2008), 22.

167 *"horrible worms, rickets, scrofula, and eye diseases"*: ibid., 42.

169 *a dancing seventeen-year-old "elf"*: ibid., 50.

169 *"true German child of the forest"*: ibid.

170 *Haeckel called Messina "the Eldorado of zoology"*: ibid., 63.

170 *"with just a few months left for his research in Italy"*: ibid.

172 *"He had no choice"*: ibid., 79.

172 *her "German Darwin-man"*: Haeckel to Darwin, August 10, 1864, in *The Correspondence of Charles Darwin*, vol. 12, 485.

172 *"The whole natural system of plants and animals"*: Haeckel (1863), quoted in Richards (2008), 94–95.

174 *"the victorious rush of an Apollonian youth"*: ibid., 83, and n. 12, quoting Furbringer (1914).

175 *what he called his "religion of monism"*: ibid., 11.

176 *a whole edifice of "natural laws"*: Haeckel (1880).

177 *"stuffed with as many lawlike proposals"*: Richards (2008), 120.

177 *"I lived then quite like a hermit"*: quoted in Gliboff (2008), 171.

177 *"It contains the foundation for all"*: Richards (2008), 117.

177 *what Schleicher called a "Darwinian" theory*: ibid., 126, 159.

181 *"the chief source of the world's knowledge of Darwinism"*: ibid., 2, 223.

182 *his subtitle,* The Developmental History of Man: ibid., 140.

184 *"superficial, inconsistent and just plain muddleheaded"*: Kelly (1981), quoted in ibid., 263.

184 *Haeckel was "Darwinian in name alone"*: Bowler (1988), 72.

184 *all the "pseudo-Darwinians" and "anti-Darwinians"*: ibid., 47, 76.

184 *"the essentially linear character of Haeckel's evolutionism"*: ibid., 87.

186 *a plant ecologist at Cornell University for whom "broad classification"*: Hagen (2012), 67.

187 *a personality that some colleagues would later call "stoic" and "intense"*: Westman and Peet (1985), 7, 10.

187 *with blurry boundaries and a "low degree of reality"*: Hagen (2012), 68.

188 *"Ecologists are familiar with divisions of the living world"*: Whittaker (1957), 536.

188 *"The kingdoms are man's classifications"*: ibid., 537.

189 *"These themes are inconsistent," he admitted*: Whittaker (1959), 223.

189 *"Recent work has made more evident the profound differences"*: Whittaker (1969), 151.

190 *lay in the idea of "ancient cellular symbioses"*: ibid.

193 *polyphyletic taxa are "unwelcome"*: Whittaker and Margulis (1978), 6.

194 *in November 1977, of "a third kingdom of life"*: Woese and Fox (1977), 5089.

195 *in correspondence with Fox and others, was "big tree"*: Carl Woese to George Fox, November 16, 1977; courtesy of George Fox.

195 *"Please give big tree the top priority"*: ibid.

196 *"For at least a century, microbiologists have attempted"*: typescript of "Big Tree," version 1, courtesy of George Fox. All other typescript versions, likewise courtesy of George Fox.

197 *"A revolution is occurring in bacterial taxonomy"*: typescript of "Big Tree," version 7.

199 *Woese wrote to Fox about several "potential points of conflict"*: Woese to Fox, August 27, 1979. Woese Archives, University of Illinois, Champaign-Urbana.

200 *The eukaryotic cell "is now recognized to be a genetic chimera"*: Fox et al. (1980), 458.

203 *its members would be "eocytes"*: Lake et al. (1984), 3786.

203 *"all your proposal does is muddy the waters"*: quoted in Sapp (2009), 247.

204 *"Your apparent need to have there be a new kingdom"*: ibid., 248.

204 *"the battle of the kingdom keepers"*: ibid., 249.

204 *dismissed the whole episode as a "ridiculous intermezzo"*: quoted in ibid., 251.

205 *"The cell is basically an historical document"*: Woese (1987), 222.

205 *"The certainty that progenotes existed"*: ibid., 263.

207 *"The progenote today is the end of an evolutionary trail"*: ibid., 264.

208 *"Someday you and I must write"*: Woese to Kandler, February 11, 1980; in the Woese Archives, University of Illinois, Champaign-Urbana.

209 *"The First Workshop on Archaebacteria"*: its *Proceedings* were published as Kandler et al. (1982).

209 *"As time goes by it becomes more and more obvious"*: Woese to Zillig, June 3, 1989; in the Woese Archives, University of Illinois, Champaign-Urbana.

210   *"I have no objections if Mark becomes a coauthor"*: Kandler to Woese, January 5, 1990; in Kandler Papers, University of Munich; as quoted in Sapp (2009), 386.

211   *The word* archaebacteria *should now disappear*: Woese et al. (1990), 4578.

PART V: Infective Heredity

215   *"very shy and aloof and difficult to get to know"*: Pollock (1970), 11.

216   *"could do more with a kerosene tin and a primus stove"*: quoted in Downie (1972), 2, from Wright (1941), 588. Downie capitalizes *Palace* but Wright did not.

217   *"there seems to be no alternative to the hypothesis of transformation"*: Griffith (1928), 154.

218   *"actually make use of the products of the dead culture"*: ibid., 150.

218   *a kind of "pabulum"*: ibid., 153.

218   *"up to the chemists" to explain*: Pollock (1970), 10.

218   *"a bombshell which fell into a fused situation"*: Olby (1994), 178.

218   *"had to be practically forced into a taxi"*: Pollock (1970), 7.

220   *"There is no consensus of opinion amongst geneticists"*: Morgan (1934), 315.

221   *"considered a 'boring molecule' or a 'stupid molecule'"*: boring: Cobb (2015), 42, 54; stupid: Judson (1979), 59, 63.

221   *"a strange impression" of holy duty seized him*: Dubos (1976), 49.

222   *His friends called him "Babe'"*: ibid., 56.

222   *began to speak of "the transforming principle"*: McCarty (1985), 85, 92.

223   *"much more impulsive and impatient"*: ibid., 101.

223   *to spray "an invisible aerosol laden with bacteria"*: ibid., 104.

224   *no longer "Babe," far from it, but "the Professor"*: Dubos (1976), 4, 62.

224   *"We were not unaware that this idea would be greeted"*: McCarty (1985), 143.

224   *"Who could have guessed it?"*: Oswald Avery to Roy Avery, May 26, 1943, quoted in Dubos (1976), 218–19.

225   *"This is it, at long last"*: McCarty (1985), 171.

227   *"In order that various genes may have the opportunity"*: Lederberg and Tatum (1946a), 558.

229   *she detected a system of "sexual compatibility"*: Lederberg, Cavalli, and Lederberg (1952), 720; see also "The True History of Fertility Factor F," Esther M. Zimmer Lederberg Memorial Website," accessed www.esthermlederberg.com/Clark_MemorialVita/HISTORY52.html, wherein Esther Lederberg asserts her priority of credit.

230   *some sort of "infective hereditary factor"*: Lederberg et al. (1952), 729.

230   *alluded likewise to "infective heredity"*: Lederberg (1952), 413.

231   *Watanabe called it "an example of 'infective heredity'"*: Watanabe (1963), 87.

235   *Its medical significance was "limited to Japan at present"*: ibid., 108.

239   *"The discovery of transferable R factors, forty years ago"*: Levy (2002), 78–79.

244   *Baker's team found "intrinsic antimicrobial resistance"*: Baker et al. (2014), 1696.

244   *the people of the Solomons were still "an antibiotic virgin population"*: Gardner (1969), 774.

245   *a person "from the innermost bush country"*: ibid., 775.

245   *"additional VRSA infections are likely to occur"*: Miller (2002), 902.

249 his *"great gift" for rubbing people the wrong way*: Anthony Tucker, "E.S. Anderson," Guardian US, last modified March 21, 2006, www.theguardian.com/society/2006/mar/22/health.science.

249 *"is their possible importance in bacterial evolution"*: Anderson (1968), 176.

249 *"the temptation is very strong to suggest"*: ibid.

250 *"This in turn could favor extremely reticulate modes"*: Jones and Sneath (1970), 69.

250 *"It may well be that gene exchange is so frequent"*: ibid.

252 gives the bacterial entity *"a huge available gene pool"*: Sonea and Panisset (1983), 112.

253 They called it a *"superorganism"*: ibid., 8, 85.

253 Ford Doolittle called it *"bold if inchoate"*: Doolittle (2004), in Cracraft and Donoghue (2004), 88–89.

256 *"could be evolutionarily significant in promoting trans-kingdom"*: Heinemann and Sprague (1989), 205, abstract.

256 *"Can Genes Jump Between Eukaryotic Species?"*: Lewin (1982).

256 an *"apparently fanciful and certainly unorthodox"* idea: ibid., 42.

256 *"massive" uploads of alien genes*: Gladyshev et al. (2008), 1210, 1213.

258 also found *"many hundreds" of foreign genes*: Eyres et al. (2015), 1.

261 *"Arguably," according to one expert, "the spread of Wolbachia represents"*: Werren (2005), 299.

266 *"Hundreds of human genes appear likely to have resulted"*: Lander et al. (2001), 860.

267 *"the most exciting news" from the Human Genome Project so far*: Andersson et al. (2001), 1.

267 called the Consortium's claim *"at least an overstatement"*: Edward R. Winstead, "Researchers Challenge Recent Claim That Humans Acquired 223 Bacterial Genes During Evolution," Genome News Network, last modified May 21, 2001, www.genomenewsnetwork.org/articles/05_01/Gene_transfer.shtml.

267 *"fresh skirmish in the genome wars"*: Nicholas Wade, "Link Between Human Genes and Bacteria Is Hotly Debated by Rival Scientific Camps," *New York Times* online, May 18, 2001.

267 *"I was immediately struck by the fact"*: quoted in ibid.

PART VI: Topiary

271 *"Dammit, one of these days"*: quoted in letter from Hugo Krubsack to Dennis Krubsack, May 13, 1975, cited in "John Krubsack," Wikipedia, https://en.wikipedia.org/wiki/John_Krubsack.

274 ignored the crossovers entirely in his *"universal phylogenetic tree"*: Woese et al. (1990), 4578, fig. 1.

276 *"shattered," according to one eminent microbiologist*: Norman Pace, quoted in Morell (1996), 1043.

276 *"This completes that basic set," he told Science*: ibid.

277 *"you can't make sense of these phylogenies"*: quoted in Pennisi (1998), 2.

277 *"Each gene has its own history"*: Robert Feldman, quoted in ibid., 3.

278 *"the last resort of the impoverished imagination"*: Doolittle remembers: email to DQ, February 5, 2017.

279 *"Extensive gene transfer," Brown and Doolittle wrote*: Brown et al. (1994), 575.

283 *"The impulse to classify organisms is ancient"*: Doolittle (1999), 2124.

284 *He called it the "current consensus" model*: ibid., 2125, fig. 2.

284 *"something quite ominous" about these E. coli results*: Martin (1999), 101.

285 *He called this one "a reticulated tree"*: Doolittle (1999), 2127, fig. 3.

288 *Horizontal gene transfer was "rampant," he told the readers*: Doolittle (2000), 94.

289 *"By swapping genes freely," he wrote*: ibid., 97.

289 *"persuading me of the importance" of HGT*: Doolittle (1999), 2128.

290 *They focused on "prokaryote" evolution, using that old word*: Gogarten et al. (2002), 2226.

291 *horizontal gene transfer might be the "principal explanatory force"*: ibid., 2234.

293 *"too fantastic for present mention in polite biological society"*: Wilson (1925), 738–39.

297 *Another . . . discussed the "mosaic" character*: Martin (1999), 99.

297 *"quite ominous" to contemplate how HGT must have played out*: ibid., 101.

298 *"the tree of 1 percent" of those genomes*: Dagan and Martin (2006), 1–2.

298 *what he began calling "the universal phylogenetic tree"*: Woese (2000), 8392.

298 *have been called his "millennial series"*: Koonin (2014), 197.

301 *"As a cell design becomes more complex and interconnected"*: Woese (2002), 8742.

303 *the tree image was "absolutely central" to Darwin's thinking*: Graham Lawton, "Axing Darwin's Tree," *New Scientist*, January 24, 2009.

303 *"There's a promiscuous exchange of genetic information"*: quoted in ibid.

304 *"the uprooting of the tree" as the start of something bigger*: ibid.

305 *"What on earth were you thinking"*: letter to the editor by Daniel Dennett, Jerry Coyne, Richard Dawkins, and Paul Myers, "Darwin Was Right," *New Scientist*, February 18, 2009.

305 *"handing the creationists a golden opportunity"*: ibid.

305 *"iconic concept of evolution" that had "fallen on hard times"*: Eric Lyons, "Startling Admission: 'Darwin Was Wrong,'" Apologetics Press. www.apologeticspress.org /APContent.aspx?category=23&article=2666.

306 *"turning out to be much more involved"*: unsigned editorial in *New Scientist*, "The Future of Life, but Not as We Know It," January 24, 2009.

306 *"None of this should give succor to creationists"*: ibid.

307 *"the history of life cannot properly be represented"*: Doolittle (1999), 2124.

308 *At what point is the "Ship of Theseus"*: Doolittle (2004), R176.

308 *It isn't just an "attractive hypothesis"*: Doolittle (2000), 97.

310 *his knack with bizarre trees, "I talk to them"*: "Axel Erlandson," Wikipedia, https:// en.wikipedia.org/wiki/Axel_Erlandson, quoting from Wilma Erlandson (2001), *My Father Talked to Trees*, 13.

PART VII: E Pluribus Human

315 *"the ecological community of commensal, symbiotic, and pathogenic"*: Lederberg (2001), 2.

317 *"a little white matter, which is as thick"*: Leeuwenhoek letter to the Royal Society, September 17, 1683, quoted in "Antony van Leeuwenhoek (1632–1723)," University of California Museum of Paleontology online, www.ucmp.berkeley.edu/history /leeuwenhoek.html.

323 *"Friendship in the kingdom"*: the photo appears in Gold (2013), 3206.

325 *"human-associated bacteria"—denizens of the human body*: Smillie et al. (2011), 242.

325 *"Taken together, these analyses indicate"*: ibid., 242.

328 *"erased the deep ancestral trace" of any uniquely valid*: ibid.

328 *called that first ping "the most important email of my life"*: Goldenfeld (2014), 248.

328 *"My telephone is 3-9369," Woese wrote*: ibid.

328 *"You may not feel too much at home with biology"*: ibid.

328 *"So began a scientific partnership and friendship"*: ibid.

329 *The dynamic process involved "non-Darwinian" mechanisms*: Vestigian et al. (2006), 10696.

330 *"the coming avalanche of genomic data"*: Goldenfeld and Woese (2007), 369.

330 *"Among microbes, HGT is pervasive and powerful"*: ibid.

331 *"self-organization" emerging from biological systems*: Kauffman (1993), xiii, 22–26.

331 *an "operating system" might have spontaneously taken form*: Goldenfeld and Woese (2007), 369.

332 *an "unearthly figure in a full neck brace"*: Lewin (2014), 273.

333 *"an unrehearsed statement, a stream of scientific consciousness"*: ibid.

333 *Harris Lewin considered himself "an improbable friend"*: ibid.

334 *That ended his "isolation" in Morrill*: ibid., 275.

334 *"I wanted to do something to honor Carl's discoveries"*: ibid., 276.

336 *"likely to have resulted from horizontal transfer"*: Lander et al. (2001), 860.

337 *"an extraordinary trove of information"*: ibid.

337 *She called them "controlling elements"*: Comfort (2001), 9.

338 *"for her discovery of mobile genetic elements"*: ibid., 251, his translation from the Swedish.

341 *"horribly disgusting, to feel numerous creatures"*: Keynes, ed. (1988), 315, and n. 1.

343 *"For Carl, winning the Crafoord Prize by himself"*: Lewin (2014), 275.

344 *declined to "an engineering discipline"*: Woese (2004), 173.

344 *might illuminate "the master plan of the living world"*: ibid.

344 *Worse, molecular biology took a "reductionist" perspective*: ibid., 174.

345 *"a biology that operates from an engineering perspective"*: Woese (2004), 173.

345 *"whose writings I encountered rather late in the game"*: Woese (2005), R112.

346 *"JAN, YOU ACCORD DARWIN SO MUCH MORE SUBSTANCE"*: Draft of "Beyond God and Darwin," in the Woese Archives, University of Illinois, Champaign-Urbana.

347 *"Book 1: Growing Up in Science"*: This CVS notebook is in the Woese Archives, University of Illinois, Champaign-Urbana.

349 *"bacteria evolve by leaps and bounds"*: Draft of "Beyond God and Darwin," in the Woese Archives, University of Illinois, Champaign-Urbana.

349 *"The cells that make up our bodies have also not arisen gradually"*: ibid.

349 *"Consider too," Sapp wrote, "that a great percentage of our own DNA"*: ibid.

352 *"We made a sort of viral archeology," Heidmann said*: Saib and Benkirane (2009), 4.

354 *genetically modified mice—"knockout" mice*: Dupressoir et al. (2005), 730.

359 clustered regularly interspaced short palindromic repeats: Morange (2015), 221.

363 *These CRISPR-associated genes (cas genes, for short)*: Jansen et al. (2002), 1565, 1569.

364 *"Modern society knows that it desperately needs"*: ibid.

366 *"I rushed to share my out-of-biology experience"*: Woese (2007), 4.

369 *"These sequences are sacred scrolls"*: Carl Woese to Harry Noller, August 22, 1974; courtesy of Harry Noller.

369 *"obstreperous, petty, and insulting"*: Carl Woese to Professor J. Health, May 17, 1988; copy to Harry Noller, which Noller shared with DQ.

370 *"Last night I discovered humor"*: Noller (2014), 230.

370 *"Carl was a profoundly creative and fiercely uncompromising scientist"*: ibid., 230–31.

375 *"Newly Discovered 'Missing Link' Shows"*: Rachel Feltman, *Washington Post*, May 6, 2015.

377 *beware of the "shitkickers" in Section 61*: Lewin (2014), 277.

377 *"Sadly, over the next few months"*: ibid.

381 *"Species are groups of actually or potentially interbreeding"*: Mayr (1942), 120.

386 *often saying "Cut" or "Hold" when he couldn't summon*: These videotapes are available for viewing, with permission, through the Carl R. Woese Institute for Genomic Biology, University of Illinois, Champaign-Urbana.

Bibliography

Acuña, Ricardo, Beatriz E. Padilla, Claudia P. Flórez-Ramos, José D. Rubio, Juan C. Herrera, Pablo Benavides, Sang-Jik Lee, Trevor H. Yeats, Ashley N. Egan, Jeffrey J. Doyle, and Jocelyn K. C. Rose. 2012. "Adaptive Horizontal Transfer of a Bacterial Gene to an Invasive Insect Pest of Coffee." *Proceedings of the National Academy of Sciences* 109 (11).

Akanni, Wasiu A., Karen Siu-Ting, Christopher J. Creevey, James O. McInerney, Mark Wilkinson, Peter G. Foster, and Davide Pisani. 2015. "Horizontal Gene Flow from Eubacteria to Archaebacteria and What It Means for Our Understanding of Eukaryogenesis." *Philosophical Transactions of the Royal Society of London (B)* 370 (1678).

Albers, Sonja-Verena, Patrick Forterre, David Prangishvili, and Christa Schleper. 2013. "The Legacy of Carl Woese and Wolfram Zillig: From Phylogeny to Landmark Discoveries." *Nature Reviews Microbiology* 11 (10).

Amunts, Alexey, Alan Brown, Jaan Toots, Sjors H. W. Scheres, and V. Ramakrishnan. 2015. "The Structure of the Human Mitochondrial Ribosome." *Science* 348 (6230).

Andam, Cheryl P., David Williams, and J. Peter Gogarten. 2010. "Natural Taxonomy in Light of Horizontal Gene Transfer." *Biology & Philosophy* 25 (4).

Andam, Cheryl P., and J. Peter Gogarten. 2011. "Biased Gene Transfer in Microbial Evolution." *Nature Reviews Microbiology* 9 (7).

Andam, Cheryl P., Gregory P. Fournier, and Johann Peter Gogarten. 2011. "Multilevel Populations and the Evolution of Antibiotic Resistance Through Horizontal Gene Transfer." *FEMS Microbiology Reviews* 35 (5).

Anderson, E. S. 1968. "The Ecology of Transferable Drug Resistance in the Enterobacteria." *Annual Review of Microbiology*, 22.

Anderson, Jan O., Asa M. Sjögren, Lesley A. M. Davis, T. Martin Embley, and Andrew J. Roger. 2003. "Phylogenetic Analyses of Diplomonad Genes Reveal Frequent Lateral Gene Transfers Affecting Eukaryotes." *Current Biology* 13.

Anderson, Norman G. 1970. "Evolutionary Significance of Virus Infection." *Nature* 227 (5265).

Anderson, O. Roger. 1983. *Radiolaria.* New York: Springer-Verlag.

Andersson, J. O. 2001. "Lateral Gene Transfer in Eukaryotes." *Cellular and Molecular Life Sciences* 62 (11).

Anderson, Jan O., W. Ford Doolittle, and Camilla L. Nesbø. 2001. "Are There Bugs in Our Genome?" *Science* 292 (5523).

Andersson, Jan O., Asa M. Sjögren, Lesley A. M. Davis, T. Martin Embley, and Andrew J. Roger. 2003. "Phylogenetic Analyses of Diplomonad Genes Reveal Frequent Lateral Gene Transfers Affecting Eukaryotes." *Current Biology* 13 (2).

Anon. 1941. "Obituary. F. Griffith, M.B., and W. M. Scott, M.D., Medical Officers, Ministry of Health." *British Medical Journal* 1 (4191).

Arbuckle, Jesse H., Maria M. Medveczky, Janos Luka, Stephen H. Hadley, Andrea Luegmayr, Dharam Ablashi, Troy C. Lund, Jakub Tolar, Kenny De Meirleir, Jose G. Montoya et al. 2010. "The Latent Human Herpesvirus-6A Genome Specifically Integrates in Telomeres of Human Chromosomes *In Vivo* and *In Vitro.*" *Proceedings of the National Academy of Sciences* 107 (12).

Archibald, J. David. 2009. "Edward Hitchcock's Pre-Darwinian (1840) 'Tree of Life.'" *Journal of the History of Biology* 42.

Archibald, J. David. 2014. *Aristotle's Ladder, Darwin's Tree: The Evolution of Visual Metaphors for Biological Order.* New York: Columbia University Press.

Archibald, John. 2014. *One Plus One Equals One: Symbiosis and the Evolution of Complex Life.* New York: Oxford University Press.

Archibald, John M. 2008. "The Eocyte Hypothesis and the Origin of Eukaryotic Cells." *Proceedings of the National Academy of Sciences* 105 (51).

———. 2015. "Gene Transfer in Complex Cells." *Nature* 524 (7566).

Arkhipova, Irina R., Mark A. Batzer, Juergen Brosius, Cedric Feschotte, John V. Moran, Jürgen Schmitz, and Jerzy Jurka. 2012. "Genomic Impact of Eukaryotic Transposable Elements." *Mobile DNA* 3 (1).

Arnold, Michael L. 2007. *Evolution Through Genetic Exchange.* New York: Oxford University Press.

———. 2009. *Reticulate Evolution and Humans: Origins and Ecology.* New York: Oxford University Press.

Arnold, Michael L., and Edward J. Larson. 2004. "Evolution's New Look." *Wilson Quarterly* (Autumn).

Arnold, Michael L., and Axel Meyer. 2006. "Natural Hybridization in Primates: One Evolutionary Mechanism." *Zoology* 109 (4).

Arnold, Michael L., Yuval Sapir, and Noland H. Martin. 2008. "Genetic Exchange and the Origin of Adaptations: Prokaryotes to Primates." *Philosophical Transactions of the Royal Society of London (B)* 363 (1505).

Avery, Oswald T., Colin M. MacLeod, and Maclyn McCarty. 1944. "Studies on the Chemical Nature of the Substance Inducing Transformation of Pneumococcal Types: Induction of Transformation by a Desoxyribonucleic Acid Fraction Isolated from Pneumococcus Type III." *Journal of Experimental Medicine* 79 (2).

Azad, Rajeev K., and Jeffrey G. Lawrence. 2011. "Towards More Robust Methods of Alien Gene Detection." *Nucleic Acids Research* 39 (9).

Baker, Kate S., Alison E. Mather, Hannah McGregor, Paul Coupland, Gemma C. Langridge, Martin Day, Ana Deheer-Graham, Julian Parkhill, Julie E. Russell, and Nicholas R. Thomson. 2014. "The Extant World War 1 Dysentery Bacillus NCTC1: A Genomic Analysis." *Lancet* 384 (9955).

Baker, Kate S., Edward Burnett, Hannah McGregor, Ana Deheer-Graham, Christine Boinett, Gemma C. Langridge, Alexander M. Wailan, Amy K. Cain, Nicholas R. Thomson, Julie E. Russell, and Julian Parkhill. 2015. "The Murray Collection of Pre-Antibiotic Era *Enterobacteriacae*: A Unique Research Resource." *Genome Medicine* 7 (97).

Balch, William E., and R. S. Wolfe. 1976. "New Approach to the Cultivation of Methanogenic Bacteria: 2-Mercaptoethanesulfonic Acid (HS-CoM)-Dependent Growth of *Methanobacterium ruminantium* in a Pressurized Atmosphere." *Applied and Environmental Microbiology* 32 (6).

Balch, William E., Linda J. Magrum, George E. Fox, Ralph S. Wolfe, and Carl R. Woese. 1977. "An Ancient Divergence Among the Bacteria." *Journal of Molecular Evolution* 9 (4).

Baldauf, Sandra L., Jeffrey D. Palmer, and W. Ford Doolittle. 1996. "The Root of the Universal Tree and the Origin of Eukaryotes Based on Elongation Factor Phylogeny." *Proceedings of the National Academy of Sciences* 93 (15).

Baltimore, David, Paul Berg, Michael Botchan, Dana Carroll, R. Alta Charo, George Church, Jacob E. Corn, George Q. Daley, Jennifer A. Doudna, Marsha Fenner et al. 2015. "A Prudent Path Forward for Genomic Engineering and Germline Gene Modification." *Science* 348 (6230).

Banfield, Jillian F., and Mark Young. 2009. "Variety—The Splice of Life—in Microbial Communities." *Science* 326 (5957).

Barker, H. A., and Robert E. Hungate. 1990. "Cornelius Bernardus van Niel, 1897–1985." From *Biographical Memoirs*, published by the National Academy of Sciences, Washington DC.

Barnett, W. Edgar, and David H. Brown. 1967. "Mitochondrial Transfer Ribonucleic Acids." *Proceedings of the National Academy of Sciences* 57 (2).

Barns, Susan M., Ruth E. Fundyga, Matthew W. Jeffries, and Norman R. Pace. 1994. "Remarkable Archaeal Diversity Detected in a Yellowstone National Park Hot Spring Environment." *Proceedings of the National Academy of Sciences* 91 (5).

Barns, Susan M., Charles F. Delwiche, Jeffrey D. Palmer, and Norman R. Pace. 1996. "Perspectives on Archaeal Diversity, Thermophily, and Monophyly from Environmental rRNA Sequences." *Proceedings of the National Academy of Sciences* 93 (17).

Barrangou, Rodolphe, Christophe Fremaux, Hélène Deveau, Melissa Richards, Patrick Boyaval, Sylvain Moineau, Dennis A. Romero, and Philippe Horvath. 2007. "CRISPR Provides Acquired Resistance Against Viruses in Prokaryotes." *Science* 315 (5819).

Barrett, Paul H., and Peter J. Gautrey, Sandra Herbert, David Kohn, Sydney Smith, editors. 1987. *Charles Darwin's Notebooks 1836–1844: Geology, Transmutation of Species, Metaphysical Enquiries*. Ithaca (NY): Cornell University Press.

Belshaw, Robert, Vini Pereira, Aris Katzourakis, Gillian Talbot, Jan Pačes, Austin Burt, and Michael Tristem. 2004. "Long-Term Reinfection of the Human Genome by Endogenous Retroviruses." *Proceedings of the National Academy of Sciences* 101 (14)

Bénit, Laurence, Nathalie de Parseval, Jean-François Casella, Isabelle Callebaut, Agnès Cordonnier, and Thierry Heidmann. 1997. "Cloning of a New Murine Endogenous Retrovirus, MuERV-L, with Strong Similarity to the Human HERV-L Element and with a *gag* Coding Sequence Closely Related to the *Fv1* Restriction Gene." *Journal of Virology* 71 (7).

Bénit, Laurence, Jean-Baptiste Lallemand, Jean-François Casella, Hervé Philippe, and Thierry Heidmann. 1999. "ERV-L Elements: A Family of Endogenous Retrovirus-like Elements Active Throughout the Evolution of Mammals." *Journal of Virology* 73 (4).

Bénit, Laurence, Philippe Dessen, and Thierry Heidmann. 2001. "Identification, Phylogeny, and Evolution of Retroviral Elements Based on Their Envelope Genes." *Journal of Virology* 75 (23).

Bergthorsson, Ulfar, Keith L. Adams, Brendan Thomason, and Jeffrey D. Palmer. 2003. "Widespread Horizontal Transfer of Mitochondrial Genes in Flowering Plants." *Nature* 424 (6945)

Bézier, Annie, Marc Annaheim, Juline Herbinière, Christoph Wetterwald, Gabor Gyapay, Sylvie Bernard-Samain, Patrick Wincker, Isabel Roditi, Manfred Heller, Maya Belghazi et al. 2009. "Polydnaviruses of Braconid Wasps Derive from an Ancestral Nudivirus." *Science* 323 (5916).

Blaise, Sandra, Nathalie de Parseval, Laurence Bénit, and Thierry Heidmann. 2003. "Genomewide Screening for Fusogenic Human Endogenous Retrovirus Envelopes Identifies Syncytin 2, a Gene Conserved on Primate Evolution." *Proceedings of the National Academy of Sciences* 100 (22).

Blaser, Martin J., MD. 2014. *Missing Microbes: How the Overuse of Antibiotics Is Fueling Our Modern Plagues.* New York: Henry Holt.

Bokulich, Nicholas A., and Charles W. Bamforth. 2013. "The Microbiology of Malting and Brewing." *Microbiology and Molecular Biology Reviews* 77 (2).

Bolotin, Alexander, Benoit Quinquis, Alexei Sorokin, and S. Dusko Ehrlich. 2005. "Clustered Regularly Interspaced Short Palindrome Repeats (CRISPRs) Have Spacers of Extrachromosomal Origin." *Microbiology* 151 (8).

Bonen, Linda, and W. Ford Doolittle. 1975. "On the Prokaryotic Nature of Red Algal Chloroplasts." *Proceedings of the National Academy of Sciences* 72 (6).

———. 1978. "Ribosomal RNA Homologies and the Evolution of the Filamentous Blue-Green Bacteria." *Journal of Molecular Evolution* 10.

Bonen, L., and W. F. Doolittle. 1976. "Partial Sequences of 16S rRNA and the Phylogeny of Blue-Green Algae and Chloroplasts." *Nature* 261 (5562).

Bonen, L., R. S. Cunningham, M. W. Gray, and W. F. Doolittle. 1977. "Wheat Embryo Mitochondrial 18S Ribosomal RNA: Evidence for Its Prokaryotic Nature." *Nucleic Acids Research* 4 (3).

Boone, David R., Yitai Liu, Zhong-Ju Zhao, David L. Balkwill, Gwendolyn R. Drake, Todd O. Stevens, and Henry C. Aldrich. 1995. "*Bacillus infernus* sp. nov., an FE(III)-and Mn(IV)-Reducing Anaerobe from the Deep Terrestrial Subsurface." *International Journal of Systematic Bacteriology* 45 (3).

Boothby, Thomas C., Jennifer R. Tenlen, Frank W. Smith, Jeremy R. Wang, Kiera A. Patanella, Erin Osborne Nishimura, Sophia C. Tintori, Qing Li, Corbin D. Jones, Mark

Yandell et al. 2015. "Evidence for Extensive Horizontal Gene Transfer from the Draft Genome of a Tardigrade." *Proceedings of the National Academy of Sciences* 112 (52).

Bordenstein, Sarah R., and Seth R. Bordenstein. 2016. "Eukaryotic Association Module in Phage WO Genomes from *Wolbachia*." *Nature Communications* 7.

Bordenstein, Seth R. 2007. "Evolutionary Genomics: Transdomain Gene Transfers." *Current Biology* 17 (21).

Bordenstein, Seth R., F. Patrick O'Hara, and John H. Werren. 2001. "*Wolbachia*-Induced Incompatibility Precedes Other Hybrid Incompatibilities in *Nasonia*." *Nature* 409 (6821).

Bosley, Katrine S., Michael Botchan, Annelien L. Bredenoord, Dana Carroll, R. Alta Charo, Emmanuelle Charpentier, Ron Cohen, Jacob Corn, Jennifer Doudna, Guoping Feng et al. 2015. "CRISPR Germline Engineering—The Community Speaks." *Nature Biotechnology* 33 (5).

Boto, Luis. 2009. "Horizontal Gene Transfer in Evolution: Facts and Challenges." *Proceedings of the Royal Society of London (B)* 277 (1683).

_____. 2014. "Horizontal Gene Transfer in the Acquisition of Novel Traits by Metazoans." *Philosophical Transactions of the Royal Society of London (B)* 281 (1777).

Bouchard, Frederic, and Philippe Huneman, editors. 2013. *From Groups to Individuals: Evolution and Emerging Individuality.* Cambridge (MA): MIT Press.

Bowler, Peter J. 1989. *Evolution: The History of an Idea.* Revised edition. Berkeley, Los Angeles, London: University of California Press. First published 1983.

_____. 1988. *The Non-Darwinian Revolution: Reinterpreting a Historical Myth.* Baltimore: Johns Hopkins University Press.

Breed, Robert S. 1928. "The Present Status of Systematic Bacteriology." *Journal of Bacteriology* 15 (3).

Brito, I. L., S. Yilmaz, K. Huang, L. Xu, S. D. Jupiter, A. P. Jenkins, W. Naisilisili, M. Tamminen, C. S. Smillie, J. R. Wortman et al. 2016. "Mobile Genes in the Human Microbiome Are Structured from Global to Individual Scales." *Nature* 535 (7612).

Brock, Thomas D., translator and editor. 1961. *Milestones in Microbiology: 1546 to 1940.* Washington (DC): ASM Press.

_____. 1967. "Life at High Temperatures." *Science* 158 (3804).

_____. 1995. "The Road to Yellowstone—And Beyond." *Annual Review of Microbiology.*

Brown, Christopher T., Laura A. Hug, Brian C. Thomas, Ital Sharon, Cindy J. Castelle, Andrea Singh, Michael J. Wilkins, Kelly C. Wrighton, Kenneth H. Williams, and Jillian F. Banfield. 2015. "Unusual Biology Across a Group Comprising More Than 15% of Domain Bacteria." *Nature* 523 (7559).

Brown, J. R., Y. Masuchi, F. T. Robb, and W. F. Doolittle. 1994. "Evolutionary Relationships of Bacterial and Archaeal Glutamine Synthetase Genes." *Journal of Molecular Evolution* 38 (6).

Brown, James R., and W. Ford Doolittle. 1995. "Root of the Universal Tree of Life Based on Ancient Aminoacyl-tRNA Synthetase Gene Duplications." *Proceedings of the National Academy of Sciences* 92 (7).

_____. 1997. "*Archaea* and the Prokaryote-to-Eukaryote Transition." *Microbiology and Molecular Biology Reviews* 61 (4).

Brownlee, George G. 2014. *Fred Sanger: Double Nobel Laureate, A Biography.* Cambridge: Cambridge University Press.

Brucker, Robert M., and Seth R. Bordenstein. 2012. "Speciation by Symbiosis." *Trends in Ecology and Evolution* 27 (8).

Buchner, Paul. 1965. *Endosymbiosis of Animals with Plant Microorganisms.* Translated by Bertha Mueller, with the collaboration of Francis H. Foeckler. New York: John Wiley & Sons.

Bulloch, William. 1979. *The History of Bacteriology.* New York: Dover Publications. First published 1938 by Oxford University Press.

Bult, Carol J., Owen White, Gary J. Olsen, Lixin Zhou, Robert D. Fleischmann, Granger G. Sutton, Judith A. Blake, Lisa M. FitzGerald, Rebecca A. Clayton, Jeannine D. Gocayne et al. (Woese and Venter). 1996. "Complete Genome Sequence of Methanogenic Archaeon, *Methanococcus jannaschii.*" *Science* 273 (5278).

Burstein, David, Christine L. Sun, Christopher T. Brown, Itai Sharon, Karthik Anantharaman, Alexander J. Probst, Brian C. Thomas, and Jillian F. Banfield. 2016. "Major Bacterial Lineages Are Essentially Devoid of CRISPR-Cas Viral Defence Systems." *Nature Communications* 7.

Bushman, Frederic. 2002. *Lateral DNA Transfer: Mechanisms and Consequences.* Cold Spring Harbor (NY): Cold Spring Harbor Laboratory Press.

Buss, Leo W. 1987. *The Evolution of Individuality.* Princeton (NJ): Princeton University Press.

Busslinger, Meinrad, Sandro Rusconi, and Max L. Birnstiel. 1982. "An Unusual Evolutionary Behaviour of a Sea Urchin Histone Gene Cluster." *EMBO Journal* 1 (1).

Campbell, Allan. 1981. "Evolutionary Significance of Accessory DNA Elements in Bacteria." *Annual Review of Microbiology* 35.

Campbell, Matthew A., James T. Van Leuven, Russell C. Meister, Kaitlin M. Carey, Chris Simon, and John P. McCutcheon. 2015. "Genome Expansion Via Lineage Splitting and Genome Reduction in the Cicada Endosymbiont *Hodgkinia.*" *Proceedings of the National Academy of Sciences* 112 (33).

Carbon, P., C. Ehresmann, B. Ehresmann, and J. P. Ebel. 1978. "The Sequence of *Escherichia coli* Ribosomal 16 S RNA Determined by New Rapid Gel Methods." *FEBS Letters* 94 (1).

Cartault, François, Patrick Munier, Edgar Benko, Isabelle Desguerre, Sylvain Hanein, Nathalie Boddaert, Simonetta Bandiera, Jeanine Vellayoudom, Pascale Krejbich-Trotot, Marc Bintner et al. 2012. "Mutation in a Primate-Conserved Retrotransposon Reveals a Noncoding RNA as a Mediator of Infantile Encephalopathy." *Proceedings of the National Academy of Sciences* 109 (13).

Cartwright, Paulyn, and Annalise M. Nawrocki. 2010. "Character Evolution in Hydrozoa (Phylum Cnidaria)." *Integrative and Comparative Biology* 50 (3).

C. elegans Sequencing Consortium. 1998. "Genome Sequence of the Nematode *C. elegans*: A Platform for Investigating Biology." *Science* 282 (5396).

Chain, P. S. G., E. Carniel, F. W. Larimer, J. Lamerdin, P. O. Stoutland, W. M. Regala, A. M. Georgescu, L. M. Vergez, M. L. Land, V. L. Motin et al. 2004. "Insights into the Evolution of *Yersinia pestis* Through Whole-Genome Comparison with *Yersinia pseudotuberculosis.*" *Proceedings of the National Academy of Sciences* 101 (38).

Chargaff, Erwin. 1978. *Heraclitean Fire: Sketches from a Life Before Nature.* New York: Rockefeller University Press.

Cho, Yangrae, Yin-Long Qiu, Peter Kuhlman, and Jeffrey D. Palmer. 1998. "Explosive Invasion of Plant Mitochondria by a Group I Intron." *Proceedings of the National Academy of Sciences* 95 (24).

Choi, B. K., B. J. Paster, F. E. Dewhirst, and U. B. Göbel. 1994. "Diversity of Cultivable and Uncultivable Oral Spirochetes from a Patient with Severe Destructive Periodontitis." *Infection and Immunity* 62 (5).

Christner, Brent C., John C. Priscu, Amanda M. Achberger, Carlo Barbante, Sasha P. Carter, Knut Christianson, Alexander B. Michaud, Jill A. Mikucki, Andrew C. Mitchell, Mark L. Skidmore, Trista J. Vick-Majors, and the WISSARD Science Team. 2014. "A Microbial Ecosystem Beneath the West Antarctic Ice Sheet." *Nature* 512 (7514).

Chung, King-Thom, and Vincent Varel. 1998. "Ralph S. Wolfe (1921—). Pioneer of Biochemistry of Methanogenesis." *Anaerobe* 4 (5).

Chuong, Edward B., Nels C. Elde, and Cedric Feschotte. 2016. "Regulatory Evolution of Innate Immunity Through Co-Option of Endogenous Retroviruses." *Science* 351 (6277).

Churchill, Frederick B. 1968. "August Weismann and a Break from Tradition." *Journal of the History of Biology* 1 (1).

———. 2015. *August Weismann: Development, Heredity, and Evolution.* Cambridge (MA): Harvard University Press.

Ciccarelli, Francesca D., Tobias Doerks, Christian von Mering, Christopher J. Creevey, Berend Snel, and Peer Bork. 2006. "Toward Automatic Reconstruction of a Highly Resolved Tree of Life." *Science* 311 (5765).

Clark, Ronald W. 1984. *Einstein, the Life and Times.* New York: Harper Perennial. First published 1971.

Clarke, Ellen. 2011. "The Problem of Biological Individuality." *Biological Theory* 5 (4).

Cobb, Matthew. 2014. "Oswald Avery, DNA, and the Transformation of Biology." *Current Biology* 24 (2).

———. 2015. *Life's Greatest Secret: The Race to Crack the Genetic Code.* New York: Basic Books.

Cohn, Ferdinand, Dr. 1939. *Bacteria, The Smallest of Living Organisms.* Translated by Charles S. Dolley, 1881. Baltimore: Johns Hopkins Press. First published 1872.

Comfort, Nathaniel C. 2001. *The Tangled Field: Barbara McClintock's Search for the Patterns of Genetic Control.* Cambridge (MA): Harvard University Press.

Cong, Le, F. Ann Ran, David Cox, Shuailiang Lin, Robert Barretto, Naomi Habib, Patrick D. Hsu, Xuebing Wu, Wenyan Jiang, Luciano A. Marraffini, and Feng Zhang. 2013. "Multiplex Genome Engineering Using CRISPR/Cas Systems." *Science* 339 (6121).

Copeland, Herbert F. 1938. "The Kingdoms of Organisms." *Quarterly Review of Biology* 13 (4).

Cordaux, Richard, and Mark A. Batzer. 2009. "The Impact of Retrotransposons on Human Genome Evolution." *Nature Reviews Genetics* 10 (10).

Cordonnier, Agnès, Jean-François Casella, and Thierry Heidmann. 1995. "Isolation of Novel Human Endogenous Retrovirus-like Elements with Foamy Virus-Related *pol* Sequence." *Journal of Virology* 69 (9).

Cornelis, Guillaume, Odile Heidmann, Sibylle Bernard-Stoecklin, Karine Reynaud, Géraldine Véron, Baptiste Mulot, Anne Dupressoir, and Thierry Heidmann. 2012. "Ancestral Capture of *syncytin-Car1*, a Fusogenic Endogenous Retroviral *Envelope*

Gene Involved in Placentation and Conserved in Carnivora." *Proceedings of the National Academy of Sciences* 109 (7).

Cornelis, Guillaume, Odile Heidmann, Séverine A. Degrelle, Cécile Vernochet, Christian Lavialle, Claire Letzelter, Sibylle Bernard-Stoecklin, Alexandre Hassanin, Baptiste Mulot, Michel Guillomot et al. 2013. "Captured Retroviral Envelope Syncytin Gene Associated with the Unique Placental Structure of Higher Ruminants." *Proceedings of the National Academy of Sciences* 110 (9).

Cornelis, Guillaume, Cécile Vernochet, Quentin Carradec, Sylvie Souquere, Baptiste Mulot, François Catzeflis, Maria A. Nilsson, Brandon R. Menzies, Marilyn B. Renfree, Gérard Pierron et al. 2015. "Retroviral Envelope Gene Captures and *syncytin* Exaptation for Placentation in Marsupials." *Proceedings of the National Academy of Sciences* 112 (5).

Costello, Elizabeth K., Keaton Stagaman, Les Dethlefsen, Brendan J. M. Bohannan, and David A. Relman. 2012. "The Application of Ecological Theory Toward an Understanding of the Human Microbiome." *Science* 336 (6086).

Cotton, James A., and James O. McInerney. 2010. "Eukaryotic Genes of Archaebacterial Origin Are More Important Than the More Numerous Eubacterial Genes, Irrespective of Function." *Proceedings of the National Academy of Sciences* 107 (40).

Cowdry, Edmund V., and Peter K. Olitsky. 1922. "Differences Between Mitochondria and Bacteria." *Journal of Experimental Medicine* 36 (5).

Cox, Cymon J., Peter G. Foster, Robert P. Hirt, Simon R. Harris, and T. Martin Embley. 2008. "The Archaebacterial Origin of Eukaryotes." *Proceedings of the National Academy of Sciences* 105 (51).

Cracraft, Joel, and Michael J. Donoghue, editors. 2004. *Assembling the Tree of Life*. New York: Oxford University Press.

Crick, Francis. 1970. "Central Dogma of Molecular Biology." *Nature* 227 (5258).

Crick, F. H. C. 1958. "On Protein Synthesis." *Symposia of the Society for Experimental Biology* 12.

Crick, F. H. C., F.R.S., Leslie Barnett, Dr. S. Brenner, and Dr. R. J. Watts-Tobin. 1961. "General Nature of the Genetic Code for Proteins." *Nature* 192 (4809).

Crisp, Alastair, Chiara Boschetti, Malcolm Perry, Alan Tunnacliffe, and Gos Micklem. 2015. "Expression of Multiple Horizontally Acquired Genes Is a Hallmark of Both Vertebrate and Invertebrate Genomes." *Genome Biology* 16 (50).

Cruz, Fernando de la, and Julian Davies. 2000. "Horizontal Gene Transfer and the Origin of Species: Lessons from Bacteria." *Trends in Microbiology* 8 (3).

Dagan, Tal, and William Martin. 2006. "The Tree of One Percent." *Genome Biology* 7 (118).

Dagan, Tal, Yael Artzy-Randrup, and William Martin. 2008. "Modular Networks and Cumulative Impact of Lateral Transfer in Prokaryote Genome Evolution." *Proceedings of the National Academy of Sciences* 105 (29).

Darwin, Charles. 1859. *On the Origin of Species by Means of Natural Selection, or the Preservation of Favoured Races in the Struggle for Life*. London: John Murray. Facsimile of the First Edition, Cambridge (MA): Harvard University Press, 1964.

_____. 1868. Letter to August Weismann, October 22, 1868. In *The Correspondence of Charles Darwin*. vol. 16. Edited by Frederick Burkhardt et al. Cambridge: Cambridge University Press. 2008.

_____. 1872. Letter to August Weismann, April 5, 1872. In *The Correspondence of Charles Darwin*. vol. 20. Edited by Frederick Burkhardt et al. Cambridge: Cambridge University Press. 2013.

Davies, Julian. 2001. "In a Map for Human Life, Count the Microbes, Too." *Science* 291 (5512).

Davies, Julian, and Dorothy Davies. 2010. "Origins and Evolution of Antibiotic Resistance." *Microbiology and Molecular Biology Reviews* 74 (3).

Davies, Roy. 2008. *The Darwin Conspiracy: Origins of a Scientific Crime*. London: Golden Square Books.

Dawes, Heather. 2004. "The Quiet Revolution." *Current Biology* 14 (15).

Dawkins, Richard. 2006. *The Selfish Gene*. 30th anniversary ed. New York: Oxford University Press. First published 1976.

Dawson, Martin H. 1930a. "The Transformation of Pneumococcal Types. I. The Conversion of R Forms of Pneumococcus into S Forms of the Homologous Type." *Journal of Experimental Medicine* 51 (1).

_____. 1930b. "The Transformation of Pneumococcal Types. II. The Interconvertiblity of Type-Specific S Pneumococci." *Journal of Experimental Medicine* 51 (1).

Dayhoff, M. O. 1964. "Computer Aids to Protein Sequence Determination." *Journal of Theoretical Biology* 8 (1).

Dayrat, Benoît. 2003. "The Roots of Phylogeny: How Did Haeckel Build His Trees?" *Systemic Biology* 52 (4).

Desmond, Adrian, and James Moore. 1991. *Darwin*. New York: Warner Books.

Dethlefsen, Les, Margaret McFall-Ngai, and David A. Relman. 2007. "An Ecological and Evolutionary Perspective on Human-Microbe Mutualism and Disease." *Nature* 449 (7164).

Dethlefsen, Les, Sue Huse, Mitchell L. Sogin, and David A. Relman. 2008. "The Pervasive Effects of an Antibiotic on the Human Gut Microbiota, as Revealed by Deep 16S rRNA Sequencing." *PLoS Biology* 6 (11).

Dodd, Matthew S., Dominic Papineau, Tor Grenne, John E. Slack, Martin Rittner, Franco Pirajno, Jonathan O'Neil, and Crispin T. S. Little. 2017. "Evidence for Early Life in Earth's Oldest Hydrothermal Vent Precipitates." *Nature* 543 (7643).

Dolan, Michael F., and Lynn Margulis. 2011. "Hans Ris. 1914–2002." From *Biographical Memoirs*, published by the National Academy of Sciences, Washington, (DC).

Dombrowski, Nina, John A. Donaho, Tony Gutierrez, Kiley W. Seitz, Andreas P. Teske, and Brett J. Baker. 2016. "Reconstructing Metabolic Pathways of Hydrocarbon-Degrading Bacteria from the Deepwater Horizon Oil Spill." *Nature Microbiology* 1 (7).

Doolittle, W. Ford. 1972. "Ribosomal Ribonucleic Acid Synthesis and Maturation in the Blue-Green Alga *Anacystis nidulans*." *Journal of Bacteriology* 111 (2).

_____. 1978. "Genes in Pieces: Were They Ever Together?" *Nature* 272 (5654).

_____. 1980. "Revolutionary Concepts in Evolutionary Cell Biology." *Trends in Biochemical Sciences* 5 (6).

_____. 1996. "At the Core of the Archaea." *Proceedings of the National Academy of Sciences* 93 (17).

_____. 1996. "Some Aspects of the Biology of Cells and Their Possible Evolutionary Significance." In *Evolution of Microbial Life*. Edited by D. McLits Roberts, P. Sharp, G. Alderson, and M.A. Collins. Cambridge: Cambridge University Press.

_____. 1998. "You Are What You Eat: A Gene Transfer Ratchet Could Account for Bacterial Genes in Eukaryotic Nuclear Genomes." *Trends in Genetics* 14 (8).

_____. 1999. "Lateral Genomics." *Trends in Cell Biology* 9 (12).

_____. 1999. "Phylogenetic Classification and the Universal Tree." *Science* 284 (5423).

_____. 2000. "Uprooting the Tree of Life." *Scientific American* 282 (2).

_____. 2004. "Q&A, W. Ford Doolittle," *Current Biology* 14 (5).

_____. 2009. "The Practice of Classification and the Theory of Evolution, and What the Demise of Charles Darwin's Tree of Life Hypothesis Means for Both of Them." *Philosophical Transactions of the Royal Society of London (B)* 364 (1527).

_____. 2010. "The Attempt on the Life of the Tree of Life: Science, Philosophy and Politics." *Biology & Philosophy* 25 (4).

_____. 2013. "Microbial Neopleomorphism." *Biology & Philosophy* 28 (2).

_____. 2015. "Is Junk DNA Bunk? A Critique of ENCODE." *Proceedings of the National Academy of Sciences* 110 (14).

Doolittle, W. Ford, and Norman R. Pace. 1970. "Synthesis of 5S Ribosomal RNA in *Escherichia coli* After Rifampicin Treatment." *Nature* 228 (5267).

_____. 1971. "Transcriptional Organization of the Ribosomal RNA Cistrons in *Escherichia coli*." *Proceedings of the National Academy of Sciences* 68 (8).

Doolittle, W. Ford, C. R. Woese, M. L. Sogin, L. Bonen, and D. Stahl. 1975. "Sequence Studies on 16S Ribosomal RNA from a Blue-Green Alga." *Journal of Molecular Evolution* 4 (4).

Doolittle, W. Ford, and Carmen Sapienza. 1980. "Selfish Genes, the Phenotype Paradigm and Genome Evolution." *Nature* 284 (5757).

Doolittle, W. Ford, and Linda Bonen. 1981. "Molecular Sequence Data Indicating an Endosymbiotic Origin for Plastids." *Annals of the New York Academy of Sciences* 361.

Doolittle, W. Ford., and James R. Brown. 1994. "Tempo, Mode, the Progenote, and the Universal Root." *Proceedings of the National Academy of Sciences* 91 (15).

Doolittle, W. Ford, Y. Boucher, C. L. Nesbø, C. J. Douady, J. O. Andersson, and A. J. Roger. 2003. "How Big Is the Iceberg of Which Organellar Genes in Nuclear Genomes Are but the Tip?" *Philosophical Transactions of the Royal Society of London (B)* 358 (1429).

Doolittle, W. Ford, and Eric Bapteste. 2007. "Pattern Pluralism and the Tree of Life Hypothesis." *Proceedings of the National Academy of Sciences* 104 (7).

Doudna, Jennifer A., and Emmanuelle Charpentier. 2014. "The New Frontier of Genome Engineering with CRISPR-Cas9." *Science* 346 (6213).

Dover, Gabriel and W. Ford Doolittle. 1980. "Modes of Genome Evolution." *Nature* 288 (December 18/25).

Downie, A. W. 1972. "Pneumococcal Transformation—A Backward View." *Journal of General Microbiology* 72 (3).

Drews, Gerhart. 1999. "Ferdinand Cohn, a Founder of Modern Microbiology." *ASM News* 65 (8).

_____. 2000. "The Roots of Microbiology and the Influence of Ferdinand Cohn on Microbiology of the 19th Century." *FEMS Microbiology Reviews* 24 (3).

Dubos, R. J. 1956. "Oswald Theodore Avery, 1877–1955." *Biographical Memoirs of the Fellows of the Royal Society* 2.

Dubos, René J. 1976. *The Professor, the Institute, and DNA.* New York: Rockefeller University Press.

Dunn, Casey W. 2005. "Complex Colony-Level Organization of the Deep-Sea Siphonophore *Bargmannia elongata* (Cnidaria, Hydrozoa) Is Directionally Asymmetric and Arises by the Subdivision of Pro-Buds." *Developmental Dynamics* 234 (4).

Dunning Hotopp, Julie C. 2011. "Horizontal Gene Transfer Between Bacteria and Animals." *Trends in Genetics* 27 (4).

_____. 2013. "Lateral Gene Transfer in Multicellular Organisms." In *Lateral Gene Transfer in Evolution.* Edited by Uri Gophna. New York: Springer.

Dunning Hotopp, Julie C., Michael E. Clark, Deodoro C. S. G. Oliveira, Jeremy M. Foster, Peter Fischer, Mónica C. Muñoz Torres, Jonathan D. Giebel, Nikhil Kumar, Nadeeza Ishmael, Shiliang Wang et al. 2007. "Widespread Lateral Gene Transfer from Intracellular Bacteria to Multicellular Eukaryotes." *Science* 317 (5845).

Dupressoir, Anne, Geoffroy Marceau, Cécile Vernochet, Laurence Bénit, Colette Kanellopoulos, Vincent Sapin, and Thierry Heidmann. 2005. "Syncytin-A and Syncytin-B, Two Fusogenic Placenta-Specific Murine Envelope Genes of Retroviral Origin Conserved in Muridae." *Proceedings of the National Academy of Sciences* 102 (3).

Dupressoir, Anne, Cécile Vernochet, Olivia Bawa, Francis Harper, Gérard Pierron, Paule Opolon, and Thierry Heidmann. 2009. "Syncytin-A Knockout Mice Demonstrate the Critical Role in Placentation of a Fusogenic, Endogenous Retrovirus-Derived, Envelope Gene." *Proceedings of the National Academy of Sciences* 106 (29).

Eck, Richard V., and Margaret O. Dayhoff. 1966. "Evolution of the Structure of Ferredoxin Based on Living Relics of Primitive Amino Acid Sequences." *Science* 152 (3720).

Ehresmann, C., P. Stiegler, P. Carbon, and J. P. Ebel. 1977. "Recent Progress in the Determination of the Primary Sequence of the 16 S RNA of *Escherichia coli.*" *FEBS Letters* 84 (2).

Eisen, Jonathan A. 2007. "Environmental Shotgun Sequencing: Its Potential and Challenges for Studying the Hidden World of Microbes." *PLoS Biology* 5 (3).

_____. 2009. "Genomic Evolvability and the Origin of Novelty: Studying the Past, Interpreting the Present, and Predicting the Future." In *Microbial Evolution and Co-Adaptation: A Tribute to the Life and Scientific Legacies of Joshua Lederberg.* Washington (DC): National Academies Press. www.nap.edu/read/12586.

Eisen, Jonathan A., and Claire M. Fraser. 2003. "Phylogenomics: Intersection of Evolution and Genomics." *Science* 300 (5626).

Elde, Nels. C., Stephanie J. Child, Michael T. Eickbush, Jacob O. Kitzman, Kelsey S. Rogers, Jay Shendure, Adam P. Geballe, and Harmit S. Malik. 2012. "Poxviruses Deploy Genomic Accordions to Adapt Rapidly Against Host Antiviral Defenses." *Cell* 150 (4).

Eliot, Theodore S. 1971. "Ivan Emmanuel Wallin. 1883–1969." *Anatomical Record* 171 (1).

Embley, T. Martin, and William Martin. 2006. "Eukaryotic Evolution, Changes and Challenges." *Nature* 440 (7084).

Embley, T. Martin, and Tom A. Williams. 2015. "Steps on the Road to Eukaryotes." *Nature* 521 (7551).

Ernster, Lars, and Gottfried Schatz. 1981. "Mitochondria: A Historical Review." *Journal of Cell Biology* 91 (3).

Espinal, P., S. Marti, and J. Vila. 2012. "Effect of Biofilm Formation on the Survival of *Acinetobacter baumannii* on Dry Surfaces." *Journal of Hospital Infection* 80 (1).

Ettema, Thijs J. G. 2016. "Mitochondria in the Second Act." *Nature* 531 (7592).

Eyres, Isobel, Chiara Boschetti, Alastair Crisp, Thomas P. Smith, Diego Fontaneto, Alan Tunnacliffe, and Timothy G. Barraclough. 2015. "Horizontal Gene Transfer in Bdelloid Rotifers Is Ancient, Ongoing and More Frequent in Species from Desiccating Habitats." *BMC Biology* 13 (90).

Fellner, P., and J. P. Ebel. 2017. "Biological Sciences: Observations on the Primary Structure of the 23S Ribosomal RNA from *E. coli*." *Nature* 225 (5238).

Feschotte, Cedric. 2008. "Transposable Elements and the Evolution of Regulatory Networks." *Nature Reviews Genetics* 9 (5).

———. 2010. "Bornavirus Enters the Genome." *Nature* 463 (7277).

———. 2015. "Transposable Elements." In *Discoveries in Modern Science: Exploration, Invention, Technology*. Edited by James Trefil. Farmington Hills (MI): Macmillan Reference USA.

Feschotte, Cedric, and Ellen J. Pritham. 2007. "DNA Transposons and the Evolution of Eukaryotic Genomes." *Annual Review of Genetics* 41.

Fitch, Walter M., and Emanuel Margoliash. 1967. "Construction of Phylogenetic Trees: A Method Based on Mutation Distances as Estimated from Cytochrome C Sequences Is of General Applicability." *Science* 155 (3760).

Fleischmann, Robert D., Mark D. Adams, Owen White, Rebecca A. Clayton, Ewen F. Kirkness, Anthony R. Kerlavage, Carol J. Bult, Jean-Francois Tomb, Brian A. Dougherty, Joseph M. Merrick et al. 1995. "Whole-Genome Random Sequencing and Assembly of *Haemophilus influenzae* Rd." *Science* 269 (5223).

Flot, Jean-François, Boris Hespeels, Xiang Li, Benjamin Noel, Irina Arkhipova, Etienne G. J. Danchin, Andreas Hejnol, Bernard Henrissat, Romain Koszul, Jean-Marc Aury et al. 2013. "Genomic Evidence for Ameiotic Evolution in the Bdelloid Rotifer *Adineta vaga*." *Nature* 500 (7463).

Forterre, Patrick, Simonetta Gribaldo, and Celine Brochier-Armanet. 2007. "Natural History of the Archaeal Domain." In *Archaea: Evolution, Physiology, and Molecular Biology*. Edited by Roger A. Garnett and Hans-Peter Klenk. Hoboken (NJ): Wiley-Blackwell.

Foster, Peter G., Cymon J. Cox, and T. Martin Embley. 2009. "The Primary Divisions of Life: A Phylogenomic Approach Employing Composition-Heterogeneous Methods." *Philosophical Transactions of the Royal Society of London (B)* 364 (1527).

Fournier, Gregory P., and E. J. Alm. 2015. "Ancestral Reconstruction of a Pre-LUCA Aminoacyl-tRNA Synthetase Ancestor Supports the Late Addition of Trp to the Genetic Code." *Journal of Molecular Evolution* 80 (3–4).

Fournier, Gregory P., Cheryl P. Andam, and Johann Peter Gogarten. 2015. "Ancient Horizontal Gene Transfer and the Last Common Ancestors." *BMC Evolutionary Biology* 15 (70).

Fox, G. E., E. Stackebrandt, R. B. Hespell, J. Gibson, J. Maniloff, T. A. Dyer, R. S. Wolfe, W. E. Balch, R. S. Tanner, L. J. Magrum et al. 1980. "The Phylogeny of Prokaryotes." *Science* 209 (4455).

Fox, George E., Linda J. Magrum, William E. Balch, Ralph S. Wolfe, and Carl R. Woese. 1977. "Classification of Methanogenic Bacteria by 16S Ribosomal RNA Characterization." *Proceedings of the National Academy of Sciences* 74 (10.)

Fox, George E., Kenneth R. Pechman, and Carl R. Woese. 1977. "Comparative Cataloging of 16S Ribosomal Ribonucleic Acid: Molecular Approach to Procaryotic Systematics." *International Journal of Systematic Bacteriology* 27 (1).

Fox, George E., Kenneth R. Luehrsen, and Carl R. Woese. 1982. "Archaebacterial 5 S Ribosomal RNA." *Zbl. Bakt. Hyg.* C3.

Fredericks, David N., editor. *The Human Microbiota: How Microbial Communities Affect Health and Disease.* Hoboken (NJ): John Wiley & Sons.

Friend, Tim. 2007. *The Third Domain: The Untold Story of Archaea and the Future of Biotechnology.* Washington, (DC): Joseph Henry Press.

Gardner, Pierce, David H. Smith, Herman Beer, and Robert C. Moellering Jr. 1969. "Recovery of Resistance (R) Factors from a Drug-Free Community." *Lancet* 2 (7624). Now labeled vol. 294.

Garrett, Roger A. 2014. "A Backward View from 16S rRNA to Archaea to the Universal Tree of Life to Progenotes: Reminiscences of Carl Woese." *RNA Biology* 11 (3).

Garrett, Roger A., and Hans-Peter Klenk, editors. 2007. *Archaea: Evolution, Physiology, and Molecular Biology.* Hoboken (NJ): Wiley-Blackwell.

Gatenby, J. Brontë. 1928. "Nature of Cytoplasmic Inclusions." *Nature* 121 (3040).

Gil, Estel, Assumpcio Bosch, David Lampe, Jose M. Lizcano, Jose C. Perales, Olivier Danos, and Miguel Chillon. 2013. "Functional Characterization of the Human *Mariner* Transposon *Hsmar2.*" *PLoS One* 8 (9).

Gilbert, Clément, John K. Pace II, and Cedric Feschotte. 2009. "Horizontal *SPIN*ing of Transposons." *Communicative & Integrative Biology* 2 (2).

Gilbert, Clément, Sarah Schaack, John K. Pace II, Paul J. Brindley, and Cedric Feschotte. 2010. "A Role for Host-Parasite Interactions in the Horizontal Transfer of Transposons Across Phyla." *Nature* 464 (7293).

Gilbert, Clément, Sharon S. Hernandez, Jaime Flores-Benabib, Eric N. Smith, and Cedric Feschotte. 2011. "Rampant Horizontal Transfer of *SPIN* Transposons in Squamate Reptiles." *Molecular Biology and Evolution* 29 (2).

Gilbert, Clément, Paul Waters, Cedric Feschotte, and Sarah Schaack. 2013. "Horizontal Transfer of OC1 Transposons in the Tasmanian Devil." *BMC Genomics* 14 (134).

Gilbert, Scott F., Jan Sapp, and Alfred I. Tauber. 2012. "A Symbiotic View of Life: We Have Never Been Individuals." *Quarterly Review of Biology* 87 (4).

Gilbert, Walter. 1986. "The RNA World." *Nature* 319 (6055).

Gladyshev, Eugene A., Matthew Meselson and Irina R. Arkhipova. 2008. "Massive Horizontal Gene Transfer in Bdelloid Rotifers." *Science* 320 (5880).

Gliboff, Sander. 2008. *H. G. Bronn, Ernst Haeckel, and the Origins of German Darwinism: A Study in Translation and Transformation.* Cambridge (MA): MIT Press.

Gitschier, Jane. 2015. "The Philosophical Approach: An Interview with Ford Doolittle." *PLoS Genetics* 11 (5).

Goffeau, A., B. G. Barrell, H. Bussey, R. W. Davis, B. Dujon, H. Feldmann, F. Galibert, J. D. Hoheisel, C. Jacq, M. Johnston et al. 1996. "Life with 6000 Genes." *Science* 274 (5287).

Gogarten, J. Peter. 1995. "The Early Evolution of Cellular Life." *TREE* 10 (4).

_____. 2003. "Gene Transfer: Gene Swapping Craze Reaches Eukaryotes." *Current Biology* 13 (2).

Gogarten, Johann Peter, Henrik Kibak, Peter Dittrich, Lincoln Taiz, Emma Jean Bowman, Barry J. Bowman, Morris F. Manolson, Ronald J. Poole, Takayasu Date, Tairo Oshima et al. 1989. "Evolution of the Vacuolar H+-ATPase: Implications for the Origin of Eukaryotes." *Proceedings of the National Academy of Sciences* 86.

Gogarten, J. Peter, Elena Hilario, and Lorraine Olendzenski. 1996. "Gene Duplications and Horizontal Gene Transfer During Early Evolution." In *Evaluation of Microbial Life*. Edited by D. McL. Roberts, P. Sharp, G. Alderson, and M.A. Collins. Cambridge: Cambridge University Press.

Gogarten, J. Peter, W. Ford Doolittle, and Jeffrey G. Lawrence. 2002. "Prokaryotic Evolution in Light of Gene Transfer." *Molecular Biology & Evolution* 19 (12).

Gogarten, J. Peter, and Jeffrey P. Townsend. 2005. "Horizontal Gene Transfer, Genome Innovation and Evolution." *Nature Reviews Microbiology* 3 (9).

Gogarten, Maria Boekels, Johann Peter Gogarten, and Lorraine Olendzenski, editors. 2009. *Horizontal Gene Transfer: Genomes in Flux*. New York: Springer/Humana Press.

Gold, Larry. 2013. "The Kingdoms of Carl Woese." *Proceedings of the National Academy of Sciences* 110 (9).

____. 2014. "Carl Woese in Schenectady: The Forgotten Years." *RNA Biology* 11 (3).

Goldenfeld, Nigel. 2014. "Looking in the Right Direction: Carl Woese and Evolutionary Biology." *RNA Biology* 11 (3).

Goldenfeld, Nigel, and Carl Woese. 2007. "Biology's Next Revolution." *Nature* 445 (7126).

Goldman, Jason G. 2012. "Ad Memoriam. Lynn Margulis (May 5, 1938–November 22, 2011)." *Studies in the History of Biology* 4 (2).

Goldner, Morris. 2007. "The Genius of Roger Stanier." *Canadian Journal of Infectious Diseases and Medical Microbiology* 18 (3).

Gontier, Nathalie. 2011. "Depicting the Tree of Life: The Philosophical and Historical Roots of Evolutionary Tree Diagrams." *Evolution: Education and Outreach* 4 (3).

Gontier, Nathalie, editor. 2015. *Reticulate Evolution: Symbiogenesis, Lateral Gene Transfer, Hybridization and Infectious Heredity*. Switzerland: Springer.

Gophna, Uri, editor. 2013. *Lateral Gene Transfer in Evolution*. New York: Springer.

Gordon, Jeffrey I. 2012. "Honor Thy Gut Symbionts Redux." *Science* 336 (6086).

Gradmann, Christoph. 2009. *Laboratory Disease: Robert Koch's Medical Bacteriology*. Translated by Elborg Forster. Baltimore: Johns Hopkins University Press.

Gray, Michael W. 2012. "Mitochondrial Evolution." *Cold Spring Harbor Perspectives in Biology* 4 (9).

____. 2014. "Organelle Evolution, Fragmented rRNAs, and Carl." *RNA Biology* 11 (3).

Gray, Michael W., and W. Ford Doolittle. 1982. "Has the Endosymbiont Hypothesis Been Proven?" *Microbiological Reviews* 46 (1).

Gray, Michael W., Gertraud Burger, and B. Franz Lang. 1999. "Mitochondrial Evolution." *Science* 283 (5407).

Green, Richard E., Johannes Krause, Adrian W. Briggs, Tomislav Maricic, Udo Stenzel, Martin Kircher, Nick Patterson, Heng Li, Weiwei Zhai, Markus His-Yang Fritz et al. 2010. "A Draft Sequence of the Neandertal Genome." *Science* 328 (5979).

Griffith, Fred. 1928. "The Significance of Pneumococcal Types." *Journal of Hygiene* 27 (2).

Groopman, Jerome. 2008. "Superbugs." *New Yorker* (August 11–18).

Grow, Edward J., Ryan A. Flynn, Shawn L. Chavez, Nicholas L. Bayless, Mark Wossidlo, Daniel J. Wesche, Lance Martin, Carol B. Ware, Catherine A. Blish, Howard Y. Chang et al. 2015. "Intrinsic Retroviral Reactivation in Human Preimplantation Embryos and Pluripotent Cells." *Nature* 522 (7555).

Guy, Lionel, and Thijs J. G. Ettema. 2011. "The Archaeal 'TACK' Superphylum and the Origin of Eukaryotes." *Trends in Microbiology* 19 (12).

Haeckel, Ernst. 2005. *Art Forms from the Ocean: The Radiolarian Atlas of 1862.* With an Introductory Essay by Olaf Breidbach. Reproduction. Munich, London, New York: Prestal.

Haeckel, Ernst Heinrich Philipp August. 2015. *The History of Creation: Or the Development of the Earth and Its Inhabitants by the Action of Natural Causes.* Translated by E. Ray Lankester. Reproduction. San Bernardino (CA): Ulan Press. First published in 1880 by New York: D. Appleton.

Hagen, Joel B. 2012. "Five Kingdoms, More or Less: Robert Whittaker and the Broad Classification of Organisms." *BioScience* 62 (1).

Hall, Ruth M., and Christina M. Collis. 1995. "Mobile Gene Cassettes and Integrons: Capture and Spread of Genes by Site-Specific Recombination." *Molecular Microbiology* 15 (4).

Handelsman, Jo. 2004. "Metagenomics: Application of Genomics to Uncultured Microorganisms." *Microbiology and Molecular Biology Review* 68 (4).

Harris, J. R. 1991. "The Evolution of Placental Mammals." *FEBS Letters* 295 (1–3).

Harris, J. Robin. 1998. "Placental Endogenous Retrovirus (ERV): Structural, Functional, and Evolutionary Significance." *BioEssays* 20 (4).

Hart, Michael W., and Richard K. Grosberg. 2009. "Caterpillars Did Not Evolve from Onychophorans by Hybridogenesis." *Proceedings of the National Academy of Sciences* 106 (47).

Hartl, Daniel L. 2001. "Discovery of the Transposable Element *Mariner.*" *Genetics* 157 (2).

Hayes, W. 1966. "Genetic Transformation: A Retrospective Appreciation." *Journal of General Microbiology* 45 (2).

Hecht, Mariana M., Nadjar Nitz, Perla F. Araujo, Alessandro O. Sousa, Ana de Cássia Rosa, Dawidson A. Gomes, Eduardo Leonardecz, and Antonio R. L. Teixeira. 2010. "Inheritance of DNA Transferred from American Trypanosomes to Human Hosts." *PLoS ONE* 5 (2).

Hedges, R. W. 1972. "The Pattern of Evolutionary Change in Bacteria." *Heredity* 28.

Hehemann, Jan-Hendrik, Gaëlle Correc, Tristan Barbeyron, William Helbert, Mirjam Czjzek, and Gurvan Michel. 2010. "Transfer of Carbohydrate-Active Enzymes from Marine Bacteria to Japanese Gut Microbiota." *Nature* 464 (7290).

Heidmann, Odile, Cécile Vernochet, Anne Dupressoir, and Thierry Heidmann. 2009. "Identification of an Endogenous Retroviral Envelope Gene with Fusogenic Activity and Placenta-Specific Expression in the Rabbit: A New 'Syncytin' in a Third Order of Mammals." *Retrovirology* 6 (107).

Heinemann, Jack A., and George F. Sprague Jr. 1989. "Bacterial Conjugative Plasmids Mobilize DNA Transfer Between Bacteria and Yeast." *Nature* 340 (6230).

Hensel, Michael, and Herbert Schmidt, editors. 2008. *Horizontal Gene Transfer in the Evolution of Pathogenesis.* New York: Cambridge University Press.

Hilario, Elena, and Johann Peter Gogarten. 1993. "Horizontal Transfer of ATPase Genes—The Tree of Life Becomes a Net of Life." *BioSystems* 31 (2–3).

Hinegardner, Ralph T., and Joseph Engelberg. 1963. "Rationale for a Universal Genetic Code." *Science* 142 (3595).

Hirt, Robert P., Cecilia Alsmark, and T. Martin Embley. 2015. "Lateral Gene Transfers and the Origins of the Eukaryote Proteome: A View from Microbial Parasites." *Current Opinion in Microbiology* 23.

Hitchcock, Edward. 1858. "Ichnology of New England. A Report of the Sandstone of the Connecticut Valley. Especially Its Fossil Footmarks." Boston: William White, Printer to the State.

_____. 1863. *Reminiscences of Amherst College, Historical, Scientific, Biographical and Autobiographical: Also, of Other and Wider Life Experiences.* Northampton (MA): Bridgman & Childs.

Hohn, Oliver, Kirsten Hanke, and Norbert Bannert. 2013. "HERV-K(HML-2), the Best Preserved Family of HERVs: Endogenization, Expression, and Implications in Health and Disease." *Frontiers in Oncology* 3 (246).

Holmes, Edward C. 2011. "The Evolution of Endogenous Viral Elements." *Cell Host & Microbe* 10 (4).

Hooper, Lora V., and Jeffrey I. Gordon. 2001. "Commensal Host-Bacterial Relationships in the Gut." *Science* 292 (5519).

Horikoshi, Koki. 1998. "Barophiles: Deep-Sea Microorganisms Adapted to an Extreme Environment." *Current Opinion in Microbiology* 1 (3).

Horvath, Philippe, and Rodolphe Barrangou. 2010. "CRISPR/Cas, the Immune System of Bacteria and Archaea." *Science* 327 (5962).

Hotchkiss, Rollin D. 1979. "The Identification of Nucleic Acids as Genetic Determinants." *Annals of the New York Academy of Sciences* 325.

Huang, Jinling. 2013. "Horizontal Gene Transfer in Eukaryotes: The Weak-Link Model." *BioEssays* 35 (10).

Huang, Jinling, Nandita Mullapudi, Cheryl A. Lancto, Marla Scott, Mitchell S. Abrahamsen, and Jessica C. Kissinger. 2004. "Phylogenomic Evidence Supports Past Endosymbiosis, Intracellular and Horizontal Gene Transfer in *Cryptosporidium parvum*." *Genome Biology* 5 (11).

Huang, Jinling, and J. Peter Gogarten. 2008. "Concerted Gene Recruitment in Early Plant Evolution." *Genome Biology* 9 (7).

Hublin, J. J. 2009. "The Origin of Neandertals." *Proceedings of the National Academy of Sciences* 106 (38).

Hug, Laura A., Brett J. Baker, Karthik Anantharaman, Christopher T. Brown, Alexander J. Probst, Cindy J. Castelle, Cristina N. Butterfield, Alex W. Hernsdorf, Yuki Amano, Kotaro Ise et al. 2016. "A New View of the Tree of Life." *Nature Microbiology* 1 (article no. 16048).

Hull, David L. 1988. *Science as a Process: An Evolutionary Account of the Social and Conceptual Development of Science.* Chicago and London: University of Chicago Press.

Hunt, Lois T. 1983. "Margaret O. Dayhoff, 1925–1983." *DNA and Cell Biology* 2 (2).

Husnik, Filip, Naruo Nikoh, Ryuichi Koga, Laura Ross, Rebecca P. Duncan, Manabu Fujie, Makiko Tanaka, Nori Satoh, Doris Bachtrog, Alex C. C. Wilson et al. 2013. "Horizontal Gene Transfer from Diverse Bacteria to an Insect Genome Enables a Tripartite Nested Mealybug Symbiosis." *Cell* 153 (7).

Ioannidis, Panagiotis, Kelly L. Johnston, David R. Riley, Nikhil Kumar, James R. White, Karen T. Olarte, Sandra Ott, Luke J. Tallon, Jeremy M. Foster, Mark J. Taylor, and Julie Dunning Hotopp. 2013. "Extensively Duplicated and Transcriptionally Active Recent Lateral Gene Transfer from a Bacterial *Wolbachia* Endosymbiont to Its Host Filarial Nematode *Brugia malayi*." *BMC Genomics* 14 (639).

Ishino, Yoshizumi, Hideo Shinagawa, Kozo Makino, Mitsuko Amemura, and Atsuo Nakata. 1987. "Nucleotide Sequence of the *iap* Gene, Responsible for Alkaline Phosphatase Isozyme Conversion in *Escherichia coli*, and Identification of the Gene Product." *Journal of Bacteriology* 169 (12).

Ivancevic, Atma M., Ali M. Walsh, R. Daniel Kortschak, and David L. Adelson. 2013. "Jumping the Fine LINE Between Species: Horizontal Transfer of Transposable Elements in Animals Catalyses Genome Evolution." *BioEssays* 35 (12).

Iwabe, Naoyuki, Kei-ichi Kuma, Masami Hasegawa, Syozo Osawa, and Takashi Miyata. 1989. "Evolutionary Relationship of Archaebacteria, Eubacteria, and Eukaryotes Inferred from Phylogenetic Trees of Duplicated Genes." *Proceedings of the National Academy of Sciences* 86 (23).

Jain, Ravi, Maria C. Rivera, and James A. Lake. 1999. "Horizontal Gene Transfer Among Genomes: The Complexity Hypothesis." *Proceedings of the National Academy of Sciences* 96 (7).

Jansen, Rudd, Jan D. A. Embden, Wim Gaastra, and Leo M. Schouls. 2002. "Identification of Genes That Are Associated with DNA." *Molecular Microbiology* 43 (6).

Jennison, A. V., and N. K. Verma. 2004. "*Shigella flexneri* Infection: Pathogenesis and Vaccine Development." *FEMS Microbiology Review* 28 (10).

Jinek, Martin, Krzysztof Chylinski, Ines Fonfara, Michael Hauer, Jennifer A. Doudna, and Emmanuelle Charpentier. 2012. "A Programmable Dual-RNA-Guided DNA Endonuclease in Adaptive Bacterial Immunity." *Science* 337 (6096).

Jones, Dorothy, and P. H. A. Sneath. 1970. "Genetic Transfer and Bacterial Taxonomy." *Bacteriological Reviews* 34 (1).

Judson, Horace Freeland. 1979. *The Eighth Day of Creation: The Makers of the Revolution in Biology*. New York: Simon & Schuster.

Kandler, Otto, and Hans Hippe. 1977. "Lack of Peptidoglycan in the Cell Walls of *Methanosarcina barkeri*." *Archives of Microbiology* 113 (1–2).

Kandler, Otto, editor 1982. *Archaebacteria*. Stuttgart (Ger.): Gustav Fischer.

Kapusta, Aurélie, Zev Kronenberg, Vincent J. Lynch, Xiaoyu Zhuo, LeeAnn Ramsay, Guillaume Bourque, Mark Yandell, and Cedric Feschotte. 2013. "Transposable Elements Are Major Contributors to the Origin, Diversification, and Regulation of Vertebrate Long Noncoding RNAs." *PLOS Genetics* 9 (4).

Katzourakis, Aris, and Robert J. Gifford. 2010. "Endogenous Viral Elements in Animal Genomes." *PLoS Genetics* 6 (11).

Kauffman, Stuart. 1993. *The Origins of Order: Self-Organization and Selection in Evolution*. New York: Oxford University Press.

_____. 1995. *At Home in the Universe: The Search for the Laws of Self-Organization and Complexity*. New York: Oxford University Press.

Kay, Lily E. (2001) "Biopower: Reflections on the Rise of Molecular Biology." In *Science, History and Social Activism: A Tribute to Everett Mendelsohn*. Edited by Garland E. Allen and Roy M. MacLeod. Boston Studies in the Philosophy of Science. Vol. 228. Dordrecht (Neth.): Springer.

Keeling, Patrick J., Gertraud Burger, Dion G. Durnford, B. Franz Lang, Robert W. Lee, Ronald E. Pearlman, Andrew J. Roger, and Michael W. Gray. 2005. "The Tree of Eukaryotes." *Trends in Ecology and Evolution* 20 (12).

Keeling, Patrick J., and Jeffrey D. Palmer. 2008. "Horizontal Gene Transfer in Eukaryotic Evolution." *Nature Reviews Genetics* 9 (8).

Keeling, Patrick J., John P. McCutcheon, and W. Ford Doolittle. 2015. "Symbiosis Becoming Permanent: Survival of the Luckiest." *Proceedings of the National Academy of Sciences* 112 (33).

Keller, Evelyn Fox. 1983. *A Feeling for the Organism: The Life and Work of Barbara McClintock.* New York: W. H. Freeman/Holt Paperback.

_____. 1986. "One Woman and Her Theory." *New Scientist* (July 3).

Keynes, R. D., editor. 1988. *Charles Darwin's Beagle Diary.* Cambridge: Cambridge University Press.

Khakhina, Liya Nikolaevna. 1992. *Concepts of Symbiogenesis: A Historical and Critical Study of the Research of Russian Botanists.* Edited by Lynn Margulis and Mark McMenamin. Translated by Stephanie Merkel and Robert Coalson. New Haven (CT): Yale University Press.

Koning, Audrey P. de, Fiona S. L. Brinkman, Steven J. M. Jones, and Patrick J. Keeling. 2000. "Lateral Gene Transfer and Metabolic Adaptation in the Human Parasite *Trichomonas vaginalis*." *Molecular Biology & Evolution* 17 (11).

Koonin, Eugene V. 2003. "Horizontal Gene Transfer: The Path to Maturity." *Molecular Microbiology* 50 (3).

_____. 2009. "On the Origin of Cells and Viruses: Primordial Virus World Scenario." *Annual Review of the New York Academy of Sciences* 1178.

_____. 2009. "Darwinian Evolution in the Light of Genomics." *Nucleic Acids Research* 37 (4).

_____. 2012. *The Logic of Chance: The Nature and Origin of Biological Evolution.* Upper Saddle River (NJ): Pearson Education.

_____. 2014. "Carl Woese's Vision of Cellular Evolution and the Domains of Life." *RNA Biology* 11 (3).

_____. 2015. "Origin of Eukaryotes from Within Archaea, Archaeal Eukaryome and Bursts of Gene Gain: Eukaryogenesis Just Made Easier?" *Philosophical Transactions of the Royal Society of London (B)* 370 (1678).

_____. 2015. "Why the Central Dogma: On the Nature of the Great Biological Exclusion Principle." *Biology Direct* 10 (52).

Koonin, Eugene V., Kira S. Makarova, and L. Aravind. 2001. "Horizontal Gene Transfer in Prokaryotes: Quantification and Classification." *Annual Review of Microbiology* 55.

Koonin, Eugene V., Tatiana G. Senkevich, and Valerian V. Dolja. 2006. "The Ancient Virus World and Evolution of Cells." *Biology Direct* 1 (29).

Koonin, Eugene V., and Yuri I. Wolf. 2009. "Is Evolution Darwinian or/and Lamarckian?" *Biology Direct* 4 (42).

_____. 2012. "Evolution of Microbes and Viruses: A Paradigm Shift in Evolutionary Biology?" *Frontiers in Cellular and Infection Microbiology* 2 (119).

Kordis, Dušan, and Franc Gubenšek. 1998. "Unusual Horizontal Transfer of a Long Interspersed Nuclear Element Between Distant Vertebrate Classes." *Proceedings of the National Academy of Sciences* 95 (18).

Kozo-Polyansky, Boris Mikhaylovich. 2010. *Symbiogenesis: A New Principle of Evolution.* Edited by Victor Fet and Lynn Margulis. Translated by Victor Fet. Introduction by Peter H. Raven. Cambridge (MA): Harvard University Press.

Kresge, Nicole, Robert D. Simoni, and Robert L. Hill. 2011. "The Molecular Genetics of Bacteriophage: The Work of Norton Zinder." In a paper series, "JBC Centennial 1905–2005, 100 Years of Biochemistry and Molecular Biology." *Journal of Biological Chemistry* 286 (25).

Kruif, Paul de. 1940. *Microbe Hunters.* New York: Pocket Books. First published in 1926.

Ku, Chuan, Shijulal Nelson-Sathi, Mayo Roettger, Sriram Garg, Einat Hazkani-Covo, and William F. Martin. 2015. "Endosymbiotic Gene Transfer from Prokaryotic Pangenomes: Inherited Chimaerism in Eukaryotes." *Proceedings of the National Academy of Sciences* 112 (33).

Kurland, Charles G. 2005. "What Tangled Web: Barriers to Rampant Horizontal Gene Transfer." *BioEssays* 27 (7).

Lacroix, Benoît, and Vitaly Citovsky. 2016. "Transfer of DNA from Bacteria to Eukaryotes." *mBio* 7 (4).

Lake, James A. 1990. "Origin of the Metazoa." *Proceedings of the National Academy of Sciences* 87 (2).

_____. 1994. "Reconstructing Evolutionary Trees from DNA and Protein Sequences: Paralinear Distances." *Proceedings of the National Academy of Sciences* 91 (4).

_____. 2015. "Eukaryotic Origins." *Philosophical Transactions of the Royal Society of London (B)* 370 (1678).

Lake, James A., Eric Henderson, Melanie Oakes, and Michael W. Clark. 1984. "Eocytes: A New Ribosome Structure Indicates a Kingdom with a Close Relationship to Eukaryotes." *Proceedings of the National Academy of Sciences* 81 (12).

Lake, James A., and Maria C. Rivera. 2004. "Deriving the Genomic Tree of Life in the Presence of Horizontal Gene Transfer: Conditioned Reconstruction." *Molecular Biology and Evolution* 21 (4).

Lake, James A., and Janet S. Sinsheimer. 2013. "The Deep Roots of the Rings of Life." *Genome Biology and Evolution* 5 (12).

Land, Miriam, Loren Hauser, Se-Ran Jun, Intawat Nookaew, Michael R. Leuze, Tae-Hyuk Ahn, Tatiana Karpinets, Ole Lund, Guruprased Kora, Trudy Wassenaar, Suresh Poudel, and David W. Ussery. 2015. "Insights from 20 Years of Bacterial Genome Sequencing." *Functional & Integrative Genomics* 15 (2).

Lander, Eric S. 2016. "The Heroes of CRISPR." *Cell* 164 (1–2).

Lander, Eric S., Lauren M. Linton, Bruce Birren, Chad Nusbaum, Michael C. Zody, Jennifer Baldwin, Keri Devon, Ken Dewar, Michael Doyle, William FitzHugh et al. 2001. "Initial Sequencing and Analysis of the Human Genome." *Nature* 409 (6822).

Lane, David J., David A. Stahl, Gary J. Olsen, Debra J. Heller, and Norman R. Pace. 1985. "Phylogenetic Analysis of the Genera *Thiobacillus* and *Thiomicrospira* by 5S rRNA Sequences." *Journal of Bacteriology* 163 (1).

Lane, David J., Bernadette Pace, Gary J. Olsen, David A. Stahl, Mitchell L. Sogin, and Norman R. Pace. 1985. "Rapid Determination of 16S Ribosomal RNA Sequences for Phylogenetic Analyses." *Proceedings of the National Academy of Sciences* 82 (20).

Lane, Nick. 2015. *The Vital Question: Energy, Evolution, and the Origins of Complex Life.* New York: W. W. Norton.

_____. 2015. "The Unseen World: Reflections on Leeuwenhoek (1677) 'Concerning Little Animals.'" *Philosophical Transactions of the Royal Society of London (B)* 370 (1666).

Lane, Nick, and William Martin. 2010. "The Energetics of Genome Complexity." *Nature* 467 (7318).

Lang, B. Franz, Gertraud Burger, Charles J. O'Kelly, Robert Cedergren, G. Brian Golding, Claude Lemieux, David Sankoff, Monique Turmel, and Michael W. Gray. 1997. "An Ancestral Mitochondrial DNA Resembling a Eubacterial Genome in Miniature." *Nature* 387 (6632).

Lapierre, Pascal, Erica Lasek-Nesselquist, and Johann Peter Gogarten. 2012. "The Impact of HGT on Phylogenomic Reconstruction Methods." *Briefings in Bioinformatics* 15 (1).

Lavialle, Christian, Guillaume Cornelis, Anne Dupressoir, Cécile Esnault, Odile Heidmann, Cécile Vernochet, and Thierry Heidmann. 2013. "Paleovirology of '*Syncytins*,' Retroviral *Env* Genes Exapted for a Role in Placentation." *Philosophical Transactions of the Royal Society of London (B)* 368 (1626).

Lawrence, Jeffrey G. 2004. "Why Genomics Is More Than Genomes." *Genome Biology* 5 (12).

Lawrence, Jeffrey G., and John R. Roth. 1996. "Selfish Operons: Horizontal Transfer May Drive the Evolution of Gene Clusters." *Genetics* 143 (4).

Lawrence, Jeffrey G., and Howard Ochman. 1997. "Amelioration of Bacterial Genomes: Rates of Change and Exchange." *Journal of Molecular Evolution* 44 (4).

_____. 1998. "Molecular Archaeology of the *Escherichia coli* Genome." *Proceedings of the National Academy of Sciences* 95 (16).

Lawrence, Jeffrey G., and Heather Hendrickson. 2003. "Lateral Gene Transfer: When Will Adolescence End?" *Molecular Microbiology* 50 (3).

Lawrence, Jeffery G., and Adam C. Retchless. 2009. "The Interplay of Homologous Recombination and Horizontal Gene Transfer in Bacterial Speciation." In *Horizontal Gene Transfer: Genomes in Flux.* Edited by Maria Boekels Gogarten, Johann Peter Gogarten, and Lorraine Olendzenski. New York: Springer/Humanz Press.

Lawrence, Jeffrey G., and Adam C. Retchless. 2010. "The Myth of Bacterial Species and Speciation." *Biology & Philosophy* 25 (4).

Lawrence, Philip J. 1972. "Edward Hitchcock: The Christian Geologist." *Proceedings of the American Philosophical Society* 116 (1).

Leahy, Sinead C., William J. Kelly, Eric Altermann, Ron S. Ronimus, Carl J. Yeoman, Diana M. Pacheco, Dong Li, Zhanhao Kong, Sharla McTavish, Carrie Sang et al. 2010. "The Genome Sequence of the Rumen Methanogen *Methanobrevibacter ruminantium* Reveals New Possibilities for Controlling Ruminant Methane Emissions." *PLoS ONE* 5 (1).

Lederberg, Joshua. 1952. "Cell Genetics and Hereditary Symbiosis." *Physiological Reviews* 32 (4).

_____. 1987. "Genetic Recombination in Bacteria: A Discovery Account." *Annual Review of Genetics* 21.

_____. 1992. "Bacterial Variation Since Pasteur: Rummaging in the Attic: Antiquarian Ideas of Transmissible Heredity, 1880–1940." *ASM News* 58 (5).

_____. 1994. "The Transformation of Genetics by DNA: An Anniversary Celebration of Avery, MacLeod and McCarty (1944)." *Genetics* 136 (2).

Lederberg, Joshua, and E. L. Tatum. 1946a. "Gene Recombination in *Escherichia coli.*" *Nature* 158 (4016).

_____. 1946b. "Novel Genotypes in Mixed Cultures of Biochemical Mutants of Bacteria." *Cold Spring Harbor Symposia on Quantitative Biology* 11.

Lederberg, Joshua, and Norton Zinder. 1948. "Concentration of Biochemical Mutants of Bacteria with Penicillin." *Journal of the American Chemical Society* 70 (12).

Lederberg, Joshua, Luigi L. Cavalli, and Esther M. Lederberg. 1952. "Sex Compatibility in *Escherichia coli.*" *Genetics* 37 (6).

Lederberg, Joshua, and E. L. Tatum. 1953. "Sex in Bacteria: Genetic Studies, 1945–1952." *Science* 118 (3059).

Lederberg, Joshua, and Esther M. Lederberg. 1956. "V. Infection and Heredity." In *Cellular Mechanisms in Differentiation and Growth*. Edited by Dorothea Rudnick. Princeton (NJ): Princeton University Press.

Levine, Donald P. 2006. "Vancomycin: A History." *Clinical Infectious Diseases* 42 (Suppl. 1).

Levy, Stuart B., MD. 2002. *The Antibiotic Paradox: How the Misuse of Antibiotics Destroys Their Curative Powers*. 2nd ed.: Cambridge (MA): Perseus. First published 2001.

Levy, Stuart B., and Paula Norman. 1970. "Segregation of Transferable R Factors into *Escherichia coli* Minicells." *Nature* 227 (5258).

Levy, Stuart B., George B. FitzGerald, and Ann B. Macone. 1976a. "Changes in Intestinal Flora of Farm Personnel After Introduction of a Tetracycline-Supplemented Feed on a Farm." *New England Journal of Medicine* 295 (11).

_____. 1976b. "Spread of Antibiotic-Resistant Plasmids from Chicken to Chicken and from Chicken to Man." *Nature* 260 (5546).

Levy, Stuart B., and Laura McMurry. 1978. "Plasmid-Determined Tetracycline Resistance Involves New Transport Systems for Tetracycline." *Nature* 276 (5683).

Lewin, Harris A. 2014. "Memories of Carl from an Improbable Friend." *RNA Biology* 11 (3).

Lewin, Roger. 1982. "Can Genes Jump Between Eukaryotic Species?" *Science* 217 (4554).

Ley, Ruth E., Peter J. Turnbaugh, Samuel Klein, and Jeffrey I. Gordon. 2006. "Human Gut Microbes Associated with Obesity." *Nature* 444 (7122).

Liu, Li, Xiaowei Chen, Geir Skogerbø, Peng Zhang, Runsheng Chen, Shunmin He, and Da-Wei Huang. 2012. "The Human Microbiome: A Hot Spot for Microbial Horizontal Gene Transfer." *Genomics* 100 (5).

Liu, Yi-Yun, Yang Wang, Timothy R. Walsh, Ling-Xian Yi, Rong Zhang, James Spencer, Yohei Doi, Guobao Tian, Baolei Dong, Xianhui Huang et al. 2015. "Emergence of Plasmid-Mediated Colistin Resistance Mechanism MCR-1 in Animals and Human Beings in China: A Microbiological and Molecular Biological Study." *Lancet Infectious Diseases* 16 (2).

Löwer, Roswitha, Johannes Löwer, and Reinhard Kurth. 1996. "The Viruses in All of Us: Characteristics and Biological Significance of Human Endogenous Retrovirus Sequences." *Proceedings of the National Academy of Sciences* 93 (11).

Luehrsen, Kenneth R. 2014. "Remembering Carl Woese." *RNA Biology* 11 (3).

Lynch, Vincent J. 2016. "A Copy-and-Paste Gene Regulatory Network." *Science* 351 (6277).

Lynch, Vincent J., Robert D. Leclerc, Gemma May, and Günter P. Wagner. 2011. "Transposon-Mediated Rewiring of Gene Regulatory Networks Contributed to the Evolution of Pregnancy in Mammals." *Nature Genetics* 43 (11).

Ma, Hong, Nuria Marti-Gutierrez, Sang-Wook Park, Jun Wu, Yeonmi Lee, Keiichiro Suzuki, Amy Koski, Dongmei Ji, Tomonari Hayama, Riffat Ahmed et al. 2017. "Correction of a Pathogenic Gene Mutation in Human Embryos." *Nature* 548 (7668).

Ma, Li-Jun, H. Charlotte van der Does, Katherine A. Borkovich, Jeffrey L. Coleman, Marie-Josée Daboussi, Antonio Di Pietro, Marie Dufresne, Michael Freitag, Manfred Grabherr, Bernard Henrissat et al. 2010. "Comparative Genomics Reveals Mobile Pathogenicity Chromosomes in *Fusarium*." *Nature* 464 (7287).

Magiorkinis, Gkikas, Robert J. Gifford, Aris Katzourakis, Joris De Ranter, and Robert Belshaw. 2012. "*Env*-less Endogenous Retroviruses Are Genomic Superspreaders." *Proceedings of the National Academy of Sciences* 109 (19).

Makarova, Kira S., L. Aravind, Nick V. Grishin, Igor B. Rogozin, and Eugene V. Koonin. 2002. "A DNA Repair System Specific for Thermophilic Archaea and Bacteria Predicted by Genomic Context Analysis." *Nucleic Acids Research* 30 (2).

Mallet, James, Nora Besansky, and Matthew W. Hahn. 2016. "How Reticulated Are Species?" *BioEssays* 37.

Mangeney, Marianne, Martial Renard, Géraldine Schlecht-Louf, Isabelle Bouallaga, Odile Heidmann, Claire Letzelter, Aurélien Richaud, Bertrand Ducos, and Thierry Heidmann. 2007. "Placental Syncytins: Genetic Disjunction Between the Fusogenic and Immunosuppressive Activity of Retroviral Envelope Proteins." *Proceedings of the National Academy of Sciences* 104 (51).

Margulis, Lynn. 1970. *Origin of Eukaryotic Cells: Evidence and Research Implications for a Theory of Origin and Evolution of Microbial, Plant, and Animal Cells on the Precambrian Earth.* New Haven (CT): Yale University Press.

_____. 1980. "Undulipodia, Flagella and Cilia." *BioSystems* 12 (1–2).

_____. 1993. *Symbiosis in Cell Evolution: Microbial Communities in the Archean and Proterozoic Eons.* 2nd ed.: New York: W. H. Freeman. First published 1981.

_____. 1998. *Symbiotic Planet: A New Look at Evolution.* New York: Basic Books.

Margulis, Lynn, and Dorion Sagan. 1986. *Micro-Cosmos: Four Billion Years of Microbial Evolution.* Foreword by Dr. Lewis Thomas. New York: Summit Books.

_____. 2003. *Acquiring Genomes: A Theory of the Origins of Species.* New York: Basic Books.

Margulis, Lynn, and Karlene V. Schwartz. 1988. *Five Kingdoms: An Illustrated Guide to the Phyla of Life on Earth.* Foreword by Stephen Jay Gould. 2nd ed. New York: W. H. Freeman. First published 1982.

Marraffini, Luciano A. 2013. "CRISPR-Cas Immunity Against Phages: Its Effects on the Evolution and Survival of Bacterial Pathogens." *PLOS Pathogens* 9 (12).

Marraffini, Luciano A., and Erik J. Sontheimer. 2008. "CRISPR Interference Limits Horizontal Gene Transfer in *Staphylococci* by Targeting DNA." *Science* 322 (5909).

Marshall, K. C., editor. 1986. *Advances in Microbial Ecology.* vol. 9. New York: Plenum Press.

Martin, Joseph P., Jr., and Irwin Fridovich. 1981. "Evidence for a Natural Gene Transfer from the Ponyfish to Its Bioluminescent Bacterial Symbiont *Photobacter leiognathi.*" *Journal of Biological Chemistry* 256 (12).

Martin, Mark. 2014. "Casting a Long Shadow in the Classroom: An Educator's Perspective of the Contributions of Carl Woese." *RNA Biology* 11 (3).

Martin, William. 1999. "Mosaic Bacterial Chromosomes: A Challenge En Route to a Tree of Genomes." *BioEssays* 21.

———. 2005. "Archaebacteria (Archaea) and the Origin of the Eukaryotic Nucleus." *Current Opinion in Microbiology* 8 (6).

Martin, William F. 1996. "Is Something Wrong with the Tree of Life?" *BioEssays* 18 (7).

Martin, William, and Rüdiger Cerff. 1986. "Prokaryotic Features of a Nucleus-Encoded Enzyme: cDNA Sequences for Chloroplast and Cystosolic Glyceraldehyde-3-Phosphate Dehydrogenases from Mustard (*Sinapis alba*)." *European Journal of Biochemistry* 159 (2).

Martin, William, and Reinhold G. Herrmann. 1998. "Gene Transfer from Organelles to the Nucleus: How Much, What Happens, and Why?" *Plant Physiology* 118 (1).

Martin, William, and Miklós Müller. 1998. "The Hydrogen Hypothesis for the First Eukaryote." *Nature* 392 (6671).

Martin, William, and Klaus V. Kowallik. 1999. "Annotated English Translation of Mereschkowsky's 1905 Paper 'Uber Natur und Ursprung der Chromatophoren im Pflanzenreiche.'" *European Journal of Phycology*, 34.

Martin, William, and Eugene V. Koonin. 2006. "Introns and the Origin of Nucleus-Cytosol Compartmentalization." *Nature* 440 (7080).

Martin, William, John Baross, Deborah Kelley, and Michael J. Russell. 2008. "Hydrothermal Vents and the Origin of Life." *Nature Reviews Microbiology* 6 (11).

Martin, William, Mayo Roettger, Thorsten Kloesges, Thorsten Thiergart, Christian Woehle, Sven Gould, and Tal Dagan. 2012. "Modern Endosymbiotic Theory: Getting Lateral Gene Transfer into the Equation." *Journal of Endocytobiosis and Cell Research* 23.

Marx, Jean. 2007. "New Bacterial Defense Against Phage Invaders Identified." *Science* 315 (5819).

Mather, Alison E., Kate S. Baker, Hannah McGregor, Paul Coupland, Pamela L. Mather, Ana Deheer-Graham, Julian Parkhill, Philippa Bracegirdle, Julie E. Russell, and Nicholas R. Thomson. 2014. "Bacillary Dysentery from World War 1 and NCTC1, the First Bacterial Isolate in the National Collection." *Lancet* 384 (9955).

Mayr, Ernst. 1982. *The Growth of Biological Thought: Diversity, Evolution, and Inheritance.* Cambridge (MA): Belknap Press of Harvard University Press.

McCarroll, Robert, Gary J. Olsen, Yvonne D. Stahl, Carl R. Woese, and Mitchell L. Sogin. 1983. "Nucleotide Sequence of the *Dictyostelium discoideum* Small-Subunit Ribosomal Ribonucleic Acid Inferred from the Gene Sequence: Evolutionary Implications." *Biochemistry* 22 (25).

McCarty, Maclyn. 1985. *The Transforming Principle: Discovering That Genes Are Made of DNA.* New York: W. W. Norton.

McCutcheon, John P. 2013. "Genome Evolution: A Bacterium with a Napoleon Complex." *Current Biology* 23 (15).

McCutcheon, John P., and Nancy A. Moran. 2007. "Parallel Genomic Evolution and Metabolic Interdependence in an Ancient Symbiosis." *Proceedings of the National Academy of Sciences* 104 (49).

_____. 2010. "Functional Convergence in Reduced Genomes of Bacterial Symbionts Spanning 200 My of Evolution." *Genome Biology and Evolution* 2.

_____. 2012. "Extreme Genome Reduction in Symbiotic Bacteria." *Nature Reviews Microbiology* 10 (1).

McCutcheon, John P., and Carol D. von Dohlen. 2011. "An Interdependent Metabolic Patchwork in the Nested Symbiosis of Mealybugs." *Current Biology* 21 (16).

McCutcheon, John P., and Patrick J. Keeling. 2014. "Endosymbiosis: Protein Targeting Further Erodes the Organelle/Symbiont Distinction." *Current Biology* 24 (14).

McInerney, James O. 2016. "A Four Billion Year Old Metabolism." *Nature Microbiology* 1 (9).

McInerney, James O., Mary J. O'Connell, and Davide Pisani. 2014. "The Hybrid Nature of the Eukaryota and a Consilient View of Life on Earth." *Nature Reviews Microbiology* 12 (6).

McInerney, James O., Davide Pisani, and Mary J. O'Connell. 2015. "The Ring of Life Hypothesis for Eukaryote Origins Is Supported by Multiple Kinds of Data." *Philosophical Transactions of the Royal Society of London (B)* 370 (1678).

McKenna, Maryn. 2010. *Superbug: The Fatal Menace of MRSA*. New York: Free Press.

Mérejkovsky, Const. de, Prof. Dr. 1920. "La Plante Considérée Comme Un Complexe Symbiotique." *Bull. Soc. Sc. Nat. Ouest*. Series 3 (6).

Mereschkowsky, M. C. 1878. "On a New Genus of Sponge." *Annals and Magazine of Natural History*, Series 5, 1 (1).

Metcalf, Jason A., Lisa J. Funkhouser-Jones, Kristen Brileya, Anna-Louise Reysenbach, and Seth R. Bordenstein. 2014. "Antibacterial Gene Transfer Across the Tree of Life." *eLife*. https://elifesciences.org/articles/04266.

Mi, Sha, Xinhua Lee, Xiang-ping Li, Geertruida M. Veldman, Heather Finnerty, Lisa Racie, Edward LaVallie, Xiang-Yang Tang, Philippe Edouard, Steve Howes et al. 2000. "Syncytin Is a Captive Retroviral Envelope Protein Involved in Human Placental Morphogenesis." *Nature* 403 (6771).

Miller, D., V. Urdaneta, and A. Weltman. 2002. "Vancomycin-Resistant *Staphylococcus aureus*—Pennsylvania, 2002." *Morbidity and Mortality Weekly Report* 51 (40).

Mirsky, Alfred E. 1968. "The Discovery of DNA." *Scientific American* 218 (6).

Modi, Sheetal R., James J. Collins, and David A. Relman. 2014. "Antibiotics and the Gut Microbiota." *Journal of Clinical Investigation* 124 (10).

Mojica, Francisco J. M., César Díez-Villaseñor, Elena Soria, and Guadalupe Juez. 2000. "MicroCorrespondence." *Molecular Microbiology* 36 (1).

Mojica, Francisco J. M., César Díez-Villaseñor, Jesús García-Martínez, and Elena Soria. 2005. "Intervening Sequences of Regularly Spaced Prokaryotic Repeats Derive from Foreign Genetic Elements." *Journal of Molecular Evolution* 60 (2).

Moore, Peter B. 2014. "Carl Woese: A Structural Biologist's Perspective." *RNA Biology* 11 (3).

Moran, Nancy A., and Paul Baumann. 2000. "Bacterial Endosymbionts in Animals." *Current Opinion in Microbiology* 3 (3).

Moran, Nancy A., John P. McCutcheon, and Atsushi Nakabachi. 2008. "Genomics and Evolution of Heritable Bacterial Symbionts." *Annual Review of Genetics* 42.

Moran, Nancy A., and Tyler Jarvik. 2010. "Lateral Transfer of Genes from Fungi Underlies Carotenoid Production in Aphids." *Science* 328 (5978).

Morange, Michel. 2015. "CRISPR-Cas: The Discovery of an Immune System in Prokaryotes." *Journal of Biosciences* 40 (2).

Morell, Virginia. 1996. "Life's Last Domain." *Science* 273 (5278).

_____. 1997. "Microbiology's Scarred Revolutionary." *Science* 276 (5313).

Morens, David M., Jeffery K. Taubenberger, and Anthony S. Fauci. 2008. "Predominant Role of Bacterial Pneumonia as a Cause of Death in Pandemic Influenza: Implications for Pandemic Influenza Preparedness." *Journal of Infectious Diseases* 198 (7).

Morgan, Gregory J. 1998. "Emile Zuckerkandl, Linus Pauling, and the Molecular Evolutionary Clock, 1959–1965." *Journal of the History of Biology* 31 (2).

Morris, J. Gareth. 1983. "Roger Yate Stanier, 1916–1982." *Journal of General Microbiology* 129.

Mukherjee, Siddhartha. 2016. *The Gene: An Intimate History.* New York: Scribner.

Nair, Prashant. 2012. "Woese and Fox: Life, Rearranged." *Proceedings of the National Academy of Sciences* 109 (4).

Nelson, Karen E., Rebecca A. Clayton, Steven R. Gill, Michelle L. Gwinn, Robert J. Dodson, Daniel H. Haft, Erin K. Hickey, Jeremy D. Peterson, William C. Nelson, Karen A. Ketchum et al. 1999. "Evidence for Lateral Gene Transfer Between Archaea and Bacteria from Genome Sequence of *Thermotoga maritima*." *Nature* 399 (6734).

Nelson-Sathi, Shijulal, Filipa L. Sousa, Mayo Roettger, Nabor Lozada-Chávez, Thorsten Thiergart, Arnold Janssen, David Bryant, Giddy Landan, Peter Schönheit, Bettina Siebers, James O. McInerney, and William F. Martin. 2015. "Origins of Major Archaeal Clades Correspond to Gene Acquisitions from Bacteria." *Nature* 517 (7532).

Ng, Hooi Jun, Mario López-Pérez, Hayden K. Webb, Daniela Gomez, Tomoo Sawabe, Jason Ryan, Mikhail Vyssotski, Chantal Bizet, François Malherbe, Valery V. Mikhailov, Russell J. Crawford, and Elena P. Ivanova. 2014. "*Marinobacter salarius* sp. nov. and *Marinobacter similis* sp. nov., Isolated from Sea Water." *PLoS One* 9 (9).

Nikoh, Naruo, Kohjiro Tanaka, Fukashi Shibata, Natsuko Kondo, Masahiro Hizume, Masakazu Shimada, and Takema Fukatsu. 2008. "*Wolbachia* Genome Integrated in an Insect Chromosome: Evolution and Fate of Laterally Transferred Endosymbiont Genes." *Genome Research* 18 (2).

Noble, W. D., Zarina Virani, and Rosemary G. A. Cree. 1992. "Co-Transfer of Vancomycin and Other Resistance Genes from *Enterococcus faecalis* NCTC 12201 to *Staphylococcus aureus*." *FEMS Microbiology Letters* 72 (2).

Noller, Harry F. 2013. "By Ribosome Possessed." *Journal of Biological Chemistry* 288 (34).

_____. 2013. "Carl Woese (1928–2012): Discoverer of Life's Third Domain, the Archaea." *Nature* 493 (7434).

_____. 2014. "Secondary Structure Adventures with Carl Woese." *RNA Biology* 11 (3).

Nowak, Rachel. 1995. "Bacterial Genome Sequence Bagged." *Science* 269 (5223).

Ochman, Howard, Jeffrey G. Lawrence, and Eduardo A. Groisman. 2000. "Lateral Gene Transfer and the Nature of Bacterial Innovation." *Nature* 405 (6784).

Olby, Robert. 1994. *The Path to the Double Helix: The Discovery of DNA.* Seattle: University of Washington Press.

Olby, Robert, and Erich Posner. 1967. "An Early Reference to Genetic Coding." *Nature* 215 (5100).

Olsen, Gary J., David J. Lane, Stephen J. Giovannoni, and Norman R. Pace. 1986. "Microbial Ecology and Evolution: A Ribosomal RNA Approach." *Annual Review of Microbiology* 40.

Olsen, Gary J., Carl R. Woese, and Ross Overbeek. 1994. "The Winds of (Evolutionary) Change: Breathing New Life into Microbiology." *Journal of Bacteriology* 176 (1).

Olsen, Gary J., and Carl R. Woese. 1997. "Archaeal Genomics: An Overview." *Cell* 89 (7).

O'Malley, Maureen A. 2009. "What *Did* Darwin Say About Microbes, and How Did Microbiology Respond?" *Trends in Microbiology* 17 (8).

O'Malley, Maureen A., and John Dupré. 2007. "Size Doesn't Matter: Towards a More Inclusive Philosophy of Biology." *Biology & Philosophy* 22 (2).

O'Malley, Maureen A., William Martin, and John Dupré. 2010. "The Tree of Life: Introduction to an Evolutionary Debate." *Biology & Philosophy* 25 (4).

O'Malley, Maureen A., and Eugene V. Koonin. 2011. "How Stands the Tree of Life a Century and a Half After *The Origin*?" *Biology Direct* 6 (32).

Orgel, L. E., and F. H. C. Crick. 1980. "Selfish DNA: The Ultimate Parasite." *Nature* 284 (5757).

Orgel, L. E., F. H. C. Crick, and C. Sapienza. 1980. "Selfish DNA." *Nature* 288 (5792).

Pääbo, Svante. 2003. "The Mosaic That Is Our Genome." *Nature* 421 (6921).

_____. 2014. *Neanderthal Man: In Search of Lost Genomes*. New York: Basic Books.

Pace, John K., II, and Cedric Feschotte. 2007. "The Evolutionary History of Human DNA Transposons: Evidence for Intense Activity in the Primate Lineage." *Genome Research* 17 (4).

Pace, John K., II, Clément Gilbert, Marlene S. Clark, and Cedric Feschotte. 2008. "Repeated Horizontal Transfer of a DNA Transposon in Mammals and Other Tetrapods." *Proceedings of the National Academy of Sciences* 105 (44).

Pace, N. R., D. H. L. Bishop, and S. Spiegelman. 1967. "The Kinetics of Product Appearance and Template Involvement in the *In Vitro* Replication of Viral RNA." *Proceedings of the National Academy of Sciences* 58 (2).

Pace, Norman R. 1997. "A Molecular View of Microbial Diversity and the Biosphere." *Science* 276 (5313).

_____. 2006. "Time For a Change." *Nature* 441 (7091).

_____. 2008. "The Molecular Tree of Life Changes How We See, Teach Microbial Diversity." *Microbe* (1).

_____. 2009. "Mapping the Tree of Life: Progress and Prospects." *Microbiology and Molecular Biology Reviews* 73 (4).

Pace, Norman R., Gary J. Olsen, and Carl R. Woese. 1986. "Ribosomal RNA Phylogeny and the Primary Lines of Evolutionary Descent." *Cell* 45 (3).

Pace, Norman R., David A. Stahl, David J. Lane, and Gary J. Olsen. 1986. "The Analysis of Natural Microbial Populations by Ribosomal RNA Sequences." In *Advances of Microbial Ecology*. Vol. 9. Edited by K. C. Marshall. New York: Plenum Press.

Pace, Norman, Jan Sapp, and Nigel Goldenfeld. 2012. "Phylogeny and Beyond: Scientific, Historical, and Conceptual Significance of the First Tree of Life." *Proceedings of the National Academy of Sciences* 109 (4).

Packard, Alpheus S., M.S., LL.D. 2009. *Lamarck, the Founder of Evolution: His Life and Work*. With translations of his writings on Organic Evolution. Repr. ed. LaVergne (TN): Kessinger. First published in 1901 by London: Longmans, Green.

Palmer, Kelli L., and Michael S. Gilmore. 2010. "Multidrug-Resistant *Enterococci* Lack CRISPR-cas." *mBio* 1 (4).

Panchen, Alec L. 1992. *Classification, Evolution, and the Nature of Biology.* Cambridge: Cambridge University Press.

Papagianni, Dimitra, and Michael A. Morse. 2013. *The Neanderthals Rediscovered: How Modern Science Is Rewriting Their Story.* London: Thames & Hudson.

Parker, Philip M., editor. 2009. *Bacteriology: Webster's Timeline History 1812–2006.* San Diego: ICON Group International.

Patterson, Nick, Daniel J. Richter, Sante Gnerre, Eric S. Lander, and David Reich. 2006. "Genetic Evidence for Complex Speciation of Humans and Chimpanzees." *Nature* 441 (7097).

Pedersen, Rolf B., Hans Tore Rapp, Ingunn H. Thorseth, Marvin D. Lilley, Fernando J. A. S. Barriga, Tamara Baumberger, Kristin Flesland, Rita Fonseca, Gretchen L. Früh-Green, and Steffen L. Jorgensen. 2010. "Discovery of a Black Smoker Vent Field and Vent Fauna at the Arctic Mid-Ocean Ridge." *Nature Communications* 1 (126).

Pennings, Jeroen L. A., Jan T. Keltjens, and Godfried D. Vogels. 1998. "Isolation and Characterization of *Methanobacterium thermoautotrohpicum* ΔH Mutants Unable to Grow Under Hydrogen-Deprived Conditions." *Journal of Bacteriology* 180 (10).

Pennisi, Elizabeth. 1998. "Genome Data Shake Tree of Life." *Science* 280 (5364).

_____. 1999. "Is It Time to Uproot the Tree of Life?" *Science* 284 (5418).

_____. 2002. "Bacteria Shared Photosynthesis Genes." *Science* 298 (5598).

_____. 2004. "The Birth of the Nucleus." *Science* 305 (5685).

Peterson, Jane, Susan Garges, Maria Giovanni, Pamela McInnes, Lu Wang, Jeffrey A. Schloss, Vivien Bonazzi, Jean E. McEwen, Kris A. Wetterstrand, Carolyn Deal et al. 2009. "The NIH Human Microbiome Project." *Genome Research* 19 (12).

Pietsch, Theodore W. 2012. *Trees of Life: A Visual History of Evolution.* Baltimore: Johns Hopkins University Press.

Pisani, Davide, James A. Cotton, and James O. McInerney. 2007. "Supertrees Disentangle the Chimerical Origin of Eukaryotic Genomes." *Molecular Biology and Evolution* 24 (8).

Piskurek, Oliver, and Norihiro Okada. 2007. "Poxviruses as Possible Vectors for Horizontal Transfer of Retroposons from Reptiles to Mammals." *Proceedings of the National Academy of Sciences* 104 (29).

Pittis, Alexandros A., and Toni Gabaldón. 2016. "Late Acquisition of Mitochondria by a Host with Chimaeric Prokaryotic Ancestry." *Nature* 531 (7592).

Pollock, M. R. 1970. "The Discovery of DNA: An Ironic Tale of Chance, Prejudice and Insight." *Journal of General Microbiology* 63.

Pollock, Robert. 2015. "Eugenics Lurk in the Shadow of CRISPR." *Science* 348 (6237).

Poundstone, William. 1999. *Carl Sagan: A Life in the Cosmos.* New York: Henry Holt.

Pourcel, C., G. Salvignol, and G. Vergnaud. 2005. "CRISPR Elements in *Yersinia pestis* Acquire New Repeats by Preferential Uptake of Bacteriophage DNA, and Provide Additional Tools for Evolutionary Studies." *Microbiology* 151 (Pt. 3).

Proctor, Lita M., Shaila Chhibba, Jean McEwen, Jane Peterson, Chris Wellington, Carl Baker, Maria Giovanni, Pamela McInnes, and R. Dwayne Lunsford. 2013. "The NIH Human Microbiome Project." In *The Human Microbiota: How Microbial Communities Affect Health and Disease.* Edited by David N. Fredricks. Hoboken (NJ): John Wiley & Sons.

Ragan, Mark A. 2009. "Trees and Networks Before and After Darwin." *Biology Direct* 4 (43).

Ragan, Mark A., James O. McInerney, and James A. Lake. 2009. "The Network of Life: Genome Beginnings and Evolution." *Philosophical Transactions of the Royal Society of London (B)* 364 (1527).

Raoult, Didier, and Eugene V. Koonin. 2012. "Microbial Genomics Challenge Darwin." *Frontiers in Cellular and Infection Microbiology* 2 (127).

Raven, Peter H. 1970. "A Multiple Origin for Plastids and Mitochondria." *Science* 169 (3946).

Reames, Richard. 2007. *Arborsculpture: Solutions for a Small Planet*. Williams (OR): Arborsmith Studios.

Reames, Richard, and Barbara Hahn Delbol. 1995. *How to Grow a Chair: The Art of Tree Trunk Topiary*. Williams (OR): Arborsmith Studios.

Reanney, Darryl. 1976. "Extrachromosomal Elements as Possible Agents of Adaptation and Development." *Bacteriological Reviews* 40 (3).

Redelsperger, François, Guillaume Cornelis, Cécile Vernochet, Bud C. Tennant, François Catzeflis, Baptiste Mulot, Odile Heidmann, Thierry Heidmann, and Anne Dupressoir. 2014. "Capture of *syncytin-Mar1*, a Fusogenic Endogenous Retroviral Envelope Gene Involved in Placentation in the Rodentia Squirrel-Related Clade." *Journal of Virology* 88 (14).

Relman, David A., MD. 2011. "Microbial Genomics and Infectious Diseases." *New England Journal of Medicine* 365 (4).

Relman, David A. 2012. "Learning About Who We Are." *Nature* 486 (7402).

_____. 2013. "Metagenomics, Infectious Disease Diagnostics, and Outbreak Investigations: Sequence First, Ask Questions Later?" *Journal of the American Medical Association* 309 (14).

_____. 2015. "The Human Microbiome and the Future Practice of Medicine." *Journal of the American Medical Association* 314 (11).

Retchless, Adam C., and Jeffrey G. Lawrence. 2007. "Temporal Fragmentation of Speciation in Bacteria." *Science* 317 (5841).

_____. 2012. "Ecological Adaptation in Bacteria: Speciation Driven by Codon Selection." *Molecular Biology and Evolution* 29 (12).

Richards, Robert J. 2008. *The Tragic Sense of Life: Ernst Haeckel and the Struggle over Evolutionary Thought*. Chicago and London: University of Chicago Press.

Ridley, Matt. 2000. *Genome: The Autobiography of a Species in 23 Chapters*. New York: HarperCollins. First published 1999.

_____. 2006. *Francis Crick: Discoverer of the Genetic Code*. New York: Atlas Books/HarperCollins.

Riley, David B., Karsten B. Sieber, Kelly M. Robinson, James Robert White, Ashwinkumar Ganesan, Syrus Nourbakhsh and Julie C. Dunning Hotopp. 2013. "Bacteria-Human Somatic Cell Lateral Gene Transfer Is Enriched in Cancer Samples." *PLoS Computational Biology* 9 (6).

Ris, Hans, and Walter Plaut. 1962. "Ultrastructure of DNA-Containing Areas in the Chloroplast of *Chlamydomonas*." *Journal of Cell Biology* 13 (3).

Rivera, Maria C., Ravi Jain, Jonathan E. Moore, and James A. Lake. 1998. "Genomic Evidence for Two Functionally Distinct Gene Classes." *Proceedings of the National Academy of Sciences* 95 (11).

Rivera, Maria C., and James A. Lake. 2004. "The Ring of Life Provides Evidence for a Genome Fusion Origin of Eukaryotes." *Nature* 431 (7005).

Roberts, D. McL., P. Sharp, G. Alderson, and M.A. Collins, editors. 1996. *Evolution of Microbial Life*. Cambridge: Cambridge University Press.

Roberts, Elijah, Anurag Sethi, Jonathan Montoya, Carl R. Woese, and Zaida Luthey-Schulten. 2008. "Molecular Signatures of Ribosomal Evolution." *Proceedings of the National Academy of Sciences* 105 (37).

Robinson, Kelly M., Karsten B. Sieber, and Julie C. Dunning Hotopp. 2013. "A Review of Bacteria-Animal Lateral Gene Transfer May Inform Our Understanding of Diseases Like Cancer." *PLOS Genetics* 9 (10).

Rogers, Matthew B., Russell F. Watkins, James T. Harper, Dion G. Durnford, Michael W. Gray, and Patrick J. Keeling. 2007. "A Complex and Punctate Distribution of Three Eukaryotic Genes Derived by Lateral Gene Transfer." *BMC Evolutionary Biology* 7 (89).

Rokas, Antonis, and Peter W. H. Holland. 2000. "Rare Genomic Changes as a Tool for Phylogenetics." *Trends in Ecology & Evolution* 15 (11).

Rokas, Antonis, and Sean B. Carroll. 2006. "Bushes in the Tree of Life." *PLoS Biology* 4 (11).

Romer, Alfred Sherwood. 1966. *Vertebrate Paleontology*. 3rd ed.: Chicago and London: University of Chicago Press. First published 1933.

Sagan, Dorion, editor. 2012. *Lynn Margulis: The Life and Legacy of a Scientific Rebel*. White River Junction (VT): Chelsea Green.

Sagan, Lynn. 1967. "On the Origin of Mitosing Cells." *Journal of Theoretical Biology* 14 (3).

Salzberg, Steven L., Owen White, Jeremy Peterson, and Jonathan A. Eisen. 2001. "Microbial Genes in the Human Genome: Lateral Transfer or Gene Loss?" *Science* 292 (5523).

Sanger, Fred. 2001. "The Early Days of DNA Sequences." *Nature Medicine* 7 (3).

Sanger, F., G. G. Brownlee, and B. G. Barrell. 1965. "A Two-Dimensional Fractionation Procedure for Radioactive Nucleotides." *Journal of Molecular Biology* 13 (2).

Sankararaman, Sriram, Swapan Mallick, Michael Dannemann, Kay Prüfer, Janet Kelso, Svante Pääbo, Nick Patterson, and David Reich. 2014. "The Genomic Landscape of Neanderthal Ancestry in Present-Day Humans." *Nature* 507 (7492).

Sapp, Jan. 1994. *Evolution by Association: A History of Symbiosis*. New York: Oxford University Press.

_____. 2002. "Paul Buchner (1886–1978) and Hereditary Symbiosis in Insects." *International Microbiology* 5 (3).

_____. 2003. *Genesis: The Evolution of Biology*. New York: Oxford University Press.

_____. 2005. "The Prokaryote-Eukaryote Dichotomy: Meanings and Mythology." *Microbiology and Molecular Biology Reviews* 69 (2).

_____. 2009. *The New Foundations of Evolution: On the Tree of Life*. New York: Oxford University Press.

Sapp, Jan, editor. 2005. *Microbial Phylogeny and Evolution: Concepts and Controversies*. New York: Oxford University Press.

Sapp, Jan, Francisco Carrapiço, and Mikhail Zolotonosov. 2002. "Symbiogenesis: The Hidden Face of Constantin Merezhkowsky." *History and Philosophy of the Life Sciences* 24 (3–4).

Sapp, Jan, and George E. Fox. 2013. "The Singular Quest for a Universal Tree of Life." *Microbiology and Molecular Biology Reviews* 77 (4).

Sarkar, Sahotra. 2014. "Woese on the Received View of Evolution." *RNA Biology* 11 (3).

Schaack, Sarah, Clément Gilbert, and Cedric Feschotte. 2010. "Promiscuous DNA: Horizontal Transfer of Transposable Elements and Why It Matters for Eukaryotic Evolution." *Trends in Ecology and Evolution* 25 (9).

Schulz, Heide N., and Bo Barker Jörgensen. 2001. "Big Bacteria." *Annual Review of Microbiology* 55.

Schwartz, Robert M., and Margaret O. Dayhoff. 1978. "Origins of Prokaryotes, Eukaryotes, Mitochondria, and Chloroplasts." *Science* 199 (4327).

Sears, Cynthia L. 2005. "A Dynamic Partnership: Celebrating Our Gut Flora." *Anaerobe* 11 (5).

Sheridan, Cormac. 2015. "CRISPR Germline Editing Reverberates Through Biotech Industry." *Nature Biotechnology* 33 (5).

Shreeve, James. 2004. *The Genome War*. New York: Alfred A. Knopf.

Sievert, D. M., M. L. Boulton, G. Stoltman, D. Johnson, M. G. Stobierski, F. P. Downes, P. A. Somsel, J. T. Rudrik, W. Brown, W. Hafeez et al. 2002. "Staphylococcus aureus Resistant to Vancomycin—United States, 2002." *Morbidity and Mortality Weekly Report* 51 (26).

Silva, Joana C., Elgion L. Loreto, and Jonathan B. Clark. 2004. "Factors That Affect the Horizontal Transfer of Transposable Elements." *Current Issues in Molecular Biology* 6 (1).

Smillie, Chris S., Mark B. Smith, Jonathan Friedman, Otto X. Cordero, Lawrence A. David, and Eric J. Alm. 2011. "Ecology Drives a Global Network of Gene Exchange Connecting the Human Microbiome." *Nature* 480 (7376).

Smith, David Roy, and Patrick J. Keeling. 2015. "Mitochondrial and Plastid Genome Architecture: Reoccurring Themes, but Significant Differences at the Extremes." *Proceedings of the National Academy of Sciences* 112 (33).

Smith, Michael, Da-Fei Feng, and Russell F. Doolittle. 1992. "Evolution by Acquisition: The Case for Horizontal Gene Transfers." *Trends in Biochemical Science* 17 (12).

Smith, Myron L., Johann N. Bruhn, and James B. Anderson. 1992. "The Fungus *Armillaria bulbosa* Is Among the Largest and Oldest Living Organisms." *Nature* 356 (6368).

Sogin, M., B. Pace, N. R. Pace, and C. R. Woese. 1971. "Primary Structural Relationship of p16 to m16 Ribosomal RNA." *Nature New Biology* 232 (28).

Sogin, Mitchell L., Hilary G. Morrison, Julie A. Huber, David Mark Welch, Susan M. Huse, Phillip R. Neal, Jesus M. Arrieta, and Gerhard J. Herndl. 2006. "Microbial Diversity in the Deep Sea and the Underexplored 'Rare Biosphere.'" *Proceedings of the National Academy of Sciences* 103 (32).

Sogin, S. J., M. L. Sogin, and C. R. Woese. 1972. "Phylogenetic Measurement in Procaryotes by Primary Structural Characterization." *Journal of Molecular Evolution* 1 (1).

Sonea, Sorin, and Maurice Panisset. 1983. *A New Bacteriology*. Boston: Jones and Bartlett. First published as *Introduction à la Nouvelle Bactériologie* in 1980 by Les Presses de L'Université de Montréal.

Sonea, Sorin. 1988. "The Global Organism: A New View of Bacteria." *Sciences* 28 (4).

Sonnenburg, Justin L. 2010. "Genetic Pot Luck." *Nature* 464 (7290).

Sorek, Rotem, C. Martin Lawrence, and Blake Wiedenheft. 2013. "CRISPR-Mediated Adaptive Immune Systems in Bacteria and Archaea." *Annual Review of Biochemistry* 82.

Soucy, Shannon M., Jinling Huang, and Johann Peter Gogarten. 2015. "Horizontal Gene Transfer: Building the Web of Life." *Nature Reviews Genetics* 16 (8).

Spang, Anja, Jimmy H. Saw, Steffen L. Jørgensen, Katarzyna Zaremba-Niedzwiedzka, Joran Martijn, Anders E. Lind, Roel van Eijk, Christa Schleper, Lionel Guy, and Thijs J. G. Ettema. 2015. "Complex Archaea That Bridge the Gap Between Prokaryotes and Eukaryotes." *Nature* 521 (7551).

Spangenburg, Ray, and Kit Moser. 2009. *Carl Sagan: A Biography.* New York: Prometheus Books.

Specter, Michael. 2007. "Darwin's Surprise." *New Yorker* (December 3).

_____. 2015. "The Gene Hackers." *New Yorker* (November 16).

_____. 2017. "Rewriting the Code of Life." *New Yorker* (January 2).

Stahl, David A., David J. Lane, Gary J. Olsen, and Norman R. Pace. 1984. "Analysis of Hydrothermal Vent-Associated Symbionts by Ribosomal RNA Sequences." *Science* 224 (4647).

Stahl, David A., David J. Lane, Gary J. Olsen, and Norman R. Pace. 1985. "Characterization of a Yellowstone Hot Spring Microbial Community by 5S rRNA Sequences." *Applied and Environmental Microbiology* 49 (6).

Stanier, Roger Y., and C. B. van Niel. 1941. "The Main Outlines of Bacterial Classification." *Journal of Bacteriology* 42 (4).

Stanier, Roger Y., and C. B. van Niel. 1962. "The Concept of a Bacterium." *Archiv. für Mikrobiologie,* 42.

Stanier, Roger Y. 1980. "The Journey, Not the Arrival, Matters." *Annual Review of Microbiology* 34.

Stanier, Roger Y., John L. Ingraham, Mark L. Wheelis, and Paige R. Painter. 1986. *The Microbial World.* 5th ed.: Englewood Cliffs (NJ): Prentice-Hall.

Stevens, P. F. 1983. "Augustin Augier's 'Arbre Botanique' (1801), a Remarkable Early Botanical Representation of the Natural System." *Taxon* 32 (2).

Stoye, Jonathan P. 2012. "Studies of Endogenous Retroviruses Reveal a Continuing Evolutionary Saga." *Nature Reviews Microbiology* 10 (6).

Stringer, Chris. 2012. *Lone Survivors: How We Came to Be the Only Humans on Earth.* New York: St. Martin's Press.

Swithers, Kristen S., Shannon M. Soucy, and J. Peter Gogarten. 2012. "The Role of Reticulate Evolution in Creating Innovation and Complexity." *International Journal of Evolutionary Biology* 2012 (418964).

Syvanen, Michael, and Clarence I. Kado, editors. 2002. *Horizontal Gene Transfer.* 2nd ed. San Diego: Academic Press, a Division of Harcourt.

Syvanen, Michael, and Jonathan Ducore. 2010. "Whole Genome Comparisons Reveals a Possible Chimeric Origin for a Major Metazoan Assemblage." *Journal of Biological Systems* 18 (2).

Taylor, F. J. R. "Max." 2003. "The Collapse of the Two-Kingdom System, the Rise of Protistology and the Founding of the International Society for Evolutionary Protistology (ISEP)." *International Journal of Systematic and Evolutionary Biology* 53 (Pt. 6).

Thomas, Christopher M., and Kaare M. Nielsen. 2005. "Mechanisms of, and Barriers to, Horizontal Gene Transfer Between Bacteria." *Nature Reviews Microbiology* 3 (9).

Timmis, Jeremy N., Michael A. Ayliffe, Chun Y. Huang, and William Martin. 2004. "Endosymbiotic Gene Transfer: Organelle Genomes Forge Eukaryotic Chromosomes." *Nature Reviews Genetics* 5 (2).

Tranter, Martyn. 2014. "Microbes Eat Rock Under Ice." *Nature* 512 (7514).

Turnbaugh, Peter J., Ruth E. Ley, Michael A. Mahowald, Vincent Magrini, Elaine R. Mardis, and Jeffrey I. Gordon. 2006. "An Obesity-Associated Gut Microbiome with Increased Capacity for Energy Harvest." *Nature* 444 (7122).

Van Boeckel, Thomas P., Charles Brower, Marius Gilbert, Bryan T. Grenfell, Simon A. Levin, Timothy P. Robinson, Aude Teillant, and Ramanan Laxminarayan. 2015. "Global Trends in Antimicrobial Use in Food Animals." *Proceedings of the National Academy of Sciences* 112 (18).

Van der Oost, John. 2013. "New Tool for Genome Surgery." *Science* 339 (6121).

Van Leuven, James T., Russell C. Meister, Chris Simon, and John P. McCutcheon. 2014. "Sympatric Speciation in a Bacterial Endosymbiont Results in Two Genomes with the Functionality of One." *Cell* 158 (6).

Van Niel, C. B. 1946. "The Classification and Natural Relationships of Bacteria." *Cold Spring Harbor Symposia on Quantitative Biology* 11.

_____. 1949. "The 'Delft School' and the Rise of General Microbiology." *Bacteriological Reviews* 13 (3).

Venter, J. Craig. 2008. *A Life Decoded, My Genome: My Life*. London: Penguin Books.

Venter, J. Craig, Mark D. Adams, Eugene W. Myers, Peter W. Li, Richard J. Mural, Granger G. Sutton, Hamilton O. Smith, Mark Yandell, Cheryl A. Evans, Robert A. Holt et al. 2001. "The Sequence of the Human Genome." *Science* 291 (5507).

Vernot, Benjamin, and Joshua M. Akey. 2014. "Resurrecting Surviving Neandertal Lineages from Modern Human Genomes." *Science* 343 (6174).

Vetsigian, Kalin, Carl Woese, and Nigel Goldenfeld. 2006. "Collective Evolution and the Genetic Code." *Proceedings of the National Academy of Sciences* 103 (28).

Vogan, Aaron A., and Paul G. Higgs. 2011. "The Advantages and Disadvantages of Horizontal Gene Transfer and the Emergence of the First Species." *Biology Direct* 6 (1).

Wallace, Alfred Russel. 1969. *Natural Selection and Tropical Nature: Essays on Descriptive and Theoretical Biology*. Facsimile ed. Westmead (UK): Gregg International. *Natural Selection* was first published in 1870. *Tropical Nature* was first published in 1878.

Wallin, Ivan E. 1917. "The Relationships and Histogenesis of Thymus-like Structures in Ammocoetes." *American Journal of Anatomy* 22 (1).

_____. 1922. "On the Nature of Mitochondria: I. Observations on Mitochondria Staining Methods Applied to Bacteria. II. Reactions of Bacteria to Chemical Treatment." *American Journal of Anatomy* 30 (2).

_____. 1923a. "The Mitochondria Problem." *American Naturalist* 57 (650).

_____. 1923b. "On the Nature of Mitochondria: III. The Demonstration of Mitochondria by Bacteriological Methods. IV. A Comparative Study of the Morphogenesis of Root-Nodule Bacteria and Chloroplasts." *American Journal of Anatomy* 30 (4).

_____. 1923c. "On the Nature of Mitochondria: V. A Critical Analysis of Portier's 'Les Symbiotes.'" *Anatomical Record* 25 (1).

_____. 1923d. "Symbionticism and Prototaxis, Two Fundamental Biological Principles." *Anatomical Record* 26 (1).

_____. 1924. "On the Nature of Mitochondria: VII. The Independent Growth of Mitochondria in Culture Media." *American Journal of Anatomy* 33 (1).

_____. 1925. "On the Nature of Mitochondria: IX. Demonstration of the Bacterial Nature of Mitochondria." *American Journal of Anatomy* 36 (1).

Wallin, Ivan Emmanuel. 1927. *Symbionticism and the Origin of Species*. Reproduction. Baltimore: Waverly Press for Williams & Wilkins.

Walsh, Ali Morton, R. Daniel Kortschak, Michael G. Gardner, Terry Bertozzi, and David L. Adelson. 2013. "Widespread Horizontal Transfer of Retrotransposons." *Proceedings of the National Academy of Sciences* 110 (3).

Walsh, David A., and W. Ford Doolittle. 2005. "The Real 'Domains' of Life." *Current Biology* 15 (7).

Walsh, David A., Mary Ellen Boudreau, Eric Bapteste, and W. Ford Doolittle. 2007. "The Root of the Tree: Lateral Gene Transfer and the Nature of the Domains." In *Archaea: Evolution, Physiology, and Molecular Biology*. Edited by Roger A. Garrett and Hans-Peter Klenk. Hoboken (NJ): Wiley-Blackwell.

Wanger, G., T. C. Onstott, and G. Southam. 2008. "Stars of the Terrestrial Deep Subsurface: A Novel 'Star-Shaped' Bacterial Morphotype from a South African Platinum Mine." *Geobiology* 6 (3).

Watanabe, Tsutomu. 1963. "Infective Heredity of Multiple Drug Resistance in Bacteria." *Bacteriological Reviews* 27 (1).

Watanabe, Tsutomu, and Toshio Fukasawa. 1961. "Episome-Mediated Transfer of Drug Resistance in *Enterobacteriaceae*. I. Transfer of Resistance Factors by Conjugation." *Journal of Bacteriology* 81 (5).

Watanabe, Tsutomu, Chizuko Ogata, and Sachiko Sato. 1964. "Episome-Mediated Transfer of Drug Resistance in *Enterobacteriaceae*. VIII. Six-Drug-Resistance R Factor." *Journal of Bacteriology* 88 (4).

Watasé, S. 1894. "On the Nature of Cell-Organization." In *Biological Lectures Delivered at the Marine Biological Laboratory of Woods Hole, in the Summer Session of 1893*. Boston: Ginn.

Watson, J. D., and F. H. C. Crick. 1953. "Molecular Structure of Nucleic Acids." *Nature* 171 (4356).

Watson, James D. 1968. *The Double Helix: A Personal Account of the Discovery of the Structure of DNA*. New York: New American Library.

Weigel, Robert D., editor. 1970. "On the Tendency of Species to Form Varieties; and On the Perpetuation of Varieties & Species by Natural Means of Selection," by Charles Darwin and A. R. Wallace. Bloomington (IN): Scarlet Ibis Press.

Weismann, August. 2014. *The Germ-Plasm: A Theory of Heredity*. Translated by W. Newton Parker, PhD, and Harriet Rönnfeldt, BSc. Reproduction. San Bernardino (CA): Ulan Press. First published 1893 by New York: Charles Scribner's Sons.

Weismann, August. 1883. "On Heredity." In *Essays Upon Heredity*. Oxford: Clarendon Press. 1889. Electronic Scholarly Publishing, www.esp.org/books/weismann/essays /facsimile.

_____. 1885. "The Continuity of the Germ-Plasm as the Foundation of a Theory of Heredity." *In Essays Upon Heredity.* Oxford: Clarendon Press. 1889. Electronic Scholarly Publishing, www.esp.org/books/weismann/essays/facsimile.

_____. 2015. *The Effect of External Influences upon Development.* Romanes lecture, 1894. Reprint from University of Michigan Library. San Bernardino (CA): Google Books. First published 1894 by London: Oxford.

Weiss, Madeline C., Filipa L. Sousa, Natalia Mrnjavac, Sinje Neukirchen, Mayo Roettger, Shijulal Nelson-Sathi, and William F. Martin. 2016. "The Physiology and Habitat of the Last Universal Common Ancestor." *Nature Microbiology* 1 (9).

Weiss, Robin A. 2006. "The Discovery of Endogenous Retroviruses." *Retrovirology* 3 (67).

Werren, John H. 1997. "Biology of *Wolbachia*." *Annual Review of Entomology* 42.

_____. 2005. "Heritable Microorganisms and Reproductive Parasitism." In *Microbial Phylogeny and Evolution: Concepts and Controversies.* Edited by Jan Sapp. New York: Oxford University Press.

Westman, W. E., and R. K. Peet. 1985. "Robert H. Whittaker (1920–1980): The Man and His Work." In *Plant Community Ecology: Papers in Honor of Robert H. Whittaker.* Edited by R. K. Peet. Dordrecht, Netherlands: Springer.

Whittaker, R. H. 1957. "The Kingdoms of the Living World." *Ecology* 38 (3).

_____. 1959. "On The Broad Classification of Organisms." *Quarterly Review of Biology* 34.

_____. 1969. "New Concepts of Kingdoms of Organisms." *Science* 163 (3863).

Whittaker, R. H., and Lynn Margulis. 1978. "Protist Classification and the Kingdoms of Organisms." *BioSystems* 10 (1–2).

White, David G., Michael N. Alekshun, and Patrick F. McDermott, editors. 2006. *Frontiers in Antimicrobial Resistance: A Tribute to Stuart B. Levy.* Washington (DC): ASM Press.

Wiedenheft, Blake, Samuel H. Sternberg, and Jennifer A. Doudna. 2012. "RNA-Guided Genetic Silencing Systems in Bacteria and Archaea." *Nature* 482 (7385).

Wilding, E. Imogen, James R. Brown, Alexander P. Bryant, Alison F. Chalker, David J. Holmes, Karen A. Ingraham, Serban Iordanescu, Chi Y. So, Martin Rosenberg, and Michael N. Gwynn. 2000. "Identification, Evolution, and Essentiality of the Mevalonate Pathway for Isopentenyl Diphosphate Biosynthesis in Gram-Positive Cocci." *Journal of Bacteriology* 182 (15).

Williams, Tom A., Peter G. Foster, Tom M. W. Nye, Cymon J. Cox, and T. Martin Embley. 2012. "A Congruent Phylogenomic Signal Places Eukaryotes Within the Archaea." *Philosophical Transactions of the Royal Society of London (B)* 279 (1749).

Williams, Tom A., and T. Martin Embley. 2014. "Archaeal 'Dark Matter' and the Origin of Eukaryotes." *Genome Biology and Evolution* 6 (3).

_____. 2015. "Changing Ideas About Eukaryotic Origins." *Philosophical Transactions of the Royal Society of London (B)* 370 (1678).

Williams, Tom A., Sarah E. Heaps, Svetlana Cherlin, Tom M. W. Nye, Richard J. Boys, and T. Martin Embley. 2015. "New Substitution Models for Rooting Phylogenetic Trees." *Philosophical Transactions of the Royal Society of London (B)* 370 (1678).

Williamson, Donald I. 2009. "Caterpillars Evolved from Onychophorans by Hybridogenesis." *Proceedings of the National Academy of Sciences* 106 (47).

Wilson, Edmund B. 1925. *The Cell in Development and Heredity.* New York: Macmillan.

Wilson, Katherine L., and Scott C. Dawson. 2011. "Functional Evolution of Nuclear Structure." *Journal of Cell Biology* 195 (2).

Wilson, Robert A. 2013. "The Biological Notion of Individual." In *The Stanford Encyclopedia of Philosophy*. Edited by Edward N. Zalta, https://plato.stanford.edu/entries/biology-individual.

Winther, Rasmus G. 2001. "August Weismann on Germ-Plasm Variation." *Journal of the History of Biology* 34 (3).

Witzany, Günther. 2009. "Introduction: A Perspective on Natural Genetic Engineering and Natural Genome Editing." *Annals of the New York Academy of Sciences* 1178.

Woese, C. R. 1961. "Coding Ratio for the Ribonucleic Acid Viruses." *Nature* 190 (4777).

Woese, C. R. 1965a. "On the Evolution of the Genetic Code." *Proceedings of the National Academy of Sciences* 54 (6).

_____. 1965b. "Order in the Genetic Code." *Proceedings of the National Academy of Sciences* 54 (1).

Woese, Carl. 1970. "Molecular Mechanics of Translation: A Reciprocating Ratchet Mechanism." *Nature* 226 (May 30).

_____. 1998. "The Universal Ancestor." *Proceedings of the National Academy of Sciences* 95 (12).

Woese, Carl R. 1964. "Universality in the Genetic Code." *Science* 144 (22 May).

_____. 1967. *The Genetic Code*. New York: Harper & Row.

_____. 1977. "Endosymbionts and Mitochondrial Origins." *Journal of Molecular Evolution* 10 (2).

_____. 1981. "Archaebacteria." *Scientific American* 244 (6).

_____. 1982. "Archaebacteria and Cellular Origins: An Overview." In *Archaebacteria*. Edited by Otto Kandler. Stuttgart (Ger.): Gustav Fischer.

_____. 1983. "The Primary Lines of Descent and the Universal Ancestor." In *Evolution from Molecules to Man*. Edited by D. S. Bendall. Cambridge: Cambridge University Press.

_____. 1987. "Bacterial Evolution." *Microbiological Reviews* 51 (2).

_____. 1994. "Microbiology in Transition." *Proceedings of the National Academy of Sciences* 91 (5).

_____. 1998. "Default Taxonomy: Ernst Mayr's View of the Microbial World." *Proceedings of the National Academy of Sciences* 95 (19).

_____. 1998. "A Manifesto for Microbial Genomics." *Current Biology* 8 (22).

_____. 2000. "Interpreting the Universal Phylogenetic Tree." *Proceedings of the National Academy of Sciences* 97 (15).

_____. 2002. "On the Evolution of Cells." *Proceedings of the National Academy of Sciences* 99 (13).

_____. 2004. "A New Biology for a New Century." *Microbiology and Molecular Biology Reviews* 68 (2).

_____. 2005. "Q&A: Carl R. Woese." *Current Biology* 15 (4).

_____. 2007. "The Birth of the Archaea: A Personal Retrospective." In *Archaea: Evolution, Physiology, and Molecular Biology*. Edited by Roger A. Garnett and Hans-Peter Klenk. Hoboken (NJ): Wiley-Blackwell.

Woese, Dr. Carl R. 1962. "Nature of the Biological Code." *Nature* 194 (4834).

Woese, C. R., D. H. Dugre, W. C. Saxinger, and S. A. Dugre. 1966. "The Molecular Basis for the Genetic Code." *Proceedings of the National Academy of Sciences* 55 (4).

Woese, Carl R., George E. Fox, Lawrence Zablen, Tsuneko Uchida, Linda Bonen, Kenneth Pechman, Bobby J. Lewis, and David Stahl. 1975. "Conservation of Primary Structure in 16S Ribosomal RNA." *Nature* 254 (5495).

Woese, Carl R., and George E. Fox. 1977a. "Phylogenetic Structure of the Prokaryotic Domain: The Primary Kingdoms." *Proceedings of the National Academy of Sciences* 74 (11).

_____. 1977b. "The Concept of Cellular Evolution." *Journal of Molecular Evolution* 10 (1).

Woese, Carl R., Otto Kandler, and Mark L. Wheelis. 1990. "Towards a Natural System of Organisms: Proposal for the Domains Archaea, Bacteria, and Eucarya." *Proceedings of the National Academy of Sciences* 87 (12).

Woese, Carl R., and Nigel Goldenfeld. 2009. "How the Microbial World Saved Evolution from the Scylla of Molecular Biology and the Charybdis of the Modern Synthesis." *Microbiology and Molecular Biology Reviews* 73 (1).

Wolf, Yuri I., L. Aravind, Nick V. Grishin, and Eugene V. Koonin. 1999. "Evolution of Aminoacyl-tRNA Synthetases—Analysis of Unique Domain Architectures and Phylogenetic Trees Reveals a Complex History of Horizontal Gene Transfer Events." *Genome Research* 9 (8).

Wolfe, Ralph. 2014. "Early Days with Carl." *RNA Biology* 11 (3).

Wolfe, Ralph S. 1991. "My Kind of Biology." *Annual Review of Microbiology* 45.

_____. 2006. "The Archaea: A Personal Overview of the Formative Years." In *The Prokaryotes*. Edited by E. Rosenberg, E. F. DeLong, S. Lory, E. Stackebrandt, and F. Thompson. Berlin/Heidelberg: Springer.

Wright, H. D. 1941. "Obituary. William McDonald Scott, M.D., B.S.C. Edin., D.T.M.&H." *Lancet* 237 (6140).

Wu, Dongying, Sean C. Daugherty, Susan E. Van Aken, Grace H. Pai, Kisha L. Watkins, Hoda Khouri, Luke J. Tallon, Jennifer M. Zaborsky, Helen E. Dunbar, Phat L. Tran et al. 2006. "Metabolic Complementarity and Genomics of the Dual Bacterial Symbiosis of Sharpshooters." *PLoS Biology* 4 (6).

Xue, Katherine. 2014. "Superbug: An Epidemic Begins." *Harvard* (May/June).

Yang, D., Y. Oyaizu, H. Oyaizu, G. J. Olsen, and C. R. Woese. 1985. "Mitochondrial Origins." *Proceedings of the National Academy of Sciences* 82 (13).

Yarus, Michael. 2010. *Life from an RNA World: The Ancestor Within*. Cambridge (MA): Harvard University Press.

Yong, Ed. 2008. "Human Gut Bacteria Linked to Obesity." *Not Exactly Rocket Science* (blog), October 6, 2008. http://phenomena.nationalgeographic.com/2008/10/06/human-gut-bacteria-linked-to-obesity.

_____. "Genes from Chagas Parasite Can Transfer to Humans and Be Passed on to Children." *Not Exactly Rocket Science* (blog). February 14, 2010. http://phenomena.nationalgeographic.com/2010/02/14/genes-from-chagas-parasite-can-transfer-to-humans-and-be-passed-on-to-children.

_____. "Meet Your Viral Ancestors—How Bornaviruses Have Been Infiltrating Our Genomes for 40 Million Years." *Not Exactly Rocket Science* (blog). January 6, 2010. http://blogs.discovermagazine.com/notrocketscience/2010/01/06/meet-your-viral-ancestors-how-bornaviruses-have-been-infiltrating-our-genomes-for-40-million-years/#.Wir PW7aZM0o.

_____. "Dormant Viruses Can Hide in Our DNA and Be Passed from Parent to Child." *Not Exactly Rocket Science* (blog). March 27, 2010. http://blogs.discover magazine.com/notrocketscience/2010/03/27/dormant-viruses-can-hide-in-our-dna -and-be-passed-from-parent-to-child/#.WirZdLaZM0o.

_____. "Tree or Ring: The Origin of Complex Cells." *Not Exactly Rocket Science* (blog). September 13, 2010. http://blogs.discovermagazine.com/notrocketscience/2010/09/13 /tree-or-ring-the-origin-of-complex-cells/#.WiraN7aZM0o.

_____. (2016). *I Contain Multitudes: The Microbes Within Us and a Grander View of Life*. New York: HarperCollins.

Yoshida, Shosuke, Kazumi Hiraga, Toshihiko Takehana, Ikuo Taniguchi, Hironao Yamaji, Yasuhito Maeda, Kiyotsuna Toyohara, Kenji Miyamoto, Yoshiharu Kimura, and Kohei Oda. 2016. "A Bacterium That Degrades and Assimilates Poly(ethylene terephthalate)." *Science* 351 (6278).

Zablen, L. B., M. S. Kissil, C. R. Woese, and D. E. Buetow. 1975. "Phylogenetic Origin of the Chloroplast and Prokaryotic Nature of Its Ribosomal RNA." *Proceedings of the National Academy of Sciences* 72 (6).

Zambryski, P., J. Tempe, and J. Schell. 1989. "Transfer and Function of T-DNA Genes from *Agrobacterium* Ti and Ri Plasmids in Plants." *Cell* 56 (2).

Zamecnik, P. C., E. B. Keller, J. W. Littlefield, M. B. Hoagland, and R. B. Loftfield. 1956. "Mechanism of Incorporation of Labeled Amino Acids into Protein." *Journal of Cellular and Comparative Physiology* 47 (Suppl. 1).

Zaremba-Niedzwiedzka, Katarzyna, Eva F. Caceres, Jimmy H. Saw, Disa Bäckström, Lina Juzokaite, Emmelien Vancaester, Kiley W. Seitz, Karthik Anantharaman, Piotr Starnawski, Kasper U. Kjeldsen et al. 2017. "Asgard Archaea Illuminate the Origin of Eukaryotic Cellular Complexity." *Nature* 541 (7637).

Zhaxybayeva, Olga, and J. Peter Gogarten. 2004. "Cladogenesis, Coalescence, and the Evolution of the Three Domains of Life." *Trends in Genetics* 20 (4).

Zhaxybayeva, Olga, and W. Ford Doolittle. 2011. "Lateral Gene Transfer." *Current Biology* 21 (7).

Zhuo, Xiaoyu, Cedric Feschotte. 2015. "Cross-Species Transmission and Differential Fate of an Endogenous Retrovirus in Three Mammal Lineages." *PLoS Pathogens* 11 (11).

Zimmer, Carl. 2008. *Microcosm: E. Coli and the New Science of Life*. New York: Pantheon Books.

_____. 2010. "Hunting Fossil Viruses in Human DNA." *New York Times* (January 12).

Zinder, Norton D. 1953. "Infective Heredity in Bacteria." *Cold Spring Harbor Symposia in Quantitative Biology* 18.

Zinder, Norton D., and Joshua Lederberg. 1952. "Genetic Exchange in *Salmonella*." *Journal of Bacteriology* 64 (5).

Zuckerkandl, Emile, and Linus Pauling. 1965a. "Evolutionary Divergence and Convergence in Proteins." In *Evolving Genes and Proteins: A Symposium*." Edited by Vernon Bryson and Henry J. Vogel. New York: Academic Press.

_____. 1965b. "Molecules as Documents of Evolutionary History." *Journal of Theoretical Biology* 8 (2).

Illustration Credits

285 From Doolittle (1999), "Phylogenetic Classification and the Universal Tree," *Science* 284, no. 5423, fig. 3. Courtesy of the American Association for the Advancement of Science, used by permission.

299 From Martin (1999), "Mosaic Bacterial Chromosomes: A Challenge En Route to a Tree of Genomes," *BioEssays* 21, no. 2, fig. 2. Courtesy of William Martin.

Index

Page numbers in *italics* refer to illustrations.